PERCIVAL FROST
JOSEPH WOLSTENHOLME

A TREATISE

ON

SOLID GEOMETRY

Elibron Classics
www.elibron.com

Elibron Classics series.

ISBN 1-4021-7098-X (paperback)
ISBN 1-4021-3635-8 (hardcover)

This Elibron Classics Replica Edition is an unabridged facsimile of the edition published in 1863 by Macmillan and Co., Cambridge and Londor

A TREATISE

ON

SOLID GEOMETRY.

A TREATISE

ON

SOLID GEOMETRY

BY THE REV.

PERCIVAL FROST, M.A.

LATE FELLOW OF ST JOHN'S COLLEGE,
MATHEMATICAL LECTURER OF KING'S COLLEGE.

AND THE REV.

JOSEPH WOLSTENHOLME, M.A.

FELLOW AND ASSISTANT TUTOR OF CHRIST'S COLLEGE.

Cambridge and London:
MACMILLAN AND CO.
1863

Cambridge:
PRINTED BY C. J. CLAY, M.A.
AT THE UNIVERSITY PRESS.

PREFACE.

The Authors of the following Treatise have endeavoured to present before students as comprehensive a view of the subject, as certain limitations have allowed them to do. The necessity of these limitations has developed itself in the course of preparing the work during a period of four years. The study of innumerable papers, by the most celebrated mathematicians of all countries, has convinced the authors that the subject is almost inexhaustible, and that, to have treated all parts of it with anything approaching to the fulness with which they have treated the first portion, would have swelled their work in a fearful proportion to what it has already attained.

Intending, as they have done, to make the subject accessible, at least in the earlier portions, to all classes of students, they have endeavoured to explain completely all the processes which are most useful in dealing with ordinary theorems and problems connected with the straight line, and plane, and particular surfaces of the second degree, and, in doing so, their object has been to direct the student to the selection of the methods which are best adapted to the exigencies of each problem. In the more difficult portions of the subject, they have considered themselves to be addressing a higher class of students, and here they have tried to lay a good foundation on which to build, if any of their readers should wish to pursue their studies in any department of the science, beyond the limits to which the work extends.

The authors would willingly have given references to all the writers from whom they have derived information in the course of their work, but they have found this to be impossible, and they regret it the less, because it will not be supposed that they lay claim to every thing in which they have made no reference. They have, however, in a very large number of cases mentioned the names of eminent men, who have advanced the boundaries of the subject, and they hope it will be apparent, that they have appreciated the labours of such men as Cayley, Salmon, McCullagh, Roberts and Townsend; at all events they are sensible that, in many departments, the treatise lately published by Salmon on the same subject proves how far their own work is from being perfect.

They cannot conclude this work without making

acknowledgments to Mr Ferrers of Caius College, and Mr Horne of St John's College, for their kindness in examining and commenting upon the proof sheets of the earlier parts of their work, and at the same time without expressing their regret that they have not escaped a large number of errors, which it will be punishment enough to them to see tabulated in an adjoining page.

ERRATA.

Page 23, *for* minimum, *read* maximum.
„ 25, Problem 15 is wrong.
„ 42, line 17, *for* (63), *read* (61).
„ 92, „ 21, *for* in, *read* is.
„ 119, „ 20, *for* $\frac{y}{b} = \cos(\theta + \alpha)$, *read* $\frac{y}{b} = \sin(\theta + \alpha)$.
„ *ib.* „ 24, *for* $\sin\theta$, *read* $\cos\theta$.
„ *ib.* „ 27, *for* $\frac{z}{c}\sin\theta\cos\alpha$, *read* $\frac{z}{c}\sin\theta\sin\alpha$.
„ 126, „ 14, *for* r^2, *read* v^2.
„ 129, „ 10 and 13, *omit* 2 throughout.
„ 143, „ 23, *for* suppose, *read* surface.
„ 167, „ 23, *for* the others, *read* another, *and for* the three, *read* two of the three.
„ 204, „ 21, *insert* $+\mu\nu\varpi$.
„ *ib.* „ 35, *for* $n - vl$, *read* $n - vl - vk$.
„ 222, „ 29, *after* cylinder, *write* or cone.
„ 243, „ 23, *remove* $\frac{1}{2}$ *from* ϕ_{n-2} *to* $D^2\phi_n$.
„ 246, „ 8, *for* these, *read* three.
„ 269, „ 5, *for* pole, *read* polar.
„ 306, „ 8, *after* cosines, *write* multiplied by A, B, C, D respectively.

Lithograph to face page 393.

CONTENTS.

CHAPTER I.

CO-ORDINATE SYSTEMS.

CHAPTER II.

GENERAL DESCRIPTION OF LOCI OF EQUATIONS.

CHAPTER III.

PROJECTIONS OF LINES AND AREAS. DIRECTIONS OF LINES.

b

CHAPTER IV.

DISTANCES OF POINTS. EQUATIONS OF A STRAIGHT LINE.

CHAPTER V.

GENERAL EQUATION OF THE FIRST DEGREE.

CHAPTER VI.

FOUR-POINT CO-ORDINATE SYSTEM.

CHAPTER VII.

TRANSFORMATION OF CO-ORDINATES.

CHAPTER VIII.

ON CERTAIN SURFACES OF THE SECOND ORDER.

CHAPTER IX.

ON GENERATION BY LINES AND CIRCLES.

CHAPTER X.

MODULAR AND UMBILICAL GENERATION OF SURFACES OF THE SECOND DEGREE.

CHAPTER XI.

GENERAL EQUATION OF THE SECOND DEGREE.

CHAPTER XII.

DIAMETRAL SURFACES. DIAMETRAL PLANES. CONJUGATE DIAMETERS.

CHAPTER XIII.

DEGREES OF SURFACES AND CURVES. INTERSECTIONS OF SURFACES. NUMBER OF CONDITIONS WHICH SURFACES CAN SATISFY. CLASSES OF SURFACES.

CHAPTER XIV.

TANGENT LINES PLANES AND CONES. POLES AND POLAR PLANES. NORMALS.

CHAPTER XV.

METHOD OF RECIPROCAL POLARS.

CHAPTER XVI.

GENERAL THEORY OF POLARS AND TANGENT LINES.

CHAPTER XVII.

FUNCTIONAL AND DIFFERENTIAL EQUATIONS OF FAMILIES OF SURFACES.

CHAPTER XVIII.

PROPERTIES OF CONICOIDS SATISFYING GIVEN CONDITIONS. FORMS OF THE EQUATION OF A SPHERE.

CHAPTER XIX.

CURVES.

CHAPTER XX.

ENVELOPES.

CHAPTER XXI.

VOLUMES, AREAS OF SURFACES, &c.

CHAPTER XXII.

CURVATURE OF CURVES.

CHAPTER XXIII.

CURVATURE OF SURFACES AND LINES OF CURVATURE.

CHAPTER XXIV.

GEODESIC LINES.

GEOMETRY OF THREE DIMENSIONS.

CHAPTER I.

ON CO-ORDINATE SYSTEMS.

1. BEFORE entering upon the application of Algebra to the investigation of Theorems, and to the solution of Problems, in Solid Geometry, we shall premise on the part of the student a complete knowledge of all the ordinary processes adopted in the case of Plane Geometry.

By this means we shall avoid the necessity of entering upon the subject of the interpretation of the affection denoted by the sign (–) prefixed to a symbol; since it is known that, if $+a$ denote a line of length a measured in any direction from a point in a line straight or curved, $-a$ may be interpreted to denote a line of length a measured in the opposite direction from any other point in the line, without this hypothesis involving any infringement of the laws of combination of these signs, assumed as the fundamental laws of Symbolical Algebra.

2. Our first object will be to explain how the position of a point in space can be represented by algebraical symbols, and with this view we shall describe the different co-ordinate systems which it has been found convenient to adopt; each of which has its peculiar advantage, according to the nature of the theorem or problem which is the subject of examination.

Co-ordinate System of Three Planes.

3. In the co-ordinate system of three planes, three planes xOy, yOz, zOx are fixed upon as planes of reference, which may be either perpendicular to one another, or inclined at angles which are known.

The three lines in which they intersect are called *co-ordinate axes*, and the point in which they meet the *origin of co-ordinates.*

The position of a point in space is then completely determined, when its distance from each of the planes, estimated parallel to the co-ordinate axes, and the direction in which those distances are measured, are given.

The absolute distance, and the direction of measurement are included in the term *algebraical distance.*

Thus $+a$ and $-a$ are the algebraical distances of two points whose absolute distances from the plane yOz are each a, and which are measured, the first in the direction Ox, the second in the direction xO from that plane.

These algebraical distances are called the *co-ordinates* of a point in this system, and are usually denoted by the letters x, y, and z.

The point, of which these are co-ordinates, is described as the point (x, y, z).

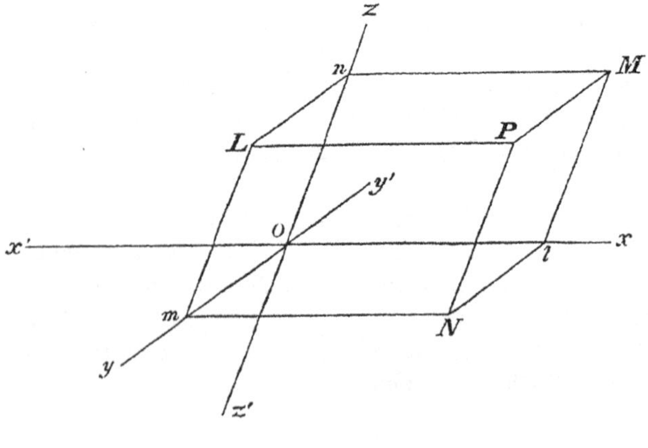

Produce xO, yO, zO backwards to x', y', z'; then, if a, b, c are absolute lengths, (a, b, c) denotes a point in the compartment $xyzO$, $(-a, b, c)$ in $x'yzO$, $(a, -b, c)$ in $xy'zO$, $(a, b, -c)$ in $xyz'O$, $(a, -b, -c)$ in $xy'z'O$, $(-a, b, -c)$ in $x'yz'O$, $(-a, -b, c)$ in $x'y'zO$, $(-a, -b, -c)$ in $x'y'z'O$.

4. If a parallelopiped be constructed, whose faces are parallel to the co-ordinate planes, the point $P(a, b, c)$ being the other extremity of the diagonal drawn from the origin, the edges LP, MP, NP will be the co-ordinates of the point P, supposed in the compartment $xyzO$.

Also, it is obvious that $x=a$ for every point in the plane face $PNlM$, or that $x=a$ is the equation of that plane, as $y=b$ and $z=c$ are the equations of the planes $PLmN$ and $PMnL$ indefinitely extended in every direction.

Thus, the point P may be considered as the intersection of the three planes, whose equations are

$$x=a, \quad y=b, \quad z=c.$$

The points l, O may be denoted by $(a, 0, 0)$ and $(0, 0, 0)$ and the points L and M by $(0, b, c)$ and $(a, 0, c)$.

I.

(1) Construct the positions of points which are represented by the equations

$$x^2 - y^2 = z^2,$$
$$x + y = 4a,$$
$$x - y = a.$$

(2)
$$x^2 + y^2 = 2z^2,$$
$$x + y = 2z,$$
$$xy = a^2.$$

(3) Shew that for every point in OP, P being (a, b, c),

$$\frac{x}{a} = \frac{y}{b} = \frac{z}{c}.$$

(4) Shew that for every point in the plane $LMlm$

$$\frac{x}{a} + \frac{y}{b} = 1.$$

Four-Plane Co-ordinate System.

5. In the co-ordinate system of four planes, four planes are fixed upon as planes of reference, which form by their intersections a pyramid or tetrahedron $ABCD$. The position of a point is determined in this system by the algebraical distances α, β, γ, δ from the four planes respectively opposite to the vertices A, B, C, D, these distances being all absolute distances when the point is within the tetrahedron.

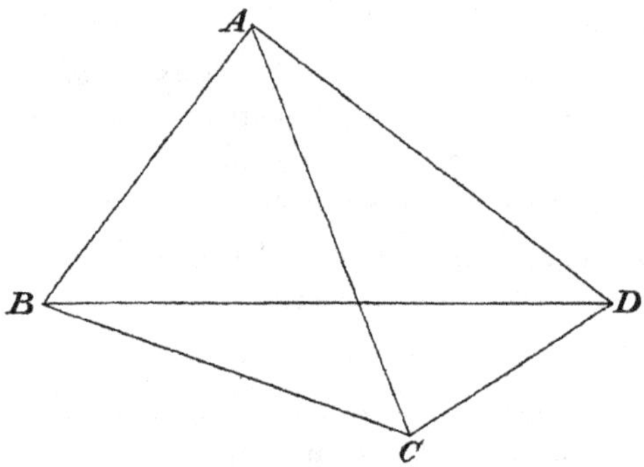

Hence, for a point in the compartment between the plane ACD and the other three produced, β will be negative and α, γ, δ positive; between BAC, CAD, and DAB, produced through A, α will be positive, and β, γ, δ all negative.

If a be positive, $\alpha = a$ is the equation of a plane parallel to BCD, at a distance a from it, on the side towards A; $\alpha = -a$ that of a plane on the opposite side at the same distance.

6. In this system of co-ordinates the following peculiarity must be observed, viz. that any three of the co-ordinates α, β, γ, δ are sufficient to determine the position of the point, since, when α, β, γ are given, three planes are determined parallel to the faces opposite to A, B, C which intersect in the point, and so determine its position completely.

Hence, when α, β, γ are given, δ ought to be known from the geometry of the figure, and we proceed to determine the relation between the co-ordinates in this system.

Relation of Co-ordinates in the Four-Plane System.

7. Let V be the volume of the tetrahedron contained by the four fixed planes, A, B, C, D the areas of the triangular faces.

If the point P whose co-ordinates are α, β, γ, δ be joined by straight lines to the angular points of the tetrahedron, four pyramids are formed, whose vertices are at P, and whose bases are the faces of the tetrahedron.

The algebraical sum of these four pyramids make up the volume of the tetrahedron; therefore, remembering that the volume of a pyramid is three times the base × the altitude,

$$A\alpha + B\beta + C\gamma + D\delta = 3V,$$

whence, when any three of the co-ordinates of a point are given, the fourth may be found.

The object of the introduction of a fourth co-ordinate, in this system, is the same as that for which trilinear co-ordinates are employed in Plane Geometry, viz. to obtain equations homogeneous with reference to the co-ordinates, and thus to arrive at symmetrical results.

By means of the equation given above, any equation which does not appear in a homogeneous form can be reduced to such a form immediately.

Thus the equation $\alpha = a$ of a plane may be reduced to the homogeneous form

$$3V\alpha = a\,(A\alpha + B\beta + C\gamma + D\delta).$$

Tetrahedral Co-ordinates.

8. The expressions involving these co-ordinates are frequently simplified by the employment, in their stead, of the tetrahedrons, which are proportional to them, namely, $\frac{1}{3}A\alpha$, $\frac{1}{3}B\beta$, $\frac{1}{3}C\gamma$, $\frac{1}{3}D\delta$, which we shall call "Tetrahedral Co-ordinates."

If α, β, γ, δ denote these co-ordinates, the relation always subsisting between them is

$$\alpha + \beta + \gamma + \delta = V.$$

Any expressions involving the former (four-plane) co-ordinates, may be at once transformed so as to involve the latter (tetrahedral) by the substitution of

$$\frac{3\alpha}{A}, \frac{3\beta}{B}, \frac{3\gamma}{C}, \frac{3\delta}{D} \text{ for } \alpha, \beta, \gamma, \delta \text{ respectively.}$$

The equation of condition is further simplified, if we take the volume of the fundamental tetrahedron as the unit of volume, or, which amounts to the same supposition, take as the co-ordinates of any point the ratios of its algebraical distances from the faces to the distances of the angular points respectively opposite to them. In this case the equation of condition becomes

$$\alpha + \beta + \gamma + \delta = 1;$$

and any given equation involving four-plane co-ordinates may be transformed into an equation referred to this system by writing $p_1\alpha$, $p_2\beta$, $p_3\gamma$, $p_4\delta$ for α, β, γ, δ respectively, p_1, p_2, p_3, p_4 being the distances of the angular points from the opposite faces.

Polar Co-ordinate System.

9. In the system of Polar Co-ordinates, a plane zOx is chosen, and in this plane a straight line Oz is drawn from a fixed point O.

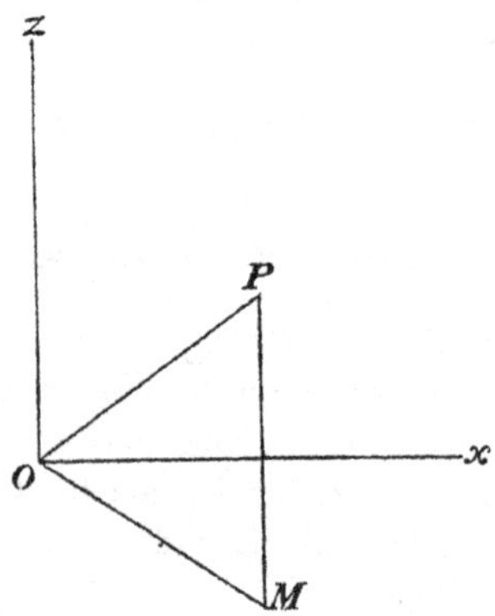

The position of a point P in space is completely determined, when its distance from the fixed point O is given, the angle through which the line OP has revolved in a plane from Oz, and the angle through which the plane zOP has revolved into its position from the fixed plane of reference zOx.

These co-ordinates are usually denoted by the symbols r, θ and ϕ, and the point P by (r, θ, ϕ).

Thus, if the longitude of a place be l, the latitude λ, and the radius of the earth a, we may take the first meridian for the plane zOx, the axis of the earth for the axis of z, and the position of the place will be expressed by

$$\left(a, \frac{\pi}{2} - \lambda, l\right).$$

If λ' be the latitude of Greenwich, its position is given by

$$\left(a, \frac{\pi}{2} - \lambda', 0\right).$$

II.

(1) Shew that for every point in a plane through the edge AB bisecting the angle between the planes CAB, DAB,

$\gamma - \delta = 0$, if the angle be the internal angle,
$\gamma + \delta = 0$, ... external ......

(2) Shew that for every point in a plane drawn through the vertex A parallel to the opposite face,

$$B\beta + C\gamma + D\delta = 0;$$

or with tetrahedral co-ordinates,

$$\beta + \gamma + \delta = 0.$$

(3) If AO be drawn perpendicular to the opposite face BCD, then for any point in AO,

$$\frac{B\beta}{\Delta COD} = \frac{C\gamma}{\Delta DOB} = \frac{D\delta}{\Delta BOC} = AO - \alpha.$$

(4) Every point in a plane through CD parallel to AB satisfies the equation in tetrahedral co-ordinates,

$$\alpha + \beta = 0.$$

(5) At any point in the straight line joining the middle points of AB and CD, the tetrahedral co-ordinates satisfy the equations

$$\alpha=\beta,\ \gamma=\delta.$$

(6) The four straight lines joining the middle points of opposite edges of the tetrahedron of reference meet in a point whose tetrahedral co-ordinates satisfy the equations

$$\alpha=\beta=\gamma=\delta=\frac{V}{4}.$$

(7) If the equations to a point O be

$$\frac{\alpha}{l}=\frac{\beta}{m}=\frac{\gamma}{n}=\frac{\delta}{p},$$

and AO, BO, CO, DO be joined and produced to A', B', C', D' such that O bisects the lines AA', &c., the tetrahedral co-ordinates of the point A' will satisfy the equations

$$\frac{2\alpha}{l-m-n-p}=\frac{\beta}{m}=\frac{\gamma}{n}=\frac{\delta}{p}=\frac{2V}{l+m+n+p},$$

and similarly for B', C', D'.

CHAPTER II.

GENERAL DESCRIPTION OF LOCI OF EQUATIONS. SURFACES. CURVES.

Locus of an equation.

10. IF an equation $F(x, y, z) = 0$ be given, in which the variables are the co-ordinates of any point, the number of solutions of this equation is generally infinite, i. e. the number of points whose co-ordinates satisfy the equation is infinite: we shall proceed to shew what is the general nature of the distribution of the points, whose co-ordinates satisfy the equation. We shall prove in the first place that no algebraical equation can be satisfied by every point of any *solid* figure, but, in the most general case, only by every point in some surface or surfaces.

11. If an equation involve only one of the co-ordinates as x, we know that such an equation $F(x) = 0$ has a finite or an infinite number of roots, a, b, c,... separated by definite intervals, and is reducible to the equations $x = a$, $x = b$,...., each of which, as $x = a$, is satisfied by every point in a plane parallel to the plane yOz, at an algebraical distance a. Hence, all the points whose co-ordinates satisfy the equation $F(x) = 0$ lie in a series of planes parallel to yOz at algebraical distances a, b, c,...

If the given equation involve two only of the variables, as $F(y, z) = 0$, on the plane yz let the curve be constructed which is the locus of $F(y, z) = 0$, and let a straight line be drawn parallel to Ox through any point in this curve, every point in this line is such that its co-ordinates satisfy the given equation, and the same is true of all points in the curve, and of no other points. Hence, all the points which satisfy the proposed equation lie in a surface generated by a straight line parallel to Ox, which passes successively through every point of the curve traced on the plane yz: such a surface is called a *cylindrical surface*, and the curve is called the *trace* on the plane of yz, and is one of an infinite number of curves, which are called *guiding curves* to the cylindrical surface.

The number of guiding curves is infinite, since, if any curve be traced upon the cylindrical surface, so as to cross every generating line, a line moving parallel to Ox, so as to traverse every portion of such a curve traced in space, would generate the entire cylindrical surface, that curve serving to guide the direction of motion of the generating line.

12. We may notice here, that, if the equation $F(y, z) = 0$ be reducible to a series of equations of such forms as

$$(y - b)^2 + (z - c)^2 = 0,$$

$$(my - nz)^2 + (z - c)^2 = 0,$$

the trace on yz is reducible to a series of points, and the locus of the equation $F(y, z) = 0$ becomes a series of straight lines parallel to Ox, passing through those points.

In such cases, the locus appears to be different in character from that of the general case, since it is a series of lines instead of being a surface.

But, it may be seen that this is only in appearance, since each of the equations whose locus is called a point represents a closed curve of infinitely small dimensions, and the lines are cylinders whose breadths are infinitely small, and the locus of the equation $F(y, z) = 0$ is as in the general case a series of surfaces: and a similar interpretation may be given in every case.

13. We shall now proceed to the general case,

$$F(x, y, z) = 0,$$

in order to examine the position of all the points which satisfy the equation; and we shall find, first, the position of those points which are at an algebraical distance g from the plane of yz, which is the same thing as examining the position of those points which lie on a plane whose equation is $x = g$.

These points are contained in the cylindrical surface whose equation is $F(g, y, z) = 0$. The trace of this surface on the plane $x = g$ is the line which contains all the points of the surface which lie on that plane; and if the series of lines be traced corresponding to different positions of the plane $x = g$ for values of g from $-\infty$ to $+\infty$, we shall evidently obtain a surface which contains all the points which satisfy the equation

$$F(x, y, z) = 0.$$

14. As an illustration of tracing surfaces, we will take the case of the surface whose equation is

$$(x + y)^2 = az.$$

If $$x = 0, \qquad y^2 = az;$$

therefore the trace on the plane of yz is a parabola whose axis is Oz and vertex O.

Similarly, that on zx is another equal parabola having the same axis and vertex.

If $$z = k, \qquad (x + y)^2 = ak,$$

which is the equation of two planes parallel to the axis Oz, equally inclined to the planes yz, zx; therefore the trace on the plane $z = k$ is two straight lines equally inclined to the planes yz, zx.

Hence, the surface is generated by straight lines, as PQ, which move parallel to the plane of xy, constantly passing through the traces on yz, zx so as to be inclined to those planes at equal angles of 135°.

The shape is therefore a cylindrical surface as in the figure.

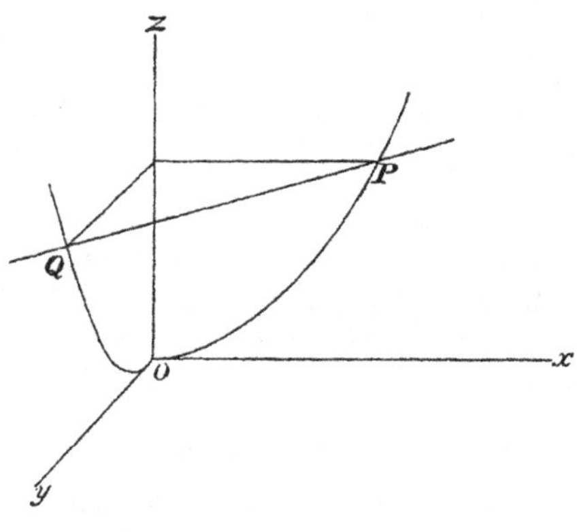

III.

Trace the surfaces represented by the equations

(1) $x^2+y^2=ax$,

(2) $z^2=ax+by$,

(3) $x^2+y^2+z^2=ax+by+cz$,

(4) $x^2+y^2=cz$,

(5) $xy=az$,

(6) $(x+z)(y+z)=ax$,

(7) $(ax+by+cz)^2=m^2(ax+by+dz)$.

15. By the introduction of constants, which admit of all values within certain limits, equations may be formed, which will represent all points within certain bounding surfaces.

For example, a sphere whose center is (a, b, c) and radius r may be represented by the equation

$$(x-a)^2+(y-b)^2+(z-c)^2=r^2$$

in a rectangular co-ordinate system, if then r be capable of receiving all values from r_1 to r_2, the equation represents all points contained between two concentric spherical surfaces, whose radii are r_1 and r_2.

Again, if a, b, c be capable of receiving all values consistent with the equation

$$a^2+b^2+c^2=d^2,$$

the above equation will represent all points contained between two spherical surfaces whose center is the origin and radii $r+d$ and $r-d$.

If c and r be constant, and a, b have all values consistent with the equation

$$a^2+b^2=d^2,$$

the equation represents all points contained within a ring, generated by a circle of radius r, revolving about the axis Oz, the center being at a distance d from that axis, and c from the plane of xy.

In the same manner, it may be shewn that

$$(x-\alpha)^2+(y-\beta)^2+(z-\gamma)^2=0,$$

in which

$$\begin{array}{llll} & \alpha \text{ may have any values between} & -a \text{ and} & +a, \\ & \beta \text{ ....} & -b \text{ ...} & +b, \\ \text{and} & \gamma \text{ ....} & -c \text{ ...} & +c \end{array}$$

is satisfied by all points which lie within a parallelopiped whose faces are parallel to the co-ordinate planes.

Locus of the Polar equation.

16. We shall examine in order the loci of equations which involve one or more of the co-ordinates.

(1) If the equation be $F(r) = 0$.

This is equivalent to a series of equations $r = a$, $r = b$,...... any one of which being satisfied the original equation is satisfied: $r = a$ is satisfied by all points at a distance a from the origin, measured in any direction; therefore the locus of $F(r) = 0$ is a series of concentric spheres, whose center is the origin.

(2) If the equation be $F(\theta) = 0$ it is equivalent to $\theta = \alpha$, $\theta = \beta$,...... any one, $\theta = \alpha$, is satisfied by every point of lines through O inclined to Oz at angles equal to α; therefore the locus of $F(\theta) = 0$ is a series of conical surfaces, whose common axis is Oz, vertex O, and vertical angles, 2α, 2β,.......

(3) If the equation be $F(\phi) = 0$, it is equivalent to $\phi = \alpha$, $\phi = \beta$,...... any one, $\phi = \alpha$, is satisfied by every point in a plane through Oz inclined at angle α to the plane zOx; therefore the locus of $F(\phi) = 0$ is a series of planes through Oz inclined to zOx at angles α, β,......

(4) If the equation involve only r and θ as $F(r, \theta) = 0$, since, for all values of ϕ the same relation exists between r and θ, the locus of the equation is the surface generated by the revolution of a curve traced on a plane through Oz, as the plane revolves about Oz as an axis.

(5) If the equation involve only ϕ and θ, as $F(\phi, \theta) = 0$, for every value of ϕ, there are a series of values of θ, corresponding to which if lines through O be drawn, every point in these

lines will be such that its co-ordinates satisfy the equation, and as ϕ changes or the plane through Oz revolves, these lines assume new positions relative to Oz, and generate during the revolution of the plane, conical surfaces, a conical surface being defined to be a surface generated by a straight line moving in any manner with the restriction that it passes through a fixed point.

(6) If the co-ordinates involved be r, ϕ as in $F(r, \phi)=0$, for each position of the plane through Oz inclined at any angle ϕ to zOx, there are values of r, which are constant for all values of θ, i.e. there are a series of concentric circles, in the plane, the co-ordinates of each point in which satisfy the equation.

The locus of the equation is therefore a surface generated by circles having their centers in O, and varying in magnitude as their planes revolve about the line Oz through which they pass.

(7) If the equation involve all the co-ordinates, as $F(r, \theta, \phi)=0$, let any value be given to ϕ, as β, then, corresponding to this value there is a plane through Oz, and if the locus of $F(r, \theta, \beta)=0$ be traced on this plane, and such curves be drawn upon all planes corresponding to values of ϕ from $-\infty$ to $+\infty$, the surface which contains all these curves is the locus of the equation.

Curves.

17. Curves in space are called generally *curves of double curvature,* because generally they do not lie entirely in one plane.

If we take three points very near to one another, these three points lie in one plane but not generally in one straight line, but a fourth point will lie generally on one side or the other of this plane, the bend first in one plane and then in another giving rise to the term *double curvature.*

Equations of curves.

18. Through every curve there can be drawn an infinite number of surfaces, the intersections of any two of which will include every point of the curve. At the same time we must observe that two surfaces, each of which contains a given curve,

may not be sufficient to determine the position of the curve, because they may intersect in other points which are not connected with the given curve.

Thus, if we take the case of a circle, it is true that it lies entirely in the intersection of a certain sphere and cylinder, but the sphere and cylinder are not sufficient to determine the circle because they may also intersect in another circle, and the circle to be considered is not defined by those surfaces: in this case it is possible to find two surfaces which do define the circle completely, as for example a plane, and either the sphere or cylinder.

19. If $F(x, y, z) = 0$, and $F_1(x, y, z) = 0$ be equations of two surfaces, these surfaces by their intersection determine a certain curve, and if another equation $\phi(x, y, z) = 0$ be obtained from those two equations, by any processes of addition or multiplication, the third equation will be satisfied by every point in the curve determined by the intersection of the first two surfaces, and we may employ this equation and either of the first two to obtain properties of the curve; although the two new equations may represent surfaces which intersect in other points than those of the curve originally proposed.

It is often convenient in practice to consider a curve as the intersection of two cylindrical surfaces, whose generating lines are parallel to two of the axes. In this way of considering curves, the equations of the surfaces are of the form

$$\phi(x, z) = 0,$$

$$\psi(y, z) = 0.$$

As a simple example of the determination of a line by two surfaces, we will take a straight line parallel to the axis of z.

$x = a$ and $y = b$ are the equations of two planes parallel to the planes of yz, zx, which intersect in a straight line parallel to Oz.

CHAPTER III.

PROJECTIONS OF LINES AND AREAS. DIRECTION-COSINES, AND DIRECTION-RATIOS.

20. Def. The *Geometrical projection* of a straight line of limited length upon any other straight line given in position, is the distance intercepted between the feet of the perpendiculars let fall from the extremities of the limited line upon the straight line on which it is to be projected.

21. *The Geometrical projection of a straight line of limited length on a given straight line is equal to the given length multiplied by the cosine of the acute angles contained between the lines.*

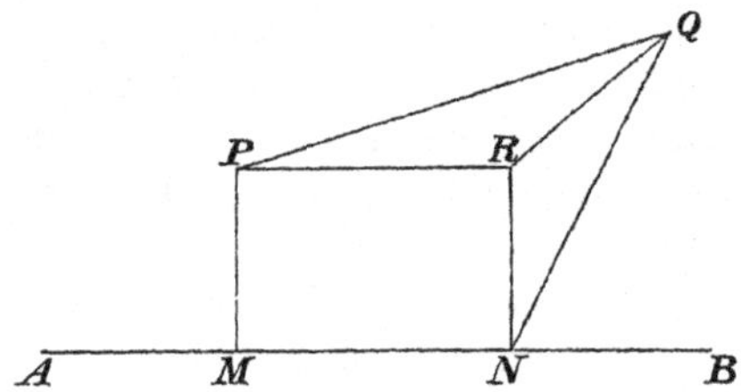

Let PQ be the line of limited length, AB the indefinite line upon which it is to be projected.

Let QRN be a plane through Q perpendicular to AB meeting it in N, PR parallel to AB meeting QRN in R.

Therefore PR being parallel to AB is perpendicular to the plane QRN, and therefore to RN and QR, and QN is perpendicular to AB; hence, if PM be drawn perpendicular to AB, MN is the projection of PQ, and QPR is the acute angle contained between PQ and AB, and since $PRNM$ is a rectangle,

$$MN = PR = PQ \cos QPR.$$

If PQ produced intersects AB, the proposition is obviously true.

22. Def. The *algebraical projection* of a line PQ upon an indefinite line AB given in position is the projection estimated in a given direction, as AB.

If α be the angle through which PQ may be supposed to have revolved from PR, drawn in the positive direction AB, the algebraical projection of $PQ = PQ \cos \alpha$.

If N lies in the opposite direction with reference to M, α is obtuse, and $PQ \cos \alpha$ is negative.

The algebraical projection of a limited straight line, upon a line given in position, measures the distance traversed in the direction of the latter line in passing from one extremity of the former to the other.

This consideration shews, that if all the sides of a closed polygon, taken in order, be projected on any straight line given in position, the sum of the algebraical projections of these sides is zero; since, in passing round the perimeter of the polygon from any point, the whole distance advanced in any direction is zero.

Hence, the algebraical projection of any side AB of a closed polygon, is the sum of the algebraical projections of the remaining sides commencing from A and terminating in B.

Note. In future, when the term projection is used, the algebraical projection is to be understood.

23. Let PQ be any line, PM, MN, NQ three straight lines drawn in any given directions so as to terminate in Q, and l, m, n the cosines of the angles which PQ makes with these directions.

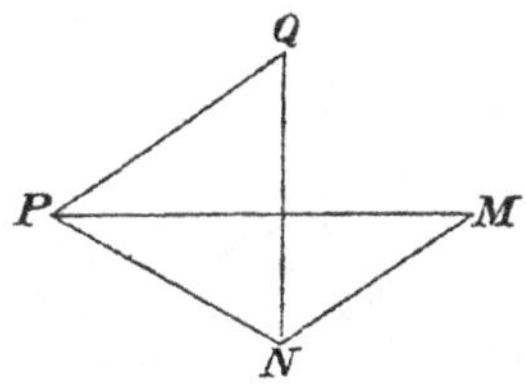

Then PQ will be the sum of the projections of PM, MN, and NQ on PQ; $\therefore\ PQ = l.PM + m.MN + n.NQ$.

Direction-cosines.

24. The direction of a straight line in space is determined when the angles which it makes with the co-ordinate axes are known.

Def. If the co-ordinate axes be perpendicular, the cosines of the inclinations to the three axes are called *direction-cosines*.

25. *To find the relation between the direction-cosines of a straight line.*

If l, m, n be the direction-cosines of PQ, and PM, MN, NQ be parallel to the co-ordinate axes,

$$PM = PQ.l, \quad MN = PQ.m, \quad NQ = PQ.n.$$

Join PN, then, since QN is perpendicular to NM, MP, and therefore to the plane PMN, PNQ is a right angle;

$$\therefore\ PQ^2 = PN^2 + NQ^2 = PM^2 + MN^2 + NQ^2;$$

$$\therefore\ 1 = l^2 + m^2 + n^2,$$

which is the relation required. Hence the three angles of inclination cannot all be assumed arbitrarily.

26. *To find the angle between two straight lines in terms of their direction-cosines.*

Let PQ, $P'Q'$ be two straight lines whose direction-cosines are (l, m, n) and (l', m', n') respectively.

Let PM, MN, NQ be drawn parallel to the axes, connecting any two points P, Q, and PP', QQ' perpendicular to $P'Q'$, and let θ be the angle between PQ and $P'Q'$.

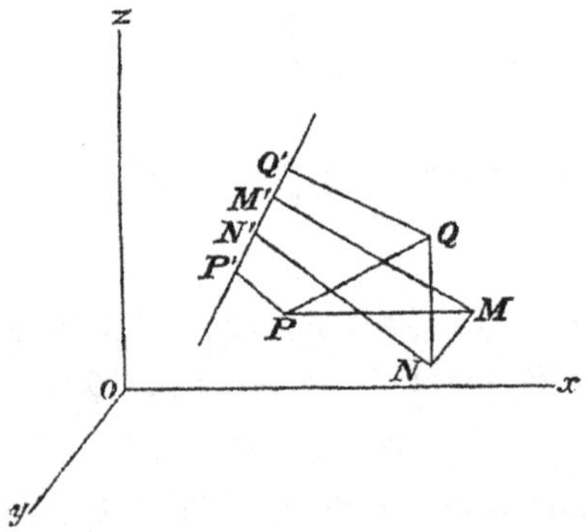

Then $P'Q'$, the projection of PQ on $P'Q'$, will be equal to the sum of the projections of PM, MN, NQ on $P'Q'$, namely, $P'M'$, $M'N'$, $N'Q'$;

$$\therefore\ PQ\cos\theta = PM.l' + MN.m' + NQ.n',$$

and since $PM = PQ.l,\ MN = PQ.m,\ NQ = PQ.n$;

$$\therefore\ \cos\theta = ll' + mm' + nn'.$$

Hence $\sin^2\theta = (l^2 + m^2 + n^2)(l'^2 + m'^2 + n'^2) - (ll' + mm' + nn')^2$

$$= (mn' - m'n)^2 + (nl' - n'l)^2 + (lm' - l'm)^2.$$

Direction-ratios.

27. Def. If the co-ordinate axes be not perpendicular to each other, the direction of a line PQ is fully determined, if the ratios of PM, MN, NQ to PQ are given, PM, MN, NQ being parallel to the axes. These ratios are called *direction-ratios.*

28. *To find the relation between the direction-ratios of a straight line.*

In the last figure, let the angles yOz, zOx, xOy be λ, μ, ν, and let α, β, γ be the angles between PQ and the axes, l, m, n the direction-ratios of PQ.

Projecting the line PQ and the bent line $PMNQ$ terminated in the same points on Ox,

$$PQ\cos\alpha = PM + MN\cos\nu + NQ\cos\mu;$$

$$\therefore\ \cos\alpha = l + m\cos\nu + n\cos\mu,$$

similarly, $\cos\beta = l\cos\nu + m + n\cos\lambda$,

and $\cos\gamma = l\cos\mu + m\cos\lambda + n$.

Also projecting $PMNQ$ on PQ,

$$PM\cos\alpha + MN\cos\beta + NQ\cos\gamma = PQ;$$

$$\therefore\ l\cos\alpha + m\cos\beta + n\cos\gamma = 1;$$

$$\therefore\ 1 = l^2 + m^2 + n^2 + 2mn\cos\lambda + 2nl\cos\mu + 2lm\cos\nu,$$

which is the relation required.

29. The following is the relation which always exists between the cosines of the angles which a straight line makes with three oblique axes.

$$\cos^2\alpha\sin^2\lambda + \cos^2\beta\sin^2\mu + \cos^2\gamma\sin^2\nu$$
$$+ 2\cos\beta\cos\gamma(\cos\mu\cos\nu - \cos\lambda) + 2\cos\gamma\cos\alpha(\cos\nu\cos\lambda - \cos\mu)$$
$$+ 2\cos\alpha\cos\beta(\cos\lambda\cos\mu - \cos\nu)$$
$$= 1 - \cos^2\lambda - \cos^2\mu - \cos^2\nu + 2\cos\lambda\cos\mu\cos\nu.$$

This may be deduced from the equations of the last article.

30. *To find the angle between two straight lines whose direction-ratios are given.*

Let (l, m, n) and (l', m', n') be the direction-ratios of two straight lines PQ, $P'Q'$, θ the angle between them, and $(\alpha', \beta', \gamma')$ the angles of inclination of $P'Q'$ to the co-ordinate axes.

Take r a limited length measured on PQ; then rl, rm, rn will be the intercepts on the axes by planes drawn through P, Q parallel to the co-ordinate planes, and $r \cos\theta$ the sum of the projections of rl, rm, rn on $P'Q'$

$$= rl\cos\alpha' + rm\cos\beta' + rn\cos\gamma'.$$

Also $\cos\alpha' = l' + m'\cos\nu + n'\cos\mu$, (Art. 28),

$$\cos\beta' = l'\cos\nu + m' + n'\cos\lambda,$$

$$\cos\gamma' = l'\cos\mu + m'\cos\lambda + n';$$

$$\therefore \cos\theta = ll' + mm' + nn'$$
$$+ (mn' + m'n)\cos\lambda + (nl' + n'l)\cos\mu$$
$$+ (lm' + l'm)\cos\nu,$$

which is the cosine of the angle required.

Projection of a Line on a Plane.

31. Def. The orthogonal projection of a line of limited length on a plane is the line intercepted between the perpendiculars drawn from the extremities of the limited line upon the plane.

32. *The orthogonal projection of a line upon a plane, is the length of the line multiplied by the cosine of the angle of inclination of the line to the plane.*

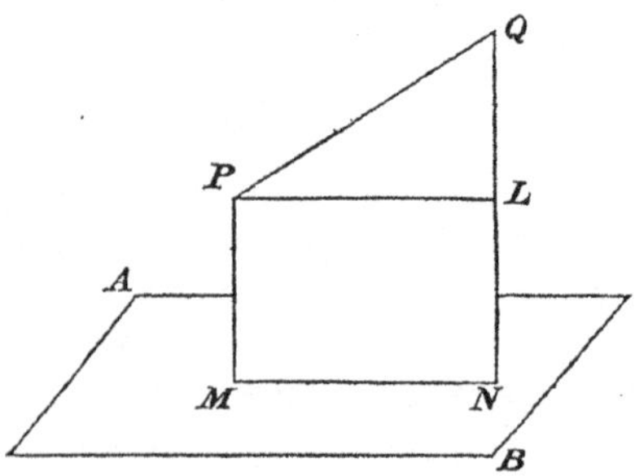

Let PQ be the given line, AB the plane, PM, QN perpendiculars upon the plane.

Since PM, QN are perpendicular to the plane AB, PM is parallel to QN, and the plane $MPQN$ is perpendicular to the plane AB; join MN, and draw PL parallel to MN;

$$\therefore \angle PLQ = \angle MNQ = \text{a right angle};$$

$$\therefore MN = PL = PQ \cos QPL,$$

and MN is the projection of PQ on AB,

$\angle QPL$ = the inclination of PQ to the plane,

whence the proposition.

Projection of a Plane Area upon a Plane.

33. DEF. The orthogonal projection of a closed plane area upon a fixed plane, is the area included within the line which is the locus of the feet of perpendiculars drawn from every point in the boundary of the plane area.

If a series of planes be taken forming a closed polyhedron, the algebraical projections of the faces upon any plane are their areas multiplied by the cosines of the angles which their normals, *drawn inwards*, make with the normal to the plane.

34. *The orthogonal projection of any plane area on a given plane is the area multiplied by the cosine of the inclination of the plane of the area to the given plane.*

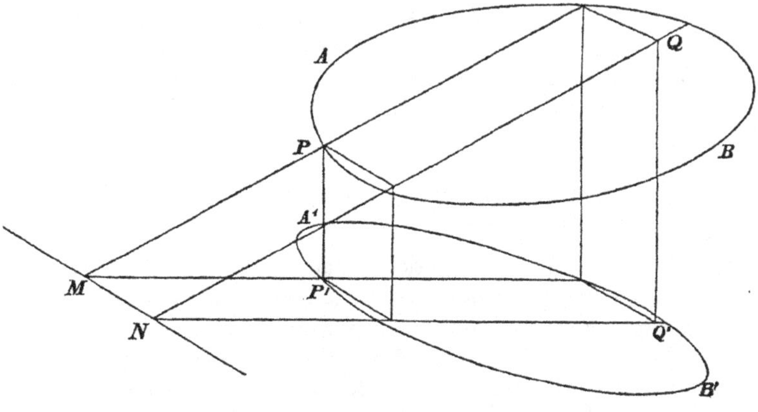

Let APB be any closed curve described upon a given plane, and $A'P'B'$ the *orthogonal projection* upon any other fixed plane, which is the locus of the feet of the perpendiculars drawn to the second plane from every point of the curve APB.

The areas APB, $A'P'B'$ may have inscribed in them any number of parallelograms, such as PQ, $P'Q'$, whose sides are in planes PMP', QNQ' drawn perpendicular to the line of intersection of the given planes, and parallel to that line, and these parallelograms are in the ratio of 1 : cosine of the inclination of the planes; therefore the sums of the parallelograms are in the same ratio.

Hence, proceeding to the limit when the breadths of these parallelograms are indefinitely diminished, the area of the projection of $APB =$ area of $APB \times$ cosine of the inclination of the planes.

35. *If the faces of any closed polyhedron be projected on any plane, the sum of the algebraical projections of the faces upon any fixed plane is zero.*

Let the polyhedron be cut by a prism having its faces perpendicular to the given plane, and let the base upon the given plane be α, and the areas of two consecutive sections of the faces of the polyhedron made by this prism be β, γ; if α be the algebraical projection of β, $-\alpha$ will be the projection of γ, the sum of which is zero; and this will be the case whatever be the number of pairs of such sections, in case the polyhedron should be re-entering, and also whatever be the form of the prism. Hence, if prisms be taken in this way so as to include the whole volume of the polyhedron, the truth of the proposition is evident.

36. *To find the area of any plane surface in terms of the areas of the projections upon any rectangular co-ordinate planes.*

Let l, m, n be the direction-cosines of a normal to the plane on which the given area A lies, A_x, A_y, A_z the areas of the projections upon the co-ordinate planes of yz, zx, xy.

Then, since l is the cosine of the angle between Ox and the normal to the plane, which is the same as the angle between the plane of A and the plane of yz, $A_x = Al$,

$$\text{and similarly, } A_y = Am, \text{ and } A_z = An;$$

$$\therefore\ A^2 = A^2(l^2 + m^2 + n^2) = A_x^2 + A_y^2 + A_z^2.$$

37. *To find the plane upon which the sum of the projections of any number of given plane areas is a minimum.*

Let A, A', A'' ... be any number of plane areas, (l, m, n), (l', m', n') ... the direction-cosines of the normals to their planes, (λ, μ, ν) those of the normal to a plane upon which they are projected; and let (A_x, A_y, A_z), (A'_x, A'_y, A'_z), ... be the areas of the projections of the given areas upon the co-ordinate planes.

Then since $l\lambda + m\mu + n\nu$ is the cosine of the angle between the plane of A, and the plane upon which it is projected, the projection of A is

$$A(l\lambda + m\mu + n\nu) = A_x\lambda + A_y\mu + A_z\nu;$$

therefore the sum of the projections of all the areas upon the plane (λ, μ, ν) is $\lambda\Sigma(A_x) + \mu\Sigma(A_y) + \nu\Sigma(A_z)$ which is to be a minimum by the variation of λ, μ, ν, subject to the condition $\lambda^2 + \mu^2 + \nu^2 = 1$;

$$\therefore \; \Sigma(A_x)\,d\lambda + \Sigma(A_y)\,d\mu + \Sigma(A_z)\,d\nu = 0,$$

and

$$\lambda d\lambda + \mu d\mu + \nu d\nu = 0,$$

must be true for an infinite number of values of $d\lambda : d\mu : d\nu$;

$$\therefore \; \frac{\lambda}{\Sigma(A_x)} = \frac{\mu}{\Sigma(A_y)} = \frac{\nu}{\Sigma(A_z)} = \frac{1}{\sqrt{[\{\Sigma(A_x)\}^2 + \{\Sigma(A_y)\}^2 + \{\Sigma(A_z)\}^2]}},$$

which determine the direction of the plane of projection in order that the sum of the projections of the areas may be a minimum.

IV.

(1) The sum of the three acute angles which a straight line forms with the co-ordinate axes is less than 180°.

(2) The sum of the acute angles which any straight line makes with rectangular co-ordinate axes can never be less than $\frac{3}{2}\sec^{-1}(-3)$.

(3) Two straight lines are drawn in the planes of xy and yz, making angles α, γ with the axes of x, z respectively; the direction-cosines of the straight line perpendicular to the two are proportional to $\tan\alpha$, -1, $\tan\gamma$.

(4) If two straight lines be inclined at an angle of 60°, and their direction-cosines be (l, m, n), (l', m', n'), there is a straight line whose direction-cosines are

$$l-l', \quad m-m', \quad n-n',$$

and this straight line is inclined at angles of 60° and 120° to the former straight lines.

(5) The direction-cosines of a straight line perpendicular to the two whose direction-cosines are proportional to l, m, n and $m+n$, $n+l$, $l+m$, are proportional to $m-n$, $n-l$, $l-m$.

(6) If the angles which a straight line forms with the co-ordinate planes be in an arithmetical progression whose difference is 45°, the line must lie in one of the co-ordinate planes.

If it form angles α, 2α, 3α, it must lie in one of the co-ordinate planes.

(7) Shew *a priori* that the rational equation connecting the direction-cosines of a straight line can only involve even powers of those quantities.

(8) The straight lines whose direction-cosines are given by the equations

$$al + bm + cn = 0,$$

$$\alpha l^2 + \beta m^2 + \gamma n^2 = 0,$$

will be perpendicular, if

$$a^2(\beta+\gamma) + b^2(\gamma+\alpha) + c^2(\alpha+\beta) = 0,$$

and parallel, if

$$\frac{a^2}{\alpha} + \frac{b^2}{\beta} + \frac{c^2}{\gamma} = 0.$$

(9) The straight lines whose direction-cosines are given by the equations

$$al + bm + cn = 0,$$

$$\frac{\alpha}{l} + \frac{\beta}{m} + \frac{\gamma}{n} = 0,$$

will be perpendicular, if

$$\frac{\alpha}{a} + \frac{\beta}{b} + \frac{\gamma}{c} = 0,$$

and parallel, if $\quad \sqrt{(a\alpha)} \pm \sqrt{(b\beta)} \pm \sqrt{(c\gamma)} = 0.$

(10) The direction-cosines of a line making equal angles with three straight lines, whose direction-cosines are

$$(l, m, n), \ (l', m', n'), \ (l'', m'', n''),$$

are proportional to

$$m(n'-n'')+m'(n''-n)+m''(n-n'),$$
$$n(l'-l'')+n'(l''-l)\;+n''(l-l'),$$
$$l(m'-m'')+l'(m''-m)+l''(m-m').$$

If the given lines are mutually at right angles, the direction-cosines are

$$\frac{l+l'+l''}{\surd 3},\quad \frac{m+m'+m''}{\surd 3},\quad \frac{n+n'+n''}{\surd 3}.$$

(11) Find the direction-cosines of the two straight lines which are equally inclined to the axis of z, and are perpendicular to each other and to the line which makes equal angles with the co-ordinate axes.

(12) If A, B, C, D be four points in a plane, A', B', C', D' their projections on any other plane, the volumes of the tetrahedrons $ABCD'$, $A'B'C'D$ are equal.

(13) Shew, by projecting upon the base, that the area of the surface of a right cone is πal, a being the radius of the base, and l the length of a slant side.

(14) The angle between two faces of a regular tetrahedron is

$$\sec^{-1} 3.$$

(15) The straight line which makes equal angles with three straight lines, also makes equal angles with three planes each containing two of the straight lines.

(16) If the edges of a tetrahedron $ABCD$ which terminate in D be a, b, c; and the respectively opposite edges a', b', c'; shew by projecting AB, BC, CD on AD, that the angle between a and a', is

$$\cos^{-1}\frac{(b^2+b'^2)\sim(c^2+c'^2)}{2aa'}.$$

Hence shew that, if two pairs of opposite edges be respectively at right angles, the third pair will also be at right angles to each other.

CHAPTER IV.

DISTANCES OF POINTS. EQUATIONS OF A STRAIGHT LINE.

Distance between two points.

38. *To find the distance between two points whose co-ordinates are given, referred to rectangular axes.*

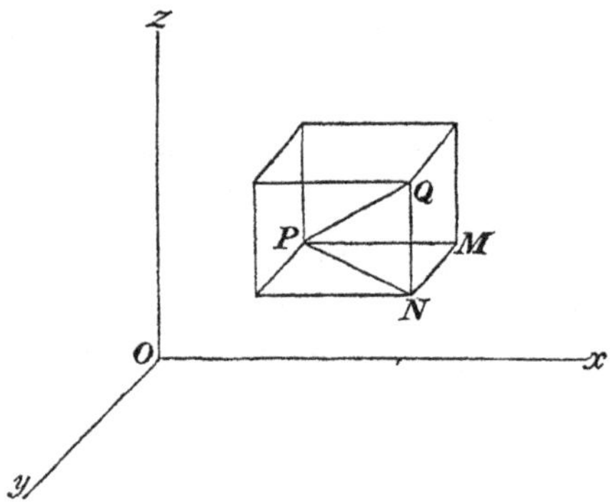

Let (x, y, z), (x', y', z') be two points P, Q whose co-ordinates are given referred to a rectangular system; and let a parallelopiped be constructed whose diagonal is PQ, and whose edges PM, MN, NQ are parallel to the co-ordinate axes Ox, Oy, Oz; and join PN.

Then, since QN is perpendicular to the plane PMN, and therefore to PN,

$$PQ^2 = PN^2 + QN^2,$$

$$\text{but } PN^2 = PM^2 + MN^2;$$

$$\therefore\ PQ^2 = PM^2 + MN^2 + NQ^2.$$

PM is the difference of the algebraical distances of Q and P from the plane yOz, and similarly for MN, NQ:

$$\therefore\ PQ^2 = (x' - x)^2 + (y' - y)^2 + (z' - z)^2.$$

If α, β, γ be the inclinations of PQ to the axes of co-ordinates,

$$x' - x = PQ \cos \alpha,$$
$$y' - y = PQ \cos \beta,$$
$$z' - z = PQ \cos \gamma;$$
$$\therefore\ 1 = \cos^2 \alpha + \cos^2 \beta + \cos^2 \gamma.$$

The double sign, which appears in the value of PQ, may be interpreted in a manner similar to that adopted in the case of the radius vector in polar co-ordinates in Plane Geometry.

If the angles α, β, γ define the direction of measurement of the distance PQ of Q from P, the opposite direction is defined by $\pi+\alpha$, $\pi+\beta$, $\pi+\gamma$, and therefore these angles with an algebraical distance $-PQ$, equally determine the position of the point Q with reference to P.

The distance of the point (x', y', z') from the origin is

$$\sqrt{(x'^2 + y'^2 + z'^2)}.$$

39. *To find the distance between two points referred to oblique axes.*

Let λ, μ, ν be the angles between the axes; and (x, y, z), (x', y', z') two points P and Q.

Let a parallelopiped be constructed whose diagonal is PQ, and edges PM, MN, NQ are parallel to the axes Ox, Oy, Oz.

Now, the projections on PM of the line PQ, and of the bent line $PMNQ$ terminated in the same points, are equal.

Therefore if α, β, γ be the angles which PQ makes with the axes,

$$\left.\begin{aligned} PQ \cos \alpha &= PM + MN \cos \nu + NQ \cos \mu, \\ \text{similarly,}\quad PQ \cos \beta &= MN + NQ \cos \lambda + PM \cos \nu, \\ \text{and}\quad PQ \cos \gamma &= NQ + PM \cos \mu + MN \cos \lambda. \end{aligned}\right\} \qquad (1),$$

Also PQ is the projection of $PMNQ$ on PQ,

$$\therefore\ PQ = PM \cos \alpha + MN \cos \beta + NQ \cos \gamma \qquad (2).$$

Therefore multiplying the equations (1) by PM, MN, NQ we have by (2),

$$PQ^2 = PM^2 + MN^2 + NQ^2 + 2\,MN.NQ \cos \lambda$$
$$+ 2\,NQ.PM \cos \mu + 2\,PM.MN \cos \nu,$$

and PM is the difference of the algebraical distances of Q and P from yOz, and therefore $= x' - x$, and similarly $MN = y' - y$, and $NQ = z' - z$;

$$\therefore PQ^2 = (x'-x)^2 + (y'-y)^2 + (z'-z)^2 + 2(y'-y)(z'-z)\cos\lambda + 2(z'-z)(x'-x)\cos\mu + 2(x'-x)(y'-y)\cos\nu,$$

whence PQ is determined as required.

40. If l, m, n be the direction-ratios of PQ,

$$PM = l.PQ, \quad MN = m.PQ, \quad NQ = n.PQ;$$

$$\therefore 1 = l^2 + m^2 + n^2 + 2mn\cos\lambda + 2nl\cos\mu + 2lm\cos\nu,$$

which is the equation connecting the direction-ratios of any line referred to oblique axes, obtained above, Art. 28.

41. *To find the distance of two points whose polar co-ordinates are given.*

Let (r, θ, ϕ) and (r', θ', ϕ') be the given points P and Q.

Join OP, OQ, QP, and let a spherical surface, whose centre is O and radius unity, intersect OP, OQ and Oz in p, q and r.

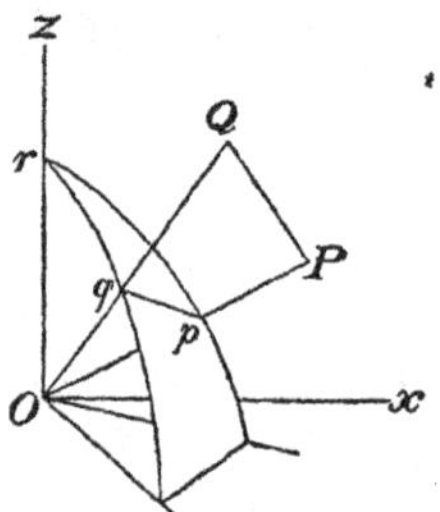

Then, $rp = \theta$, $rq = \theta'$, and $\angle qrp = \phi' - \phi$.

$$PQ^2 = OP^2 + OQ^2 - 2OP.OQ\cos pq$$
$$= r^2 + r'^2 - 2rr'\cos pq.$$

But $\cos pq = \cos pr \cos pr + \sin pr \sin qr \cos prq$

$$= \cos\theta\cos\theta' + \sin\theta\sin\theta'\cos(\phi' - \phi);$$

$$\therefore PQ^2 = r^2 + r'^2 + 2rr'\{\cos\theta\cos\theta' + \sin\theta\sin\theta'\cos(\phi' - \phi)\},$$

whence the distance PQ is determined in terms of the polar co-ordinates of P and Q.

42. The distance may be determined without Spherical Trigonometry as follows.

Draw PM, QN perpendicular to the plane of xy, join MN, OM, and ON, and draw PR perpendicular to QN;

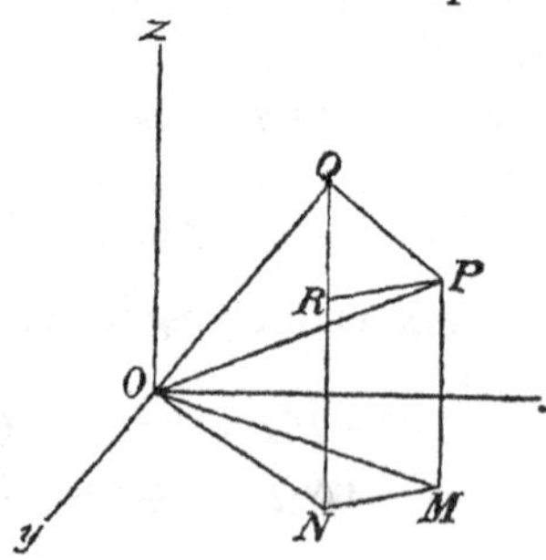

$$\therefore PQ^2 = QR^2 + PR^2$$
$$= QR^2 + MN^2,$$
$$QR = r'\cos\theta' - r\cos\theta,$$

and $MN^2 = OM^2 + ON^2 - 2OM.ON\cos MON$

$$= r^2\sin^2\theta + r'^2\sin^2\theta' - 2rr'\sin\theta\sin\theta'\cos(\phi'-\phi);$$

$$\therefore\ PQ^2 = r^2 + r'^2 - 2rr'\{\cos\theta\cos\theta' + \sin\theta\sin\theta'\cos(\phi'-\phi)\},$$

which gives the required distance.

The Straight Line.

43. The general equations of the straight line which will be employed are of two forms: one form is symmetrical and the equations are deduced from the consideration that the position of a straight line is completely determined, when one point in the line is given, and the direction in which the straight line is drawn. The symmetry of this form gives great advantages, and in all questions of a general nature, the general symmetrical equations will be almost exclusively employed. The other form is unsymmetrical, and the equations are deduced from the consideration that a straight line is the intersection of two planes, and is completely determined when the equations of the two planes are given. These equations in their simplest forms are the equations of planes parallel to two of the co-ordinate axes, and are the same as the equations of the projections of the straight line parallel to these axes upon two of the co-ordinate planes. It will be seen that in particular cases it is advantageous to use the unsymmetrical form of the equations.

44. *To find the symmetrical equations of a straight line.*

Let A be a fixed point (α, β, γ) of a straight line, P any other point (x, y, z), l, m, n, the direction-cosines of AP; and let $AP = r$.

Then the projection of PQ on the axis of x is $x - \alpha$, and it is also lr;

hence, $\dfrac{x-\alpha}{l} = r$, and similarly, $\dfrac{y-\beta}{m} = r$, $\dfrac{z-\gamma}{n} = r$.

The equations of the straight line are then

$$\frac{x-\alpha}{l} = \frac{y-\beta}{m} = \frac{z-\gamma}{n},$$

$$\text{or}\quad \frac{x-\alpha}{L} = \frac{y-\beta}{M} = \frac{z-\gamma}{N},$$

if L, M, N be any quantities proportional respectively to l, m, n.

It should be carefully remembered that, when the former equations are used, each member of the equations is equal to the distance r between the two points (x, y, z) and (α, β, γ). The equations of a straight line will be of the same form if the axes be oblique, the same interpretation being given to r, and l, m, n being the direction-ratios. The projections employed in the above proof will then be the intercepts on the axes made by planes parallel to the co-ordinate planes.

45. *To find the non-symmetrical equations of a straight line.*

If a straight line PQ be projected by straight lines parallel to the axes Oy, Ox, whether rectangular or oblique, on the two co-ordinate planes xz, yz, each projection will be a straight line as pq, $p'q'$, in those planes respectively.

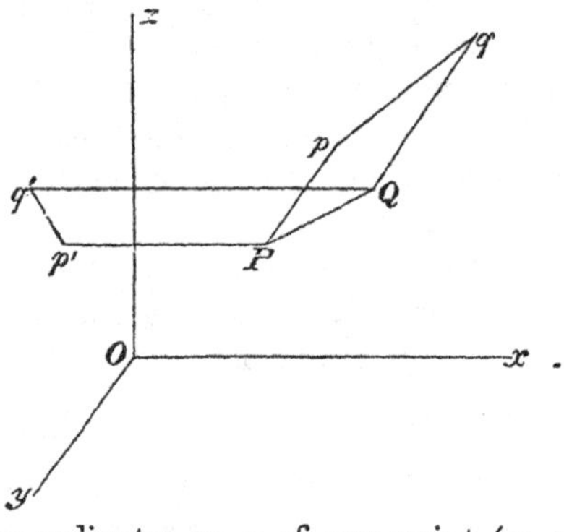

Hence the co-ordinates x, z of any point (x, y, z), in PQ being the same as those of the projection of the point in pq, satisfy an equation of the form

$$x = mz + a,$$

and the co-ordinates (y, z), similarly an equation of the form

$$y = nz + b,$$

and consequently the equations of the line may be written

$$x = mz + a, \quad y = nz + b.$$

46. *On the number of constants employed in the equations of a straight line.*

It may be noticed that the latter system of equations involves only four constants, whilst the symmetrical system involves six.

Of the three l, m, n, however, we know that they are connected by the relation $l^2+m^2+n^2=1$, (Art. 25), which renders them equivalent to only two independent constants; and if we take L, M, N, since these are only required to be proportional to l, m, n, one of them may be assumed arbitrarily, and they are still equivalent to two constants only.

Also, of the three α, β, γ, one may be assumed at pleasure; for, since the straight line cannot be parallel to all the co-ordinate planes, let it be not parallel to that of yz; then, at whatever distance α from yz we take a parallel plane, the straight line will meet this plane, and we may take the point where they meet for the point α, β, γ; that is, we may give to α any value we please, and the three α, β, γ are consequently equivalent only to two independent constants.

47. *To find the equations of a straight line parallel to a co-ordinate plane.*

If a straight line be parallel to a co-ordinate plane, as that of yz, every point in it is at a constant distance from this plane, and we have the equation $x=a$, therefore its equations will be of the form

$$x=a, \quad y=nz+b.$$

Taking the symmetrical form, since the line will be perpendicular to the axis of x, $l=0$, and therefore $L=0$, and the equations of the line assume the form

$$\frac{x-\alpha}{0}=\frac{y-\beta}{m}=\frac{z-\gamma}{n}=r,$$

$$\text{or,} \quad \frac{x-\alpha}{0}=\frac{y-\beta}{M}=\frac{z-\gamma}{N},$$

which form implies that $x=\alpha$ for every point in the line at a finite distance, since the members are not infinite for such values.

48. *To find the equations of a straight line parallel to one of the co-ordinate axes.*

If the straight line be parallel to one of the co-ordinate axes, it will be parallel to the two co-ordinate planes passing through that axis, and consequently any point in it will be at an invariable distance from each of these planes. Thus, if a straight line be parallel to the axis of x, the distances of any point in it from the planes xz, yz, will be constant, a fact expressed by the equations,

$$y = b, \quad z = c,$$

which will therefore be the equations of the line.

As before, the symmetrical form is

$$\frac{x-\alpha}{L} = \frac{y-\beta}{0} = \frac{z-\gamma}{0}.$$

49. *To find the angle between two straight lines whose equations are given.*

If the equations of a straight line be given in the form

$$\frac{x-\alpha}{L} = \frac{y-\beta}{M} = \frac{z-\gamma}{N};$$

then, if l, m, n be its direction-cosines,

$$\frac{l}{L} = \frac{m}{M} = \frac{n}{N} = \frac{\pm\sqrt{(l^2+m^2+n^2)}}{\sqrt{(L^2+M^2+N^2)}} = \frac{\pm 1}{\sqrt{(L^2+M^2+N^2)}};$$

or, the direction-cosines are

$$\frac{\pm L}{\sqrt{(L^2+M^2+N^2)}}, \quad \frac{\pm M}{\sqrt{(L^2+M^2+N^2)}}, \quad \frac{\pm N}{\sqrt{(L^2+M^2+N^2)}} \qquad (1).$$

If the equations be given in the form

$$x = mz + a, \quad y = nz + b,$$

since these may be written,

$$\frac{x-a}{m} = \frac{y-b}{n} = z,$$

the direction-cosines of the line are

$$\frac{\pm m}{\sqrt{(m^2+n^2+1)}}, \quad \frac{\pm n}{\sqrt{(m^2+n^2+1)}}, \quad \frac{\pm 1}{\sqrt{(m^2+n^2+1)}} \qquad (2).$$

In (1) and (2), the ambiguities have the same sign.

Hence, if the equations of two straight lines be

$$\frac{x-\alpha}{L}=\frac{y-\beta}{M}=\frac{z-\gamma}{N},$$

$$\frac{x-\alpha'}{L'}=\frac{y-\beta'}{M'}=\frac{z-\gamma'}{N'},$$

the angle between them is

$$\cos^{-1}\frac{LL'+MM'+NN'}{\sqrt{(L^2+M^2+N^2)}\sqrt{(L'^2+M'^2+N'^2)}},\ \text{(Art. 26)},$$

and if the equations be

$$x = mz + a,\quad y = nz + b,$$
$$x = m'z + a',\quad y = n'z + b',$$

the angle between them is

$$\cos^{-1}\frac{mm'+nn'+1}{\sqrt{(m^2+n^2+1)}\sqrt{(m'^2+n'^2+1)}}.$$

50. *To find the conditions that two straight lines whose equations are given, may be parallel.*

Hence, if the two straight lines

$$\frac{x-\alpha}{L}=\frac{y-\beta}{M}=\frac{z-\gamma}{N},$$

$$\frac{x-\alpha'}{L'}=\frac{y-\beta'}{M'}=\frac{z-\gamma'}{N'}$$

be parallel, they will have the same direction-cosines, and since L, M, N and also L', M', N' are respectively proportional to these direction-cosines,

$$\frac{L}{L'}=\frac{M}{M'}=\frac{N}{N'}$$

are the conditions of parallelism.

These conditions may be derived from the general value of the cosine of the angle between them, which will then be unity.

$$\text{Hence,}\quad 1=\frac{LL'+MM'+NN'}{\sqrt{(L^2+M^2+N^2)}\sqrt{(L'^2+M'^2+N'^2)}};$$

$$\text{or,}\quad (L^2+M^2+N^2)(L'^2+M'^2+N'^2)-(LL'+MM'+NN')^2=0,$$

$$\text{or,}\quad (LM'-L'M)^2+(MN'-M'N)^2+(NL'-N'L)^2=0,$$

which is equivalent to the conditions

$$\frac{L}{L'}=\frac{M}{M'}=\frac{N}{N'}.$$

Similarly, if the straight lines

$$x=mz+a,\quad y=nz+b,$$
$$x=m'z+a',\quad y=n'z+b',$$

be parallel,

$$m=m',\quad n=n',$$

which results follow from the consideration that if the straight lines be parallel, their projections will also be parallel.

51. *To find the condition that two straight lines whose equations are given, may be perpendicular.*

If the straight lines be perpendicular, the cosine of the angle between them is 0, and the condition in order to this, is

$$LL'+MM'+NN'=0,\quad \text{or,}\quad mm'+nn'+1=0,$$

according to the systems of equations given.

52. *To find the condition that two straight lines whose equations are given, may intersect.*

Let the equations of the two straight lines be

$$\frac{x-\alpha}{L}=\frac{y-\beta}{M}=\frac{z-\gamma}{N}, \qquad (1)$$

$$\frac{x-\alpha'}{L'}=\frac{y-\beta'}{M'}=\frac{z-\gamma'}{N'}, \qquad (2)$$

and let each member of (1) be equal to k, and of (2) equal to k'.

Then, if the lines intersect, equations (1) and (2) must be simultaneously satisfied by the co-ordinates of the point in which they intersect.

Hence,

$$\alpha-\alpha'=L'k'-Lk,$$
$$\beta-\beta'=M'k'-Mk,$$
$$\gamma-\gamma'=N'k'-Nk,$$

and eliminating k and k', we obtain the required condition,

$$(\alpha-\alpha')(MN'-M'N)+(\beta-\beta')(NL'-N'L)$$
$$+(\gamma-\gamma')(LM'-L'M)=0.$$

With the equations

$$x = mz + a, \quad y = nz + b,$$
$$x = m'z + a', \quad y = n'z + b',$$

the condition is immediately found to be

$$\frac{a - a'}{m - m'} = \frac{b - b'}{n - n'},$$

by eliminating x, y, and z.

Straight Line under given Conditions.

53. *To find the equations of a straight line passing through a given point.*

If (α, β, γ) be the given point, we have already seen that the equations

$$\frac{x - \alpha}{L} = \frac{y - \beta}{M} = \frac{z - \gamma}{N},$$

represent a straight line passing through that point.

54. *To find the equations of a straight line passing through two given points.*

If the straight line pass through the points (α, β, γ) and $(\alpha', \beta', \gamma')$, we shall have

$$\frac{\alpha' - \alpha}{L} = \frac{\beta' - \beta}{M} = \frac{\gamma' - \gamma}{N},$$

and the equations of the line become

$$\frac{x - \alpha}{\alpha' - \alpha} = \frac{y - \beta}{\beta' - \beta} = \frac{z - \gamma}{\gamma' - \gamma}.$$

55. *To find the equations of a straight line passing through a given point, and parallel to a given straight line.*

The equations of a straight line passing through a point (α, β, γ) are

$$\frac{x - \alpha}{L} = \frac{y - \beta}{M} = \frac{z - \gamma}{N},$$

and if this be parallel to a straight line whose direction-cosines are l, m, n,

$$\frac{L}{l} = \frac{M}{m} = \frac{N}{n}, \text{ (Art. 52)},$$

or, the required equations are

$$\frac{x-\alpha}{l}=\frac{y-\beta}{m}=\frac{z-\gamma}{n}.$$

56. *To find the equations of a straight line passing through a given point, and perpendicular to and intersecting a given straight line.*

Let (α, β, γ) be the given point, and the equations of the given straight line be

$$\frac{x-\alpha'}{l}=\frac{y-\beta'}{m}=\frac{z-\gamma'}{n}.$$

Then, $$\frac{x-\alpha}{L}=\frac{y-\beta}{M}=\frac{z-\gamma}{N}$$

will be the required equations of the straight line, when the ratio $L : M : N$ are determined by the equations

$$Ll+Mm+Nn=0, \quad \text{(Art. 53)},$$

$$(\alpha-\alpha')(Mn-Nm)+(\beta-\beta')(Nl-Ln)$$
$$+(\gamma-\gamma')(Lm-Ml)=0, \quad \text{(Art. 54)}.$$

57. *To find the equations of a straight line passing through a given point, parallel to a given plane, and intersecting a given straight line.*

Let (α, β, γ) be the given point, (l, m, n) the direction-cosines of a normal to the plane, which will therefore be perpendicular to the straight line whose equations are required, and let the equations of the given straight line be

$$\frac{x-\alpha'}{l'}=\frac{y-\beta'}{m'}=\frac{z-\gamma'}{n'}.$$

The required equations will then be

$$\frac{x-\alpha}{L}=\frac{y-\beta}{M}=\frac{z-\gamma}{N};$$

when $L : M : N$ are determined by the equations

$$Ll + Mm + Nn = 0,$$

$$(\alpha - \alpha')(Mn' - Nm') + (\beta - \beta')(Nl' - Ln') + (\gamma - \gamma')(Lm' - Ml') = 0.$$

58. *To find the distance from a given point to a given straight line.*

Let A be the given point $(x'y'z')$, B the point (α, β, γ) of the given straight line whose equations are

$$\frac{x-\alpha}{l} = \frac{y-\beta}{m} = \frac{z-\gamma}{n},$$

AP the perpendicular from A on the straight line; then the projections of BA on the axes of x, y, z are respectively

$$x' - \alpha, \quad y' - \beta, \quad z' - \gamma;$$

and the projections of these on the given straight line are

$$l(x' - \alpha), \quad m(y' - \beta), \quad n(z' - \gamma),$$

but the sum of these projections is the projection of BA on the straight line, or

$$BP = l(x' - \alpha) + m(y' - \beta) + n(z' - \gamma),$$

hence, $AP^2 = BA^2 - BP^2$

$$= (x'-\alpha)^2 + (y'-\beta)^2 + (z'-\gamma)^2 - \{l(x'-\alpha) + m(y'-\beta) + n(z'-\gamma)\}^2,$$

and the required distance is

$$\sqrt{[(x'-\alpha)^2 + (y'-\beta)^2 + (z'-\gamma)^2 - \{l(x'-\alpha) + m(y'-\beta) + n(z'-\gamma)\}^2]}.$$

If the equations of the line be

$$x = mz + a, \quad y = nz + b,$$

which are equivalent to

$$\frac{x-a}{m} = \frac{y-b}{n} = z,$$

the distance will be

$$\sqrt{\left[(x'-a)^2 + (y'-b)^2 + z'^2 - \frac{\{m(x'-a) + n(y'-b) + z'\}^2}{m^2 + n^2 + 1}\right]},$$

replacing α, β, γ by a, b, 0, and l, m, n by m, n, 1, in the expression already found.

59. *To shew that the shortest distance between two straight lines which do not intersect is perpendicular to each.*

Let AP, BQ be the two straight lines, and let a plane be drawn through BQ parallel to AP, and RB be the orthogonal

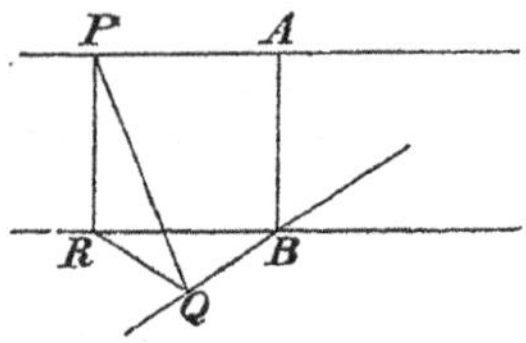

projection of AP upon this plane, B being the projection of A; therefore AB will be perpendicular to both straight lines, for it meets two parallel lines AP, BR, to one of which, BR, it is perpendicular, and it is also perpendicular to BQ.

Let P, Q be any points in AP, BQ, join PQ, draw PR perpendicular to BR, and join QR; then PQ is greater than PR, being opposite to the greater angle, and $PR = AB$; therefore AB is less than PQ, or the distance which is perpendicular to both straight lines is less than any other distance.

60. *To find the shortest distance between two straight lines whose equations are given.*

Let the equations of the two straight lines be

$$\frac{x-\alpha}{l} = \frac{y-\beta}{m} = \frac{z-\gamma}{n}, \text{ and } \frac{x-\alpha'}{l'} = \frac{y-\beta'}{m'} = \frac{z-\gamma'}{n'},$$

and let λ, μ, ν be the direction-cosines of the straight line perpendicular to each, then

$$l\lambda + m\mu + n\nu = 0,$$
$$l'\lambda + m'\mu + n'\nu = 0.$$

Hence,
$$\frac{\lambda}{mn' - m'n} = \frac{\mu}{nl' - n'l} = \frac{\nu}{lm' - l'm}$$
$$= \frac{\sqrt{(\lambda^2 + \mu^2 + \nu^2)}}{\sqrt{\{(mn' - m'n)^2 + (nl' - n'l)^2 + (lm' - l'm)^2\}}} = \frac{1}{\sin\theta},$$

θ being the angle between the lines, (Art. 26).

Now, if we suppose P, Q to be the points (α, β, γ), $(\alpha', \beta', \gamma')$, the projection of PQ on AB will be $\lambda(\alpha-\alpha')+\mu(\beta-\beta')+\nu(\gamma-\gamma')$, but this projection will be AB itself;

hence, $AB=\dfrac{(\alpha-\alpha')(mn'-m'n)+(\beta-\beta')(nl'-n'l)+(\gamma-\gamma')(lm'-l'm)}{\sin\theta}$.

The equations of the straight line on which the shortest distance lies, may be obtained in the following form,

$$l(x-\alpha)+m(y-\beta)+n(z-\gamma)=\frac{u+u'\cos\theta}{\sin^2\theta},$$

$$\text{and } l'(x-\alpha')+m'(y-\beta')+n'(z-\gamma')=\frac{u'+u\cos\theta}{\sin^2\theta};$$

$$\text{where } u=l(\alpha'-\alpha)+m(\beta'-\beta)+n(\gamma'-\gamma),$$

$$\text{and } u'=l'(\alpha-\alpha')+m'(\beta-\beta')+n'(\gamma-\gamma').$$

61. The simplest form in which the equations of two straight lines can be presented will be obtained by taking the middle point of the shortest distance between them for the origin, the line on which the shortest distance lies for one of the axes, the axis of z suppose, and the two planes equally inclined to the two straight lines as those of zx, zy. If 2α be the angle between the two straight lines, $2c$ the shortest distance between them, their equations will thus become

$$y=x\tan\alpha,\ z=c;\ \text{and } y=-x\tan\alpha,\ z=-c.$$

62. *To find the equations of a straight line referred to four-plane co-ordinates.*

If $(\alpha', \beta', \gamma', \delta')$ be four-plane co-ordinates of a fixed point in a straight line, $(\alpha, \beta, \gamma, \delta)$ those of any other point in it, ρ the distance of these points,

$$\text{then } \frac{\alpha-\alpha'}{\rho},\ \frac{\beta-\beta'}{\rho},\ \frac{\gamma-\gamma'}{\rho},\ \frac{\delta-\delta'}{\rho}$$

will be equal respectively to the cosines of the angles between the straight line and the normals to the fundamental planes, and if these be l, m, n, r, we shall have

$$\frac{\alpha-\alpha'}{l}=\frac{\beta-\beta'}{m}=\frac{\gamma-\gamma'}{n}=\frac{\delta-\delta'}{r}=\rho;$$

$$\text{or, } \frac{\alpha-\alpha'}{L}=\frac{\beta-\beta'}{M}=\frac{\gamma-\gamma'}{N}=\frac{\delta-\delta'}{R}, \qquad (1)$$

if L, M, N, R be respectively proportional to l, m, n, r.

It is obvious that, since the direction of the straight line, expressed in this system, involves the three ratios $L : M : N : R$, and in the former system depended only on two, some invariable relation must subsist between L, M, N, R. In fact, we have

$$A\alpha + B\beta + C\gamma + D\delta = 3V = A\alpha' + B\beta' + C\gamma' + D\delta',$$

$$A(\alpha - \alpha') + B(\beta - \beta') + C(\gamma - \gamma') + D(\delta - \delta') = 0;$$

$$\text{whence } AL + BM + CN + DR = 0.$$

If tetrahedral co-ordinates be employed, the relation will be

$$L + M + N + R = 0.$$

63. It should be noticed that two of the three equations (1) are sufficient to determine the straight line, the third being directly deducible. If, for example, we have

$$\frac{\alpha - \alpha'}{L} = \frac{\beta - \beta'}{M} = \frac{\gamma - \gamma'}{N},$$

each member is equal to

$$\frac{A(\alpha - \alpha') + B(\beta - \beta') + C(\gamma - \gamma')}{AL + BM + CN};$$

or to $\dfrac{-D(\delta - \delta')}{AL + BM + CN}$, which is the same as $\dfrac{\delta - \delta'}{R}$.

64. If the straight line pass through one of the angular points of the tetrahedron of reference, as the one opposite D, $(0, 0, 0, p_4)$ the equations become

$$\frac{\alpha}{L} = \frac{\beta}{M} = \frac{\gamma}{N} = \frac{\delta - p_4}{R}.$$

V.

(1) The straight line given by the equations

$$x + 2y + 3z = 0, \quad 3x + 2y + z = 0,$$

makes equal angles with the axes of x and z, and an angle $\sin^{-1}\dfrac{1}{\sqrt{3}}$ with the axis of y.

(2) The equations $\dfrac{x^3 + 1}{x + 1} = \dfrac{y^3 + 1}{y + 1} = \dfrac{z^3 + 1}{z + 1}$ denote thirteen straight lines. Shew that four are equally inclined to each other, and construct for the rest.

(3) Find the direction-cosines of the straight line determined by the equations

$$lx + my + nz = mx + ny + lz = nx + ly + mz.$$

(4) Find the equations of the straight line which passes through the origin and intersects at right angles the straight line whose equations are

$$(m+n)x + (n+l)y + (l+m)z = a,$$

$$(m-n)x + (n-l)y + (l-m)z = a;$$

and obtain the co-ordinates of the point of intersection.

(5) Find the equations of the straight line passing through the points (b, c, a) (c, a, b), and shew that it is perpendicular to the line passing through the origin and through the middle point of the line joining the two points, and also to each of the straight lines whose equations are

$$x = y = z, \quad \frac{x}{a} = \frac{y}{b} = \frac{z}{c}.$$

(6) Interpret the equation

$$(x^2 + y^2 + z^2)(l^2 + m^2 + n^2) = (lx + my + nz)^2,$$

and give a geometrical illustration.

(7) The straight lines determined by the equations

$$lx + my + nz = 0,$$

$$l(b-c)yz + m(c-a)zx + n(a-b)xy = 0,$$

are at right angles to each other.

(8) The straight lines given by the equations

$$lx + my + nz = 0, \quad \frac{a}{x} + \frac{b}{y} + \frac{c}{z} = 0,$$

will be at right angles, if $\frac{a}{l} + \frac{b}{m} + \frac{c}{n} = 0.$

(9) The equations of two straight lines are

$$\frac{x}{\sin\alpha} = \frac{y}{\cos\alpha} = \frac{z-c}{0},$$

$$\frac{x}{\sin\alpha} = \frac{y}{-\cos\alpha} = \frac{z+c}{0};$$

shew that the distance between two points on these straight lines whose distances from the axis of z are a, b respectively is

$$\sqrt{(4c^2 + a^2 + b^2 \mp 2ab\cos 2\alpha)}.$$

(10) Shew that the equations

$$\frac{\alpha + mz - ny}{l} = \frac{\beta + nx - lz}{m} = \frac{\gamma + ly - mx}{n}$$

are reducible to $\dfrac{x + n\beta - m\gamma}{l} = \dfrac{y + l\gamma - n\alpha}{m} = \dfrac{z + m\alpha - l\beta}{n}$,

l, m, and n being direction-cosines.

(11) The equations of a straight line are given in the form

$$\frac{\alpha - ny + mz}{\lambda} = \frac{\beta - lz + nx}{\mu} = \frac{\gamma - mx + ly}{\nu},$$

obtain them in the form

$$\frac{x - \dfrac{\mu\gamma - \nu\beta}{l\lambda + m\mu + n\nu}}{l} = \frac{y - \dfrac{\nu\alpha - \lambda\gamma}{l\lambda + m\mu + n\nu}}{m} = \frac{z - \dfrac{\lambda\beta - \mu\alpha}{l\lambda + m\mu + n\nu}}{n}.$$

(12) The equations of the straight line on which lies the shortest distance ($2d$) between the two straight lines,

$$\frac{y}{b} + \frac{z}{c} = 1, \quad x = 0; \quad \text{and } \frac{x}{a} - \frac{z}{c} = 1, \quad y = 0,$$

$$\text{are } \frac{y}{b} + \frac{z}{c} - \frac{x}{a} = 1 - \frac{ax}{d^2} = \frac{by}{d^2} - 1.$$

Shew that $$\frac{1}{d^2} = \frac{1}{a^2} + \frac{1}{b^2} + \frac{1}{c^2}.$$

(13) The locus of the middle points of all straight lines terminated by two fixed straight lines is a plane bisecting the shortest distance between the fixed straight lines. Take as axes the system suggested, Art. (63).

(14) The locus of the middle points of all straight lines of constant length terminated by two fixed straight lines, is an ellipse whose centre bisects the shortest distance between the fixed lines, and whose axes are equally inclined to them.

CHAPTER V.

GENERAL EQUATION OF THE FIRST DEGREE. EQUATION OF A PLANE.

65. *The locus of the general equation of the first degree is a plane.*

The general equation of the first degree is

$$Ax + By + Cz + D = 0.$$

Let (α, β, γ) $(\alpha', \beta', \gamma')$ be two points in the locus.

The equations of the straight line joining these points are

$$\frac{x-\alpha}{l} = \frac{y-\beta}{m} = \frac{z-\gamma}{n} = r;$$

$$\text{where } \frac{\alpha'-\alpha}{l} = \frac{\beta'-\beta}{m} = \frac{\gamma'-\gamma}{n},$$

and since (α, β, γ) is in the locus of the equation

$$A\alpha + B\beta + C\gamma + D = 0, \qquad (1)$$

$$\text{similarly,} \quad A\alpha' + B\beta' + C\gamma' + D = 0;$$

$$\therefore A(\alpha'-\alpha) + B(\beta'-\beta) + C(\gamma'-\gamma) = 0;$$

$$\therefore Al + Bm + Cn = 0. \qquad (2)$$

Now, the straight line meets the locus in all points for which the equation in r,

$$A(\alpha + lr) + B(\beta + mr) + C(\gamma + nr) + D = 0,$$

is satisfied, i. e. for all values of r, by (1) and (2); therefore, every point in the straight line lies in the locus, and this is true wherever the two points are chosen.

Hence, the locus is a plane.

66. The student will readily deduce the following special positions of the plane.

(1) If $D = 0$, the plane passes through the origin.

(2) If $A = 0$, the plane is parallel to the axis of x.

(3) If A and $B=0$, the plane is parallel to the plane of xy.

(4) If A, B and $D=0$, the plane is that of xy.

(5) If A, B, and $C=0$, while D remains finite, the plane is at an infinite distance. For, the point in which the axis of x meets the plane is given by the equations

$$y=0,\ z=0,\ Ax+D=0.$$

Hence, the distance from the origin being $-\dfrac{D}{A}$, if A be indefinitely diminished while D is finite, the plane cuts the axis of x at an infinite distance from the origin, and the same being true for each axis, it follows that the plane is at an infinite distance from the origin.

Equation of a Plane.

67. *To find the equation of a plane in the form*

$$lx+my+nz=p,$$

in which p *is the perpendicular from the origin upon the plane, and* l, m, n *its direction-cosines.*

A plane may be considered as the locus of a straight line which passes through a given point, and is perpendicular to a given straight line.

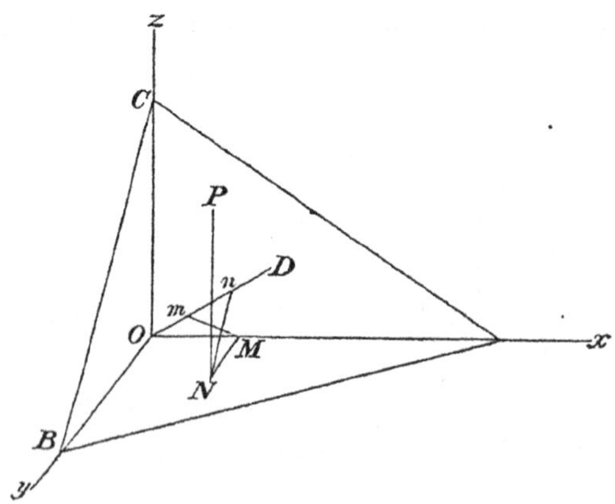

Let $OD = p$ be the perpendicular from the origin upon a plane, l, m, n its direction-cosines, (x, y, z) any point P in the plane, then, by the definition, PD is perpendicular to OD, and OD is the sum of projections of the co-ordinates of P on OD;

$$\therefore\ lx + my + nz = p,$$

which is the equation of the plane in the form required.

If the axes be rectangular,

$$l^2 + m^2 + n^2 = 1.$$

If the axes be oblique, and a, b, c be the cosines of the angles yOz, zOx, xOy,

$$l^2(1-a^2) + m^2(1-b^2) + n^2(1-c^2) + 2mn(bc-a) + 2nl(ca-b) + 2lm(ab-c) = 1 - a^2 - b^2 - c^2 + 2abc. \quad \text{(Art. 29)}.$$

68. *Interpretation of the expression*

$$p - lx - my - nz.$$

The equation $p - lx - my - nz = 0$ represents a plane, in which p is the perpendicular from the origin, and l, m, n are its direction-cosines.

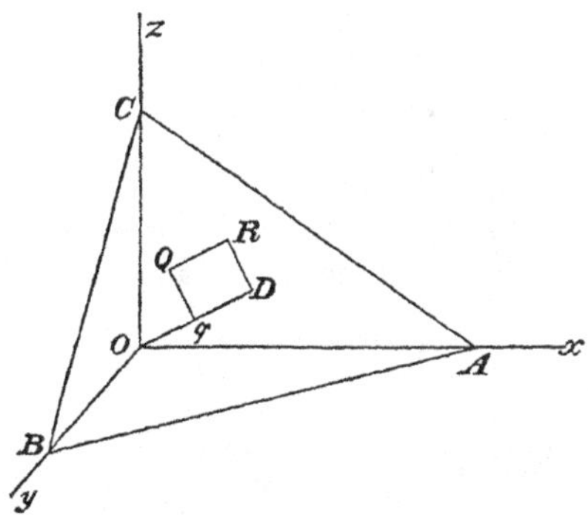

Let ABC be this plane, and suppose OD, QR to be drawn perpendicular to it, in the direction defined by (l, m, n), from the origin, and from the point $Q(x, y, z)$, and join RD which will be perpendicular to OD. Let $QR = q$, and project x, y, z and q on OD.

Then,
$$p = lx + my + nz + q;$$
$$\therefore\ q = p - lx - my - nz.$$

Hence, the expression $p - lx - my - nz$ represents the perpendicular drawn from (x, y, z) upon the plane

$$p - lx - my - nz = 0,$$

estimated positive in the direction defined by the cosines l, m and n.

69. *To find the angle between two planes whose equations are given.*

Let $Lx + My + Nz = D$, and $L'x + M'y + N'z = D'$,
be the given equations: then (L, M, N) and (L', M', N') are proportional respectively to the direction-cosines of the normals: but the angle between two planes is equal to the angle between their normals; hence the angle between the planes is

$$\cos^{-1}\left\{\frac{LL' + MM' + NN'}{\sqrt{(L^2 + M^2 + N^2)}\sqrt{(L'^2 + M'^2 + N'^2)}}\right\}.$$

The conditions of parallelism and perpendicularity are therefore respectively,

$$\frac{L}{L'} = \frac{M}{M'} = \frac{N}{N'},$$

$$\text{and } LL' + MM' + NN' = 0.$$

The student may also deduce the conditions of parallelism from the consideration that parallel planes intersect in a straight line at infinity, or directly from the parallelism of the normals.

70. *To find the angle between a straight line and plane whose equations are given.*

Let
$$\frac{x-\alpha}{L} = \frac{y-\beta}{M} = \frac{z-\gamma}{N} \quad\text{..................... (1)},$$

$$L'x + M'y + N'z = D \quad\text{..................... (2)},$$

be the given equations. The angle between a straight line and a plane is the complement of the angle between the straight line and the normal to the plane; hence the required angle is

$$\sin^{-1}\frac{LL' + MM' + NN'}{\sqrt{(L^2 + M^2 + N^2)}\sqrt{(L'^2 + M'^2 + N'^2)}}.$$

71. *To determine the perpendiculars from a point* (g, h, k) *upon a plane whose equation is* $Ax + By + Cz + D = 0$.

If we compare the equation

$$Ax + By + Cz + D = 0$$

with the equation of the plane in the form

$$lx + my + nz - p = 0;$$

then, $\frac{l}{A} = \frac{m}{B} = \frac{n}{C} = \frac{p}{-D} = \pm \frac{1}{\sqrt{(A^2 + B^2 + C^2)}}$.

where, if the ambiguous sign be so taken that p shall be an absolute length, l, m, n will be completely determined.

The perpendicular from (g, h, k) upon the plane, estimated positive when drawn in the direction defined by these cosines,

$$= p - lg - mh - nk$$

$$= \frac{Ag + Bh + Ck + D}{\mp \sqrt{(A^2 + B^2 + C^2)}},$$

that sign being chosen which is the same as that of D.

72. Hence, the connection between a system of rectangular axes and a four-plane system is established.

For, let O, the origin of co-ordinates, be supposed within the tetrahedron $ABCD$, and let l, m, n be the direction-cosines of the perpendicular from O on BCD.

Then α, one of the four-plane co-ordinates of a point (x, y, z), is the algebraical distance of that point from BCD, estimated in the same direction as the perpendicular from A on BCD.

Hence, $\alpha = p - lx - my - nz.$

73. *To find the perpendicular distance from a given point upon a given plane, the co-ordinates being oblique.*

Let the equation of the given plane be

$$Ax + By + Cz + D = 0;$$

(g, h, k) the given point, a, b, c the cosines of the angles yOz, zOx, xOy respectively, l, m, n the direction-cosines of a normal to the plane;

$$\therefore \frac{A}{l} = \frac{B}{m} = \frac{C}{n} = \rho, \text{ suppose.}$$

Now, Art. (29),

$$l^2(1-a^2)+m^2(1-b^2)+n^2(1-c^2)+2mn(bc-a)+2nl(ca-b)$$
$$+2lm(ab-c)=1-a^2-b^2-c^2+2abc,$$

whence ρ^2 is determined in terms of A, B, C.

The perpendicular on the plane being

$$-\frac{D}{\rho}-lg-mh-nk,$$

in which the sign of ρ is chosen so that $-\frac{D}{\rho}$ is positive, the length required, estimated in the direction $\left(\frac{A}{\rho}, \frac{B}{\rho}, \frac{C}{\rho}\right)$, is

$$\frac{Ag+Bh+Ck+D}{-\rho}.$$

74. *To find the distance from a given point to a given plane, measured in any given direction.*

Let the equation of the plane be

$$Ax+By+Cz+D=0,$$

and let (g, h, k) be the given point, (l, m, n) the given direction, l, m, n being direction-cosines for rectangular axes, and direction-ratios for oblique.

The equations of a line drawn through (g, h, k) in the given direction are

$$\frac{x-g}{l}=\frac{y-h}{m}=\frac{z-k}{n}=r,$$

and where this straight line meets the plane,

$$A(g+lr)+B(h+mr)+C(k+nr)+D=0;$$

$$\therefore \text{ the required distance is } -\frac{Ag+Bh+Ck+D}{Al+Bm+Cn}.$$

Hence, if the given direction be perpendicular to the plane, and the axes be rectangular,

$$\frac{l}{A}=\frac{m}{B}=\frac{n}{C}=\frac{Al+Bm+Cn}{A^2+B^2+C^2}=\frac{1}{\pm\sqrt{(A^2+B^2+C^2)}},$$

and the perpendicular distance is $\dfrac{Ag+Bh+Ck+D}{\mp\sqrt{(A^2+B^2+C^2)}}$,

the sign being chosen so that $\dfrac{D}{\mp\sqrt{(A^2+B^2+C^2)}}$ is positive.

75. *To find the equation of a plane in the form*

$$\frac{x}{a}+\frac{y}{b}+\frac{z}{c}=1.$$

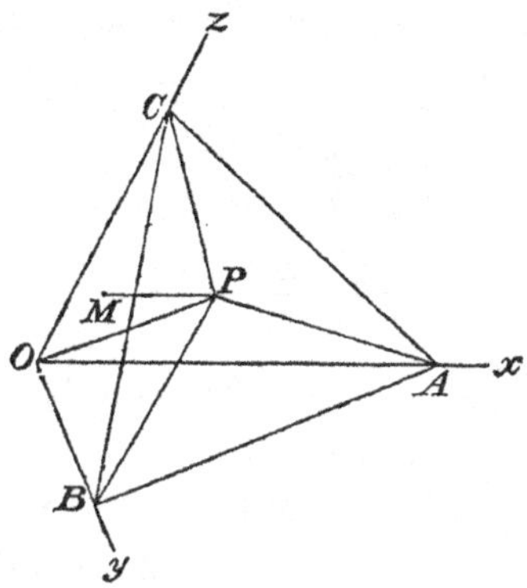

Let $OA=a$, $OB=b$, $OC=c$ be the intercepts of the axes of x, y, z by the plane ABC, and let PA, PB, PC, PO be drawn from the point $P\,(x, y, z)$ in the plane.

Draw PM parallel to xO meeting yOz in M. Since the pyramids $POBC$, $AOBC$ are on the same base,

$$\text{vol } POBC : \text{vol } OABC :: PM : AO :: x : a;$$

$$\therefore \frac{x}{a}=\frac{\text{vol } POBC}{\text{vol } OABC}.$$

Similarly, $$\frac{y}{b}=\frac{\text{vol } POCA}{\text{vol } OABC},$$

and $$\frac{z}{c}=\frac{\text{vol } POAB}{\text{vol } OABC},$$

and $$\text{vol } POBC+\text{vol } POCA+\text{vol } POAB=\text{vol } OABC;$$

$$\therefore \frac{x}{a}+\frac{y}{b}+\frac{z}{c}=1,$$

which is the equation required.

The student is recommended to investigate this equation by the employment of a figure in which P lies in another compartment, as $x'y'z$, of the co-ordinate planes, taking care to interpret the geometrical into algebraical distances.

76. If q be the perpendicular from a point Q (x, y, z) on the plane ABC estimated in the direction of p, the perpendicular from O on the plane,

$$\frac{q}{p} = \frac{\text{vol } QABC}{\text{vol } OABC}$$

$$= 1 - \frac{x}{a} - \frac{y}{b} - \frac{z}{c}.$$

77. The equation of Art. (75) may be obtained from the general equation of the first degree.

For let a, b, c be the intercepts of the axes of x, y, z,

$Ax + By + Cz - D = 0$, the equation of the plane.

Since $(a, 0, 0)$ is a point in the plane,

$$-D = Aa, \text{ and similarly, } = Bb = Cc.$$

Hence, the equation of the plane is $\frac{x}{a} + \frac{y}{b} + \frac{z}{c} = 1$.

78. *To find the equation of the plane in the form*

$$z = mx + ny + c.$$

Consider the plane as a surface generated by a straight line which moves subject to the conditions that it always intersects one given straight line and is parallel to another.

Let the equations of the line which it intersects be

$$z = mx + c, \quad y = 0; \tag{1}$$

and those of the line to which it is parallel,

$$z = ny, \quad x = 0. \tag{2}$$

Then the equations of the moving line will be of the form

$$z = ny + \beta, \quad x = \alpha. \tag{3}$$

Since (3) intersects (1), $\beta = m\alpha + c$;

therefore for every point in the plane, $z - ny = mx + c$;

$$\text{or, } z = mx + ny + c.$$

In this form of the equation, c is the intercept of the axis of z cut off by the plane, m, n are the tangents of the angles made respectively with the axes of x and y by the traces on the planes of zx, yz, if the co-ordinates be rectangular; and the ratios of the sines of the angles made with the axes in those planes, if the co-ordinates be oblique.

79. *To find the polar equation of a plane.*

Let (c, α, β) be the polar co-ordinates of the foot of the perpendicular from the origin on the plane; (r, θ, ϕ) those of any point in the plane, then if ψ be the angle between the lines joining these points to the origin,

$$c = r \cos \psi,$$

and $\cos \psi = \cos \theta \cos \alpha + \sin \theta \sin \alpha \cos (\phi - \beta)$, Art. (41),

whence $\dfrac{c}{r} = \cos \theta \cos \alpha + \sin \theta \sin \alpha \cos (\phi - \beta)$;

the most convenient form of the equation of a plane when referred to polar co-ordinates.

Equation of a Plane referred to a Four-Plane Co-ordinate System.

80. *The general equation of the first degree represents a plane.*

Let $l\alpha + m\beta + n\gamma + r\delta = 0$ be the equation of the first degree, $(\alpha', \beta', \gamma', \delta')$, $(\alpha'', \beta'', \gamma'', \delta'')$, any two points in the locus;

$$\therefore\ l(\alpha'' - \alpha') + m(\beta'' - \beta') + n(\gamma'' - \gamma') + r(\delta'' - \delta') = 0,$$

but the equations of the line joining the two points are

$$\frac{\alpha - \alpha'}{\alpha'' - \alpha'} = \frac{\beta - \beta'}{\beta'' - \beta'} = \frac{\gamma - \gamma'}{\gamma'' - \gamma'} = \frac{\delta - \delta'}{\delta'' - \delta'}.$$

Therefore, for any point in the straight line joining them

$$l(\alpha - \alpha') + m(\beta - \beta') + n(\gamma - \gamma') + r(\delta - \delta') = 0,$$

$$\text{and } l\alpha' + m\beta' + n\gamma' + r\delta' = 0;$$

$$\therefore\ l\alpha + m\beta + n\gamma + r\delta = 0.$$

Hence, every point in the straight line lies in the locus.

The locus is therefore such, that any two points being taken the straight line joining them lies entirely in the surface, which is therefore a plane superficies.

81. *Interpretation of the constants in the equation of a plane,*

$$l\alpha + m\beta + n\gamma + r\delta = 0.$$

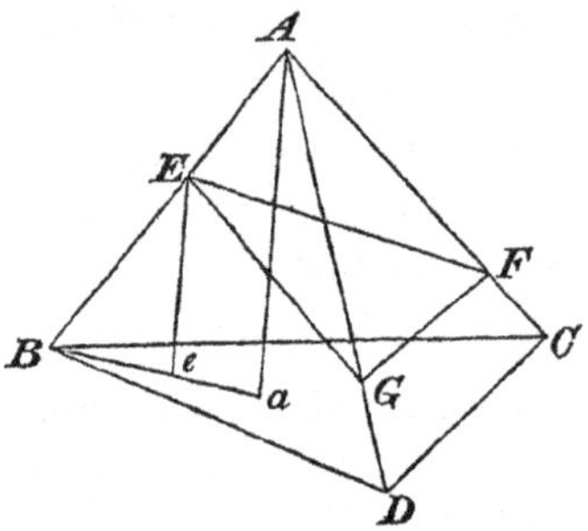

Let E be the point $(\alpha', \beta', 0, 0)$, in which the plane cuts AB;

$$\therefore\ l\alpha' + m\beta' = 0.$$

Draw Ee, Aa perpendicular to BCD;

$$\therefore\ \frac{Ee}{Aa} = \frac{BE}{AB}; \quad \text{or,} \quad \frac{\alpha'}{p_1} = \frac{BE}{AB}.$$

Similarly, $$\frac{\beta'}{p_2} = \frac{AE}{AB},$$

and if α_1, β_1 be the perpendiculars from A, B upon the given plane, estimated in the same direction;

$$\frac{\alpha_1}{AE} = \frac{-\beta_1}{BE};$$

$$\therefore\ \frac{\alpha_1\alpha'}{p_1} + \frac{\beta_1\beta'}{p_2} = 0;$$

$$\therefore\ \frac{l}{\frac{\alpha_1}{p_1}} = \frac{m}{\frac{\beta_1}{p_2}} = \frac{n}{\frac{\gamma_1}{p_3}} = \frac{r}{\frac{\delta_1}{p_3}},$$

and the equation of the plane is

$$\frac{\alpha_1}{p_1}\alpha + \frac{\beta_1}{p_2}\beta + \frac{\gamma_1}{p_3}\gamma + \frac{\delta_1}{p_4}\delta = 0.$$

In tetrahedral co-ordinates, the equation becomes

$$\alpha_1\alpha + \beta_1\beta + \gamma_1\gamma + \delta_1\delta = 0. \qquad \text{Art. (8).}$$

If the plane be at an infinite distance, $\alpha_1 = \beta_1 = \gamma_1 = \delta_1$; therefore, in four-plane co-ordinates, $\frac{\alpha}{p_1} + \frac{\beta}{p_2} + \frac{\gamma}{p_3} + \frac{\delta}{p_4} = 0$, or, in tetrahedral co-ordinates, $\alpha + \beta + \gamma + \delta = 0$, is the equation of a plane at an infinite distance.

82. *To find the condition of parallelism of two planes.*

Let the equations of two planes be

$$l\alpha + m\beta + n\gamma + r\delta = 0, \quad l'\alpha + m'\beta + n'\gamma + r'\delta = 0.$$

These intersect in a line which lies on the plane at infinity, whose equation is

$$\alpha + \beta + \gamma + \delta = 0,$$

using tetrahedral co-ordinates. Hence, these planes intersecting in a straight line are satisfied simultaneously by an infinite number of values of α, β, γ, δ: therefore, employing indeterminate multipliers,

$$\lambda l + \lambda' l' + 1 = 0,$$
$$\lambda m + \lambda' m' + 1 = 0,$$
$$\lambda n + \lambda' n' + 1 = 0,$$
$$\lambda r + \lambda' r' + 1 = 0,$$

from which we obtain the required conditions of parallelism,

$$\frac{l - m}{l' - m'} = \frac{m - n}{m' - n'} = \frac{n - r}{n' - r'}.$$

Planes under Particular Conditions.

83. *Equation of a plane passing through a given point.*

Let (a, b, c) be the co-ordinates of the given point,

$$lx + my + nz = p,$$

the equation of the plane, then since (a, b, c) is a point in this plane

$$la + mb + nc = p,$$

or, eliminating p,

$$l(x-a)+m(y-b)+n(z-c)=0,$$

is the general equation of a plane passing through the point (a, b, c).

84. *Equation of a plane passing through a point determined by the intersection of three given planes.*

If the point be given by the equations of three planes,

$$u=0, \quad v=0, \quad w=0,$$

passing through it and not intersecting in one straight line, then $lu+mv+nw=0$ will be the general equation of a plane passing through that point, for it is satisfied by the values of x, y, z which are given by the equations

$$u=0, \quad v=0, \quad w=0,$$

taken simultaneously, and therefore passes through the intersection of these planes, which is the given point; and since this equation is of the first degree, and involves two arbitrary constants, namely, the ratios $l:m:n$, it is the general equation of a plane passing through the given point.

If the three planes, $u=0$, $v=0$, $w=0$, do intersect in a straight line, then these equations, and therefore the equation $lu+mv+nw=0$, will be simultaneously satisfied for all points lying in that straight line. Hence, $lu+mv+nw=0$, cannot be the *general* equation of a plane passing through a given point. The position of a point is not, in this case, completely determined by the given equations, but only the fact that it lies on a certain straight line.

85. *Equation of a plane passing through two given points.*

Let (a, b, c), (a', b', c') be the given points: the equation of a plane passing through (a, b, c) is

$$l(x-a)+m(y-b)+n(z-c)=0.$$

If this plane pass also through (a', b', c'), we shall have

$$l(a'-a)+m(b'-b)+n(c'-c)=0,$$

and eliminating n, we obtain the equation

$$l\{(z-a)(c'-c)-(z-c)(a'+a)\}+m\{(y-b)(c'-c)-(z-c)(b'-b)\}=0,$$

or, altering the arbitrary constant,

$$\lambda\left(\frac{x-a}{a'-a}-\frac{z-c'}{c'-c}\right)+\mu\left(\frac{y-b}{b'-b}-\frac{z-c}{c'-c}\right)=0,$$

for the general equation of a plane passing through two given points.

This equation may be written symmetrically,

$$\lambda\frac{x-a}{a'-a}+\mu\frac{y-b}{b'-b}+\nu\frac{z-c}{c'-c}=0;$$

λ, μ, ν being connected by the equation

$$\lambda+\mu+\nu=0.$$

86. *Equation of a plane passing through two given points, lying on given planes.*

If the straight line passing through the two points be given by the equations $u=0$, $v=0$, the equation $lu+mv=0$ will represent a plane passing through the points of intersection of $u=0$ and $v=0$, and therefore through the given points, and since this equation involves an arbitrary constant $(l:m)$, it will be the *general* equation of a plane passing through the two given points.

Or, the equation $lu+mv+nw=0$, if l, m, n be connected by a relation of the form $al+bm+cn=0$, will represent a plane passing through the two points given by the equations

$$u=0,\quad v=0,\quad w=0,$$
$$u=a,\quad v=b,\quad w=c.$$

87. *Equation of a plane passing through three given points.*

Let (a,b,c), (a',b',c'), (a'',b'',c'') be the three given points,

$$l(x-a)+m(y-b)+n(z-c)=0, \tag{1}$$

the equation of a plane passing through (a, b, c).

If this plane also pass through (a', b', c') and (a'', b'', c''), we shall have

$$l(a'-a)+m(b'-b)+n(c'-c)=0, \tag{2}$$
$$l(a''-a)+m(b''-b)+n(c''-c)=0, \tag{3}$$

and eliminating l, m, n between (1), (2), and (3), we obtain

$$\begin{aligned}&(x-a)\{b(c'-c'')+b'(c''-c)+b''(c-c')\}\\&+(y-b)\{c(a'-a'')+c'(a''-a)+c''(a-a')\}\\&+(z-c)\{a(b'-b'')+a'(b''-b)+a''(b-b')\}=0, \qquad (4)\end{aligned}$$

as the equation of the plane passing through three given points.

The coefficients of x, y, z in this equation are the projections on the co-ordinate planes of the triangle formed by the three given points, call these A_x, A_y, A_z; then xA_x will be equal to three times the volume of the pyramid whose base is A_x, and vertex the point (x, y, z).

Hence, equation (4) asserts that the algebraical sum of the pyramids whose bases are the projections of any triangle on the co-ordinate planes, and common vertex any point in the plane of the triangle, is constant for all positions of this point.

The equation here obtained becomes nugatory if

$$b(c'-c'')+b'(c''-c)+b''(c-c')=0,$$
$$c(a'-a'')+c'(a''-a)+c''(a-a')=0,$$
$$\text{and } a(b'-b'')+a'(b''-b)+a''(b-b')=0;$$

which are equivalent to

$$(b-b')(c''-c')-(c-c')(b''-b')=0,$$
$$(c-c')(a''-a')-(a-a')(c''-c')=0,$$
$$(a-a')(b''-b')-(b-b')(a''-a')=0;$$
$$\text{or, to } \frac{a-a'}{a'-a''}=\frac{b-b'}{b'-b''}=\frac{c-c'}{c'-c''},$$

which are the conditions that the three given points should lie in a straight line.

88. *To find the equation of a plane passing through a given point, and parallel to a given plane.*

If (a, b, c) be the given point, and l, m, n the direction-cosines of a normal to the given plane, the equation of the proposed plane will be

$$l(x-a)+m(y-b)+n(z-c)=0.$$

89. *To find the equation of a plane passing through two given points, and parallel to a given straight line.*

Let (a, b, c), (a', b', c') be the given points, l, m, n the direction-cosines of the given straight line; then the equation of any plane passing through (a, b, c), (a', b', c'), is

$$\lambda \frac{x-a}{a'-a} + \mu \frac{y-b}{b'-b} + \nu \frac{z-c}{c'-c} = 0, \quad (1)$$

λ, μ, ν being connected by the equation

$$\lambda + \mu + \nu = 0, \quad (2)$$

and if this plane is parallel to the line whose direction-cosines are l, m, n, its normal is perpendicular to this line, hence

$$\frac{l\lambda}{a'-a} + \frac{m\mu}{b'-b} + \frac{n\nu}{c'-c} = 0; \quad (3)$$

and eliminating λ, μ, ν between (1), (2), and (3), we obtain the required equation,

$$\frac{x-a}{a'-a}\left(\frac{m}{b'-b} - \frac{n}{c'-c}\right) + \frac{y-b}{b'-b}\left(\frac{n}{c'-c} - \frac{l}{a'-a}\right) + \frac{z-c}{c'-c}\left(\frac{l}{a'-a} - \frac{m}{b'-b}\right) = 0.$$

This equation becomes identical if $\frac{l}{a'-a} = \frac{m}{b'-b} = \frac{n}{c'-c}$, which are the conditions that the given straight line may be parallel to the line joining the two given points. The equations (2) and (3) are in this case coincident, or every plane passing through the two points will necessarily be parallel to the given straight line, as is otherwise evident. The required equation will then be the equation of any plane passing through the two given points.

90. *To find the equation of a plane passing through a given point, and parallel to two given straight lines.*

If the direction-cosines of the two straight lines be l, m, n and l', m', n', and the co-ordinates of the given point a, b, c, the equation of the plane will be

$$(mn'-m'n)(x-a) + (nl'-n'l)(y-b) + (lm'-l'm)(z-c) = 0. \text{ (Art. 60).}$$

If $\frac{l}{l'} = \frac{m}{m'} = \frac{n}{n'}$, this equation is satisfied for all values of x, y, z; or, if the given straight lines are parallel, there are an infinite number of planes satisfying the given conditions, the direction of the normal to the required plane being indeterminate.

91. *To find the equation of a plane equidistant from two given straight lines, not in the same plane.*

Let the equations of the two given straight lines be

$$\frac{x-\alpha}{l}=\frac{y-\beta}{m}=\frac{z-\gamma}{n}=r, \qquad (1)$$

$$\frac{x-\alpha'}{l'}=\frac{y-\beta'}{m'}=\frac{z-\gamma'}{n'}=r', \qquad (2)$$

(x_1, y_1, z_1) a point in (1), (x_2, y_2, z_2) a point in (2), (X, Y, Z) the middle point of the line joining (x_1, y_1, z_1) and (x_2, y_2, z_2).

Then, $2X=x_1+x_2=\alpha+\alpha'+lr+l'r'$,

$$2Y=y_1+y_2=\beta+\beta'+mr+m'r',$$

$$2Z=z_1+z_2=\gamma+\gamma'+nr+n'r',$$

and eliminating r, and r', we obtain for the locus of (X, Y, Z), the equation

$$(2X-\alpha-\alpha')(mn'-m'n)+(2Y-\beta-\beta')(nl'-n'l) + (2Z-\gamma-\gamma')(lm'-l'm)=0. \qquad (3)$$

The plane represented by this equation bisects all lines joining any point of (1) to any point of (2), and therefore bisects the shortest distance between them; and since the direction-cosines of the normal to (3) are proportional to

$$mn'-m'n, \quad nl'-n'l, \quad lm'-l'm,$$

the normal is parallel to the shortest distance between the lines, (Art. 60). Hence this plane bisects at right angles the shortest distance between the lines.

VI.

(1) Find the equation of the plane passing through the points (a, b, c), (b, c, a), (c, a, b), and the equations of the planes each of which passes through two of the points, and is perpendicular to the former plane.

(2) The equation of a plane passing through the origin, and containing a straight line

$$\frac{x-a}{l}=\frac{y-\beta}{m}=\frac{z-\gamma}{n},$$

is $$\frac{x}{l}\left(\frac{\beta}{m}-\frac{\gamma}{n}\right)+\frac{y}{m}\left(\frac{\gamma}{n}-\frac{a}{l}\right)+\frac{z}{n}\left(\frac{a}{l}-\frac{\beta}{m}\right)=0.$$

Hence, find the equations of the straight line passing through the origin, and intersecting two given straight lines; and examine the case in which the straight lines are parallel.

(3) The equation of a plane passing through the origin, and containing the straight line whose equations are

$$x+2y+3z+4=2x+3y+4z+1=3x+4y+z+2 \text{ is } x+y-2z=0.$$

(4) The equation of a plane passing through the origin, and containing the straight line

$$\frac{a+mz-ny}{l}=\frac{\beta+nx-lz}{m}=\frac{\gamma+ly-mx}{n},$$

is $$(l^2+m^2+n^2)(ax+\beta y+\gamma z)=(la+m\beta+n\gamma)(lx+my+nz).$$

(5) Shew that the co-ordinates of a point, which divide the distance between the points (a, β, γ), (a', β', γ'), in the ratio $\lambda' : \lambda$,

$$\text{are } \frac{\lambda a+\lambda' a'}{\lambda+\lambda'},\ \frac{\lambda\beta+\lambda'\beta'}{\lambda+\lambda'},\ \text{and } \frac{\lambda\gamma+\lambda'\gamma'}{\lambda+\gamma'}.$$

(1) Hence, shew that the locus of a point dividing the distance between any two points on the two straight lines

$$\frac{x-a}{l}=\frac{y-\beta}{m}=\frac{z-\gamma}{n},\quad \frac{x-a'}{l'}=\frac{y-\beta'}{m'}=\frac{z-\gamma'}{n'},$$

in the ratio $\lambda' : \lambda$, is the plane whose equation is

$$(mn'-m'n)\left(x-\frac{\lambda a+\lambda' a'}{\lambda+\lambda'}\right)+\text{\&c.}=0.$$

(2) Also, that the equation $Ax+By+Cz=D$ represents a plane, according to Euclid's definition.

(6) Shew that if the straight lines

$$\frac{x}{a}=\frac{y}{\beta}=\frac{z}{\gamma},\quad \frac{x}{aa}=\frac{y}{b\beta}=\frac{z}{c\gamma},\quad \frac{x}{l}=\frac{y}{m}=\frac{z}{n},$$

lie in one plane, then $\frac{l}{a}(b-c)+\frac{m}{\beta}(c-a)+\frac{n}{\gamma}(a-b)=0.$

(7) Find the equation of the plane which passes through the two parallel lines

$$\frac{x-a}{l}=\frac{y-\beta}{m}=\frac{z-\gamma}{n};\quad \frac{x-a'}{l}=\frac{y-\beta'}{m}=\frac{z-\gamma'}{n};$$

and explain the result when $\frac{a-a'}{l}=\frac{\beta-\beta'}{m}=\frac{\gamma-\gamma'}{n}$.

(8) The equation of a plane passing through the two straight lines

$$\frac{x-a}{a'}=\frac{y-b}{b'}=\frac{z-c}{c'}, \quad \frac{x-a'}{a}=\frac{y-b'}{b}=\frac{z-c'}{c},$$

is $$(bc'-b'c)\,x+(ca'-c'a)\,y+(ab'-a'b)\,z=0.$$

Give a geometrical interpretation of the equations.

(9) Shew that the three planes

$$lx+my+nz=0, \quad (m+n)\,x+(n+l)\,y+(l+m)\,z=0, \quad x+y+z=0$$

intersect in one straight line

$$\frac{x}{m-n}=\frac{y}{n-l}=\frac{z}{l-m}.$$

(10) Determine the conditions necessary in order that the planes

$$ax+c'y+b'z=0, \quad c'x+by+a'z=0, \quad b'x+a'y+cz=0,$$

may have a common line of intersection, and shew that the equations of that line are

$$x\,(aa'-b'c')=y\,(bb'-c'a')=z\,(cc'-a'b').$$

Find the conditions necessary in order that the three planes may be coincident.

(11) If A, A'; B, B'; C, C' are fixed points in any three fixed straight lines passing through a point; the intersections of the planes ABC, $A'B'C'$; $A'BC$, $AB'C'$; $AB'C$, $A'BC'$; and ABC', $A'B'C$ are four straight lines lying in a plane dividing the fixed lines harmonically.

(12) The equation of the planes which pass through the straight line

$$\frac{x}{l}=\frac{y}{m}=\frac{z}{n},$$

and make an angle α with the plane $l'x+m'y+n'z=0$, is

$$\{l'(ny-mz)+m'(lz-nx)+n'(mx-ly)\}^2$$
$$=\cos^2\alpha\,(l'^2+m'^2+n'^2)\,\{(ny-mz)^2+(lz-nx)^2+(mx-ly)^2\}.$$

What limitation is there to the value of α? Shew that for the limiting values the two planes coincide.

(13) The equation of any plane containing the straight line

$$\frac{x-\alpha}{l}=\frac{y-\beta}{m}=\frac{z-\gamma}{n} \quad \text{is} \quad \frac{\lambda(x-\alpha)}{l}+\frac{\mu\,(y-\beta)}{m}+\frac{\nu\,(z-\gamma)}{n}=0,$$

λ, μ, ν being connected by the equations $\lambda+\mu+\nu=0$. Hence, find the equation of a plane containing one given straight line, and parallel to another.

CHAPTER VI.

FOUR-POINT CO-ORDINATE SYSTEM.

92. IN the *Four-Plane Co-ordinate System*, the position of a point is given by its algebraical distances from four fundamental planes, given in position, which do not pass through one point, so that they form the plane faces of a tetrahedron of finite volume.

The position of a plane is given by a relation between the four-plane co-ordinates, which exists for every point which lies in that plane.

In the *Four-Point Co-ordinate System*, the position of a plane is given by its distances from four fundamental points, given in position, which do not all lie in one plane, so that they form the angular points of a tetrahedron of finite volume.

These distances are called *Point Co-ordinates* of the plane.

An infinite number of planes can be drawn through any given point, and it can be shewn that the point co-ordinates of each of these planes satisfy a linear equation; this equation gives the position of the point, and is called the equation of the point.

93. *To find the distance, estimated in any direction, of a point, whose position relative to fixed points is known, from a plane whose distances from the fixed points are given.*

Let L, M be two points, and let a point G be taken in the straight line joining them, such that $l \, . \, LG = m \, . \, MG$.

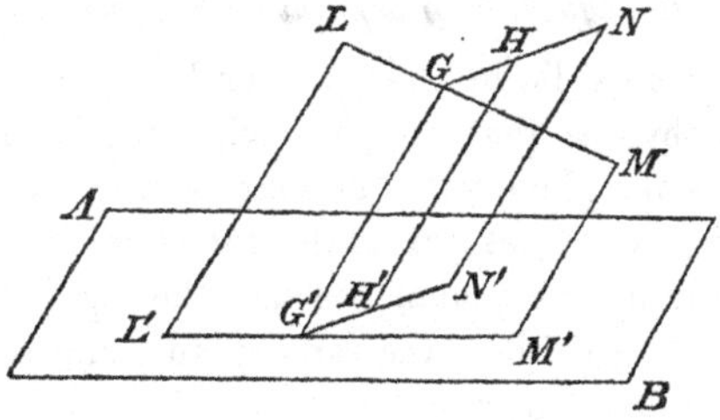

Let LL', MM', GG' be parallel lines, drawn in a given

direction, meeting a given plane AB in L', G', M', then it is evident that

$$\frac{LL' - GG'}{LG} = \frac{GG' - MM'}{MG},$$

$$\text{and } l.(LL' - GG') = m.(GG' - MM');$$

$$\therefore\ l.LL' + m.MM' = (l+m)\,GG'. \qquad (1)$$

If N be any other point and H be taken in GN so that $(l+m)\,GH = n.HN$, and if HH', NN' be drawn parallel to LL',

$$\text{then } n.NN' + (l+m)\,GG' = (l+m+n)\,HH';$$

$$\therefore\ l.LL' + m.MM' + nNN' = (l+m+n)\,HH'. \qquad (2)$$

The equations (1) and (2) give the distance of a point, whose position relative to two or three fixed points is known from the plane AB.

94. If four fundamental points be taken L, M, N, R, and K be taken in KR, such that $r.KR = (l+m+n)\,K$,
and RB', KK' be drawn parallel to LL' meeting the plane AB in R', K',

$$\text{then } r.RR' + (l+m+n)\,HH' = (l+m+n+r)\,KK';$$

$$\therefore\ l.LL' + m.MM' + n.NN' + r.RR' = (l+m+n+r)\,KK'. \qquad (3)$$

The distance of the point K from the plane AB is given by (3) when its position relative to the four fundamental points $LMNR$ is given.

The position of the point is given by l, m, n, r, and it may be denoted by (l, m, n, r).

95. *To find the equation of a point in four-point co-ordinates.*

If the four points lie in a plane, then by the construction of Art. (94), it is obvious that the point K, being in the line NR, will lie in the same plane with the four points. This accounts for the restriction with respect to the fundamental points, that they shall not lie in one plane, because the equation obtained would then denote a point in the same plane, and could not be the equation of any point in space.

If A, B, C, D be any four points which do not lie in a plane, α, β, γ, δ the perpendicular distances of a plane from these points, estimated in a given direction, (l, m, n, r) a point P, with reference to these fundamental points, and π the perpendicular from P upon the plane;

$$\text{then } l\alpha + m\beta + n\gamma + r\delta = (l + m + n + r)\,\pi, \quad \text{(Art. 94)},$$

therefore for every plane passing through P,

$$l\alpha + m\beta + n\gamma + r\delta = 0, \text{ which is the equation required.}$$

Hence, upon the same principle, that, in the four-plane co-ordinates, $\alpha, \beta, \gamma, \delta$ being the co-ordinates of any point,

$$l\alpha + m\beta + n\gamma + r\delta = 0$$

is the equation of that plane, in which a series of points lie, whose co-ordinates satisfy the equation: so, $\alpha, \beta, \gamma, \delta$ being the co-ordinates of a plane in the four-point system of co-ordinates, $l\alpha + m\beta + n\gamma + r\delta = 0$ is the equation of a point, through which all planes pass whose co-ordinates satisfy the equation.

96. If the tetrahedron be formed whose angular points are A, B, C, D; $p_1, 0, 0, 0$ are the co-ordinates of the face BCD; and $\alpha = 0$ is the equation of the point A.

97. *To interpret the constants in the equation of a point.*

Let $l\alpha + m\beta + n\gamma + r\delta = 0$ be the equation of a point P, $\alpha_1, \beta_1, \gamma_1, \delta_1$ the perpendicular distances of P from the faces opposite to A, B, C, D;

$$\therefore\ lp_1 = (l + m + n + r)\,\alpha_1. \quad \text{(Art. 94)}.$$

Hence,
$$\frac{p_1 l}{\alpha_1} = \frac{p_2 m}{\beta_1} = \frac{p_3 n}{\gamma_1} = \frac{p_4 r}{\delta_1},$$

and the equation of a point whose four-plane co-ordinates are $\alpha_1, \beta_1, \gamma_1, \delta_1$, may be written

$$\frac{\alpha_1}{p_1}\alpha + \frac{\beta_1}{p_2}\beta + \frac{\gamma_1}{p_3}\gamma + \frac{\delta_1}{p_4}\delta = 0.$$

If $\alpha_1, \beta_1, \gamma_1, \delta_1$ be tetrahedral co-ordinates, this equation becomes

$$\alpha_1\alpha + \beta_1\beta + \gamma_1\gamma + \delta_1\delta = 0.$$

98. *To find the equation of a point which divides the straight line joining two points, whose equations are given, in a given ratio.*

Let the equations of the two points P, P' be

$$l\alpha + m\beta + n\gamma + r\delta = 0, \text{ and } l'\alpha + m'\beta + n'\gamma + r'\delta = 0,$$

and let Q be a point in PP', such that

$$PQ : QP' :: \mu : \lambda;$$

therefore for every plane through Q, whose distances from P and P' are π and π', $\lambda\pi + \mu\pi' = 0$, (Art. 95);

$$\therefore \lambda . \frac{l\alpha + m\beta + n\gamma + r\delta}{l + m + n + r} + \mu . \frac{l'\alpha + m\beta + n'\gamma + r'\delta}{l' + m' + n' + r'} = 0, \text{ (Art. 94)},$$

which is therefore the equation of Q.

99. *To shew that the straight lines joining the middle points of opposite edges of a tetrahedron intersect and bisect each other.*

The equation of the middle point of AB is $\alpha + \beta = 0$, and of the middle point of CD is $\gamma + \delta = 0$: therefore the equation of the middle point of the line joining these is $\alpha + \beta + \gamma + \delta = 0$, which for the same reason bisects the lines joining the middle points of the other opposite edges.

100. The student is recommended to examine carefully the processes employed in the following applications of Point-Co-ordinates.

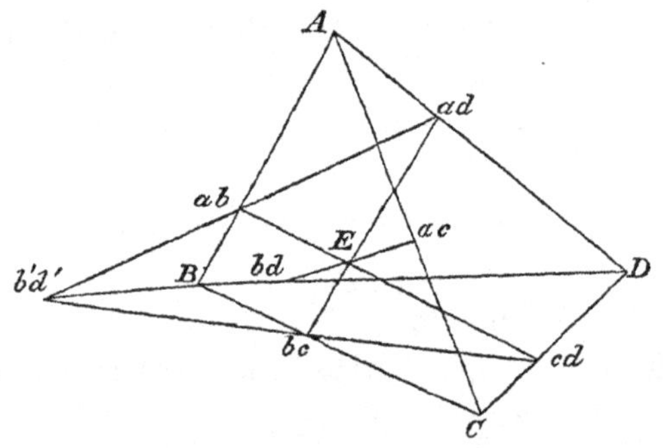

Let $l\alpha + m\beta + n\gamma + r\delta = 0$ be the equation of any point E, and let planes be drawn through this point and each of the edges; and let (ab) denote the point in which the plane ECD meets AB, and similarly for the other edges.

The point E lies in the straight line joining (ab), for which $l\alpha + m\beta = 0$, and (cd), for which $m\gamma + r\delta = 0$, since its equation is of the form

$$L(l\alpha + m\beta) + M(m\gamma + r\delta) = 0.$$

$$\begin{aligned}
\text{Since, for } (ab),\quad & l\alpha + m\beta = 0, \\
(ad),\quad & l\alpha + r\delta = 0, \\
(bc),\quad & m\beta + n\gamma = 0, \\
(cd),\quad & n\gamma + r\delta = 0,
\end{aligned}$$

the straight lines joining these pairs of points meet BD in a point $b'd'$ whose equation is $m\beta = r\delta$, and the equation of bd is $m\beta + r\delta = 0$; therefore $b'd'$, bd divide BD harmonically.

Similarly, the line joining (ab), (ac); and (bd), (cd) intersect BC in $b'c'$, for which $m\beta = n\gamma$; and the straight line (ab) (cd) meets the plane passing through A and the points $b'c'$, $b'd'$, in the point whose equation is

$$2l\alpha + 2m\beta - n\gamma - r\delta = 0,$$

since this equation may be written

$$2l\alpha + (m\beta - n\gamma) + (m\beta - r\delta) = 0.$$

Again, the equation $2l\alpha + m\beta + n\gamma = 0$ being of the form

$$L\alpha + M\delta + N(l\alpha + m\beta + n\gamma + r\delta) = 0,$$

represents a point in plane AED, and, being of the form $L(l\alpha + m\beta) + M(l\alpha + n\gamma) = 0$, lies on the line joining (ab), (ac), and obviously lies in the plane ABC.

Let the straight line AE, BE meet the opposite faces in A', B', the equations of these points are

$$m\beta + n\gamma + r\delta = 0, \quad \text{and} \quad l\alpha + n\gamma + r\delta = 0,$$

and therefore $A'B'$ intersects AB in the point $l\alpha - m\beta = 0$, the same point in which (ac) (bc), (ad) (bd), meet AB.

101. *To find the equation of a point at an infinite distance.*

Let the equation of the point be

$$l\alpha + m\beta + n\gamma + r\delta = 0;$$

since the point is at an infinite distance, the distance

$$\frac{l\alpha + m\beta + n\gamma + r\delta}{l + m + n + r}$$

from any plane, whose co-ordinates α, β, γ, δ are finite, is infinite;

$$\therefore\ l + m + n + r = 0.$$

Then, since l, m, n, r are proportional to the tetrahedral co-ordinates of the point, (Art. 99), if α_1, β_1, γ_1, δ_1 be these co-ordinates, we have

$$\alpha_1 + \beta_1 + \gamma_1 + \delta_1 = 0,$$

which is the same result as was given in (Art. 81).

102. *To find the relations between the constants in the equation of a point, in order that the point may lie in the different portions of space cut off by the indefinite planes which form the faces of the fundamental tetrahedron.*

Let the equation of a point P be

$$l\alpha + m\beta + n\gamma + r\delta = 0,$$

and let $$l + m + n + r = s.$$

If AP or AP produced meet the opposite face in Q, the equation of Q is

$$m\beta + n\gamma + r\delta = 0;$$

and if ϵ be the perpendicular from Q on any plane through P, whose co-ordinates are α, β, γ, δ,

$$(m + n + r)\,\epsilon + l\alpha = 0, \qquad \text{(Art. 93)},$$

α and ϵ being estimated in the same direction.

If P be between A and Q, α and ϵ have opposite signs;

$\therefore \dfrac{m+n+r}{l}$ is positive, or $\dfrac{s}{l}$ is between 1 and ∞.

If P be in AQ produced, α and ϵ are of the same sign, α is greater than ϵ;

$\therefore \dfrac{m+n+r}{l}$ is less than -1, or $\dfrac{s}{l}$ is between 0 and $-\infty$.

If P be in QA produced, α and ϵ are of the same sign, α is less than ϵ;

$\therefore \dfrac{m+n+r}{l}$ is between 0 and -1, or $\dfrac{s}{l}$ is between 1 and 0.

I. For points within the fundamental tetrahedron,

$\dfrac{s}{l}, \dfrac{s}{m}, \dfrac{s}{n}, \dfrac{s}{r}$ are all between 1 and ∞;

$\therefore$ l, m, n, r are all of the same sign.

II. For points between BCD, and ABC, ACD, ADC produced,

$$\frac{s}{l} \text{ is negative, } \frac{s}{m},\ \frac{s}{n},\ \frac{s}{r} \text{ between 1 and } \infty.$$

III. For points within the solid angle formed by BCA, CDA, DBA produced,

$$\frac{s}{l} \text{ is between 0 and 1, } \frac{s}{m},\ \frac{s}{n},\ \frac{s}{r} \text{ negative.}$$

IV. For points between ACD, BCD, BAC, ABD produced,

$$\frac{s}{m} \text{ and } \frac{s}{l} \text{ are negative, } \frac{s}{n} \text{ and } \frac{s}{r} \text{ are positive.}$$

And similarly for the nine other compartments.

103. The results of the preceding article may be obtained, if we assume a knowledge of Tetrahedral Co-ordinates, by observing that the tetrahedral co-ordinates of the point are

$$\frac{l}{s},\ \frac{m}{s},\ \frac{n}{s},\ \frac{r}{s}. \qquad \text{(Art. 97).}$$

104. *To find the distance between two points whose equations are given.*

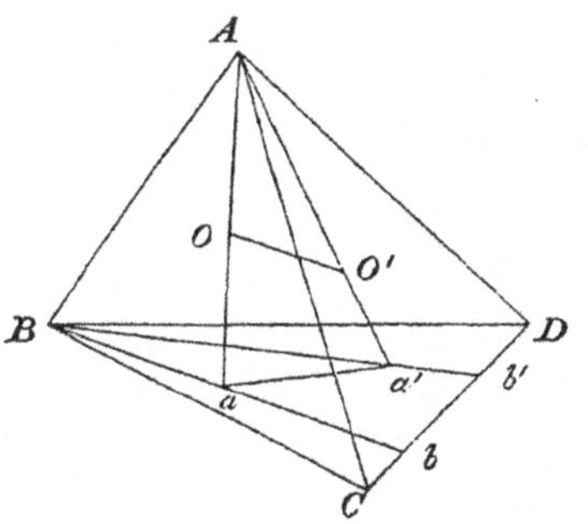

Let the equations of the two points O, O' be

$$l\alpha + m\beta + n\gamma + r\delta = 0,$$

$$l'\alpha + m'\beta + n'\gamma + r'\delta = 0,$$

and let $l + m + n + r = s$, $l' + m' + n' + r' = s'$.

Then, if AO, AO' meet BCD in a, a', and Ba, Ba' meet CD in b, b', we have from the triangles OAO', aAa',

$$\frac{AO^2+AO'^2-OO'^2}{2AO \cdot AO'} = \cos OAO' = \frac{Aa^2+Aa'^2-aa'^2}{2Aa \cdot Aa'}.$$

But $$\frac{Oa}{l}=\frac{Aa}{s}=\frac{AO}{s-l}, \text{ and } \frac{O'a'}{l'}=\frac{Aa'}{s'}=\frac{AO'}{s'-l'};$$

$$\therefore\ OO'^2 = \left(1-\frac{l}{s}\right)^2 Aa^2 + \left(1-\frac{l'}{s'}\right)^2 Aa'^2$$

$$-\left(1-\frac{l}{s}\right)\left(1-\frac{l'}{s'}\right)(Aa^2+Aa'^2-aa'^2)$$

$$=\left(1-\frac{l}{s}\right)\left(\frac{l'}{s'}-\frac{l}{s}\right)Aa^2$$

$$+\left(1-\frac{l'}{s'}\right)\left(\frac{l}{s}-\frac{l'}{s'}\right)Aa'^2+\left(1-\frac{l}{s}\right)\left(1-\frac{l'}{s'}\right)aa'^2.$$

From which form it is manifest that the final result will be an expression in terms of the squares of the edges, and we shall obtain the result by investigating the coefficient of AB^2, and deducing the rest by symmetry.

Now the only terms which can involve AB^2 are contained in Aa^2 and Aa'^2, and, as before,

$$\frac{AB^2+Ba^2-Aa^2}{2AB \cdot Ba} = \cos ABa = \frac{AB^2+Bb^2-Ab^2}{2AB \cdot Bb},$$

also $$\frac{ab}{m}=\frac{Bb}{s-l}=\frac{Ba}{s-l-m};$$

$$\therefore\ Aa^2 = AB^2 - \left(1-\frac{m}{s-l}\right)AB^2 + \ldots\ldots$$

$$=\frac{m}{s-l}AB^2+\ldots\ldots$$

writing down only the terms in which AB^2 appears.

Similarly, $$Aa'^2 = \frac{m'}{s'-l'}AB^2+\ldots\ldots$$

$$\therefore\ OO'^2 = \frac{m}{s}\left(\frac{l'}{s'} - \frac{l}{s}\right) AB^2 + \frac{m'}{s'}\left(\frac{l}{s} - \frac{l'}{s'}\right) AB^2 + \ldots\ldots$$

$$= \left(\frac{l'}{s'} - \frac{l}{s}\right)\left(\frac{m}{s} - \frac{m'}{s'}\right) AB^2 + \ldots\ldots$$

which gives the distance required.

105. If $\alpha, \beta, \gamma, \delta$ and $\alpha', \beta', \gamma', \delta'$ be tetrahedral co-ordinates of the points O, O' we deduce the expression for the square of the distance,

$$OO'^2 = (\alpha' - \alpha)(\beta - \beta') AB^2 + \ldots\ldots$$

which gives the distance between two points, when their tetrahedral co-ordinates are given.

*106. *To find the relation between the co-ordinates of a plane.*

Let $\alpha, \beta, \gamma, \delta$ be the co-ordinates of a plane, λ, μ, ν, ρ the cosines of the angles between the direction in which the co-ordi-

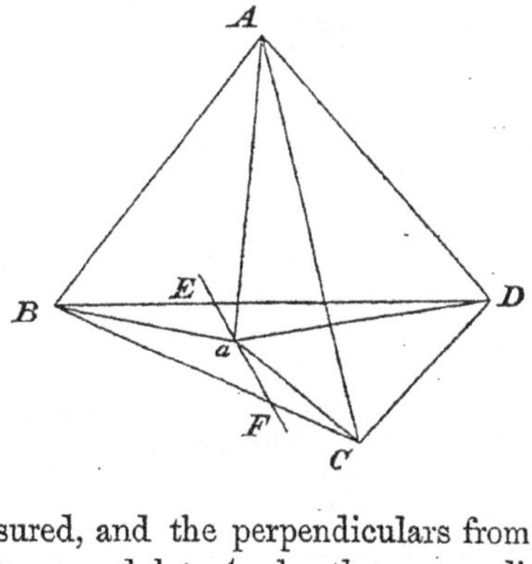

nates are measured, and the perpendiculars from A, B, C, D on the opposite faces, and let Aa be the perpendicular on BCD, $\varpi_b, \varpi_c, \varpi_d$ those from B, C, D on the plane drawn through a parallel to the given plane, EF the trace of this plane on BCD.

Project Aa on the normal to the given plane;

$$\left.\begin{aligned}\therefore\ p_1\lambda &= \alpha - \beta + \varpi_b \\ &= \alpha - \gamma + \varpi_c \\ &= \alpha - \delta + \varpi_d\end{aligned}\right\}. \qquad (1)$$

Now since ϖ_b, ϖ_c, ϖ_d are proportional to the perpendiculars on EF, from B, C, D,

$$\varpi_b \, . \, aCD + \varpi_c \, . \, aDB + \varpi_d \, . \, aBC = 0 ;*$$

$$\text{and } aCD + aDB + aBC = BCD;$$

$\therefore$ by (1), $p_1 \lambda \, . \, BCD = \alpha \, . \, BCD - \beta \, . \, aCD - \gamma \, . \, aDB - \delta \, . \, aBC$;

also, $p_1 \, . \, BCD = p_2 \, . \, ACD = p_3 \, . \, ADB = p_4 \, . \, ABC$;

$$\therefore \lambda = \frac{\alpha}{p_1} - \frac{\beta}{p_2} \cos(CD) - \frac{\gamma}{p_3} \cos(DB) - \frac{\delta}{p_1} \cos(BC),$$

denoting by (CD) the angle between the faces which meet in CD, and similarly for the rest.

Similar values may be obtained in the same way for μ, ν, and ρ.

If P be the foot of the perpendicular from A on the given plane, the perpendiculars from P on the faces of the tetrahedron will be

$$a\lambda - p_1, \quad a\mu, \quad a\nu, \quad a\rho,$$

$$\therefore \alpha \frac{a\lambda - p_1}{p_1} + \beta \frac{a\mu}{p_2} + \gamma \frac{a\nu}{p_1} + \delta \, . \, \frac{a\rho}{p_4} = 0 ; \qquad \text{Art. (81)}$$

$$\therefore \frac{\alpha}{p_1} \lambda + \frac{\beta}{p_2} \mu + \frac{\gamma}{p_3} \nu + \frac{\delta}{p_4} \rho = 1,$$

$$\text{or, } \frac{\alpha^2}{p_1^2} + \frac{\beta^2}{p_2^2} + \frac{\gamma^2}{p_3^2} + \frac{\delta^2}{p_4^2}$$

$$- \frac{2\alpha\beta}{p_1 p_2} \cos(CD) - \frac{2\alpha\gamma}{p_1 p_3} \cos(DB) - \frac{2\alpha\delta}{p_1 p_4} \cos(BC)$$

$$- \frac{2\gamma\delta}{p_3 p_4} \cos(AB) - \frac{2\delta\beta}{p_4 p_2} \cos(AC) - \frac{2\beta\gamma}{p_2 p_3} \cos(AD) = 1,$$

which is the relation required.

---

* The equation of a straight line, referred to trilinear co-ordinates, is

$$\alpha \frac{\varpi_1}{p_1} + \beta \frac{\varpi_2}{p_2} + \gamma \frac{\varpi_3}{p_3} = 0,$$

where ϖ_1, ϖ_2, ϖ_3 are the perpendiculars, estimated in the same direction, from the angular points of the fundamental triangle upon the straight line, and p_1, p_2, p_3 those upon the sides. Now, taking EF for the straight line, and BCD for the fundamental triangle, for a, a point in that line, $\frac{\alpha}{p_1} : \frac{\beta}{p_2} : \frac{\gamma}{p_3} :: aCD : aDB : aBC$,

$$\therefore \varpi_1 \, . \, aCD + \varpi_2 \, . \, aDB + \varpi_3 \, . \, aBC = 0,$$

whence the equation in the text.

107. *To find the relations between the co-ordinates of a plane at an infinite distance.*

If the plane be at an infinite distance, the difference between any two of its co-ordinates will vanish compared with either of them; whence, if α, β, γ, δ be the co-ordinates of the plane, we must have

$$\alpha = \beta = \gamma = \delta,$$

the relations required.

108. Since each of the co-ordinates is of infinite magnitude, the expression for λ in the last article will give us

$$0 = \frac{1}{p_1} - \frac{1}{p_2}\cos(CD) - \frac{1}{p_3}\cos(DB) - \frac{1}{p_4}\cos(BC),$$

which is the equation arising from the projection of the faces meeting in A upon BCD.

From this and the three similar equations, we may eliminate p_1, p_2, p_3, and p_4, and obtain the relation which subsists between the cosines of the inclinations of the faces.

109. *To find the perpendicular from a given point upon a given plane, referred to Tetrahedral Co-ordinates.*

Let $(\alpha', \beta', \gamma', \delta')$ be the co-ordinates of the point,

$$l\alpha + m\beta + n\gamma + r\delta = 0,$$

the equation of the plane.

If α_1, β_1, γ_1, δ_1 be the point-co-ordinates of the plane, the equation of the plane may be written

$$\alpha\alpha_1 + \beta\beta_1 + \gamma\gamma_1 + \delta\delta_1 = 0, \qquad \text{Art. (81)},$$

and the perpendicular required is equal to

$$\alpha'\alpha_1 + \beta'\beta_1 + \gamma'\gamma_1 + \delta'\delta_1. \qquad \text{Art. (94)}.$$

Now $$\frac{\alpha_1}{l} = \frac{\beta_1}{m} = \frac{\gamma_1}{n} = \frac{\delta_1}{r} = \frac{\pm\sqrt{\left\{\frac{\alpha_1^2}{p_1^2} + \ldots\ldots - \frac{2\alpha_1\beta_1}{p_1p_2}\cos(CD) - \ldots\ldots\right\}}}{\sqrt{\left\{\frac{l^2}{p_1^2} + \ldots\ldots - \frac{2lm}{p_1p_2}\cos(CD) - \ldots\ldots\right\}}},$$

therefore the perpendicular required is

$$\frac{l\alpha' + m\beta' + n\gamma' + r\delta'}{\pm\sqrt{\left\{\frac{l^2}{p_1^2} + \ldots\ldots - \frac{2lm}{p_1p_2}\cos(CD) - \ldots\ldots\right\}}}.$$

It may be shewn, without much difficulty, that the denominator of this expression cannot vanish for real values of $l : m : n : r$, except when $l = m = n = r$, in which case, since $\alpha_1 = \beta_1 = \gamma_1 = \delta_1$, the plane is at an infinite distance.

110. *To find the cosines of the angles which a plane, whose co-ordinates are given, makes with the faces of the tetrahedron.*

If λ, μ, ν, ρ be these cosines, α, β, γ, δ the co-ordinates of the given plane, we have deduced in the previous article

$$\lambda = \frac{\alpha}{p_1} - \frac{\beta}{p_2}\cos(CD) - \frac{\gamma}{p_3}\cos(DB) = \frac{\delta}{p_4}\cos(BC),$$

and similarly for μ, ν, ρ.

The relation between λ, μ, ν and ρ is formed immediately from the consideration that the equation

$$\frac{\lambda\alpha}{p_1} + \frac{\mu\beta}{p_2} + \frac{\nu\gamma}{p_3} + \frac{\rho\delta}{p_4} = 1,$$

when the plane moves parallel to itself to an infinite distance, becomes

$$\frac{\lambda}{p_1} + \frac{\mu}{p_2} + \frac{\nu}{p_3} + \frac{\rho}{p_4} = 0.$$

If the plane be the face BCD,

$$\lambda = 1, \quad \mu = -\cos CD, \quad \nu = -\cos DB, \text{ and } \rho = -\cos BC;$$

$$\therefore \frac{1}{p_1} - \frac{1}{p_2}\cos CD - \frac{1}{p_3}\cos DB - \frac{1}{p_4}\cos BC = 0,$$

as before, Art. (108).

111. *To find the condition that two planes whose co-ordinates are given may be parallel.*

If $(\alpha, \beta, \gamma, \delta)$ $(\alpha_1', \beta_1', \gamma_1', \delta')$ be the two planes,

$$\alpha - \alpha' = \beta - \beta' = \gamma - \gamma' = \delta - \delta',$$

since the perpendicular distance between the planes is constant.

112. *To find the equation of a line.*

Let the equations of two points in the line be

$$u = l\alpha + m\beta + n\gamma + \nu\delta = 0,$$

$$v = l'\alpha + m'\beta + n'\gamma + \nu'\delta = 0,$$

then, for every point in the line joining them, $\lambda u + \mu v = 0$, for some value of $\lambda : \mu$.

Hence, if

$$L\alpha + M\beta + N\gamma + R\delta = 0$$

be an equation in which L, M, N, R involve only one variable in the first degree, this is the equation of any point lying in a certain straight line, and may therefore be considered as the equation of a straight line.

113. *To find the equation of a plane.*

Let $u = 0$, $v = 0$, $w = 0$ be the equations of three points in the plane, not in the same straight line.

Then $\lambda u + \mu v + \nu w = 0$ for any point in that plane with certain values of $\lambda : \mu : \nu$.

Hence, if

$$L\alpha + M\beta + N\gamma + R\delta = 0$$

be an equation in which L, M, N, R involve any two variables in the first degree, the equation is that of any point lying in a certain plane, and may therefore be considered as the equation of the plane.

VIII.

(1) The equation of the center of gravity of the face ABC is

$$\alpha + \beta + \gamma = 0.$$

Hence, shew that the lines joining the vertices with the centers of gravity of the opposite faces meet in a point.

(2) The equation of the center of the circle circumscribing the triangle ABC is

$$\alpha \sin 2A + \beta \sin 2B + \gamma \sin 2C = 0.$$

(3) The co-ordinates of the plane passing through the centers of gravity of the faces ACD, ADB, and ABC are given by the equations

$$\frac{\alpha}{2} = \beta = \gamma = \delta = \frac{p_1}{3}.$$

(4) If P be any point on BD, Q, R points in AC, such that

$$AQ : QC :: DP : PB :: CR : RA,$$

then PQ and PR will intersect the lines joining the middle points of BC, AD, and AB, CD respectively, and divide them in the same ratio as AC.

(5) If through the middle points of the edges BC, CD, DB straight lines be drawn parallel respectively to the opposite edges, these straight lines will meet in a point; and the line joining this point with A will pass through the center of gravity of the pyramid.

(6) The equation of the center of gravity of the surface of the tetrahedron is

$$(A+B+C+D)(\alpha+\beta+\gamma+\delta) = A\alpha + B\beta + C\gamma + D\delta.$$

(7) Shew that the equation of the centers of the eight spheres which touch the faces or the faces produced of the fundamental tetrahedron are

$$A\alpha \pm B\beta \pm C\gamma \pm D\delta = 0.$$

(8) The center of the inscribed sphere lies on the line joining the centers of gravity of the volume and of the surface of the tetrahedron, and divides it in the ratio 3 : 1.

(9) The points B, C, D are joined to the centers of gravity of the opposite faces, and the joining lines produced to points b, c, d, so that B, b, &c., are equidistant from the corresponding faces, prove that the co-ordinates of the plane bcd are given by the equations

$$-2\alpha = \beta = \gamma = \delta,$$

and that this plane divides the edges AB, AC, AD in the ratio 1 : 2.

(10) If points be taken in the lines joining B, C, D to the centers of gravity of the opposite faces, dividing them in the ratio $m : n$, the plane containing these points divides the edges AB, AC, AD in the ratio $m : 2m+3n$.

(11) If through any point P straight lines AP, BP, CP, DP be drawn meeting the opposite faces in a, b, c, d, the straight lines AB, ab intersect, and their point of intersection and the point in which Cd meets AB divide AB harmonically.

(12) The straight lines joining D to the intersection of AB, ab, and A to the intersection of DB, db, will intersect in a point lying on Bc.

CHAPTER VII.

TRANSFORMATION OF CO-ORDINATES.

114. The investigation of the properties of a surface represented by a given equation, is often rendered more convenient by referring it to a different system of co-ordinate axes, in the choice of which we must be guided by the nature of the investigation proposed. We proceed to obtain formulæ by means of which such transformation may be effected.

115. *To change the origin of co-ordinates from one point to another, without altering the direction of the axes.*

Let α, β, γ be the co-ordinates of the new origin O referred to the primary system, x, y, z, x', y', z' co-ordinates of any the same point P referred to the first and second systems respectively. Then the algebraic distance of P from the plane of yz, measured parallel to the axis of x, is equal to its algebraic distance from the plane of $y'z'$, together with the algebraic distance of that plane from the plane of yz. But these distances are x, x', α respectively.

Hence $$x = x' + \alpha,$$
and similarly $$y = y' + \beta,$$
$$z = z' + \gamma,$$
are the formulæ required.

116. Since the formulæ thus obtained involve three arbitrary constants, we can generally by this transformation make the coefficients of three terms in the resulting equation vanish, but as the coefficients of the terms of highest dimensions are unaltered, none of the three terms, so eliminated, can be of the same dimensions as the degree of the equation. Thus, in an equation of the second degree, we can generally destroy the terms of one dimension in x, y, and z; in an equation of the third degree, three of the terms of two dimensions, and so on with equations of higher orders. If, however, the terms, whose coefficients we destroy,

differ by more than one dimension from the degree of the equation, the equations for determining α, β, γ in order to effect this result, will rise to the second, third, or higher orders, but in a surface of the nth degree, the coefficients of the terms of $n-1$ dimensions in the transformed equation will involve α, β, γ only in the first power.

For, if $f(x, y, z) = 0$ be the equation of the nth degree, the transformed equation will be $f(x' + \alpha, y' + \beta, z' + \gamma) = 0$,

$$\text{or } f(\alpha, \beta, \gamma) + \left(x' \frac{df}{d\alpha} + y' \frac{df}{d\beta} + z' \frac{df}{d\gamma}\right) + \dots\dots\dots$$

$$+ \left(x'^{n-1} \frac{d^{n-1}f}{d\alpha^{n-1}} + \dots\right) + \left(x'^{n} \frac{d^{n}f}{d\alpha^{n}} + \dots\right) = 0,$$

and the coefficients of x'^{n-1}, $x'^{n-2}y'$, &c..... in the transformed equation will be

$$\frac{d^{n-1}f}{d\alpha^{n-1}}, \quad (n-1)\frac{d^{n-2}f}{d\alpha^{n-2}d\beta}, \quad \&c\dots\dots$$

each of which, since $f(\alpha, \beta, \gamma)$ is a function of n dimensions, will be of the first degree only in α, β, γ. Hence, if we equate three of these quantities to zero, we obtain three equations of the first degree in α, β, γ, which, if they be independent and consistent will give the point to which the origin must be transferred in order to destroy the three corresponding terms of the equation.

117. *To transform from one system of co-ordinates to another system having the same origin, both systems being rectangular.*

Let Ox, Oy, Oz be the first system; Ox', Oy', Oz' the second; a_1, b_1, c_1; a_2, b_2, c_2; a_3, b_3, c_3; the direction-cosines of Ox', Oy', Oz', referred to Ox, Oy, Oz; x, y, z; x', y', z'; co-ordinates of the same point in the two systems.

Then the algebraic distance of the point from the plane of yz is x; but measured successively parallel to Ox', Oy', Oz', this same distance is $a_1x' + a_2y' + a_3z'$. Hence

and similarly,

$$\left.\begin{aligned} x &= a_1x' + a_2y' + a_3z', \\ y &= b_1x' + b_2y' + b_3z', \\ z &= c_1x' + c_2y' + c_3z', \end{aligned}\right\} \qquad (1)$$

the formulæ required.

The nine constants introduced in these results are connected by six equations of condition, expressing that the two systems of co-ordinates are rectangular, for since Ox, Oy, Oz are mutually at right angles, we have the system of equations

$$\left.\begin{aligned} a_1^2 + b_1^2 + c_1^2 = 1, \\ a_2^2 + b_2^2 + c_2^2 = 1, \\ a_3^2 + b_3^2 + c_3^2 = 1, \end{aligned}\right\} \qquad \text{(A)}$$

and by reason of Ox', Oy', Oz' being also at right angles, the system

$$\left.\begin{aligned} a_2a_3 + b_2b_3 + c_2c_3 = 0, \\ a_3a_1 + b_3b_1 + c_3c_1 = 0, \\ a_1a_2 + b_1b_2 + c_1c_2 = 0. \end{aligned}\right\} \qquad \text{(B)}$$

The number of disposable constants in this transformation is therefore only three.

The relations (A), (B) subsisting among the nine constants involved in these formulæ may also be written

$$\left.\begin{aligned} a_1^2 + a_2^2 + a_3^2 = 1, \\ b_1^2 + b_2^2 + b_3^2 = 1, \\ c_1^2 + c_2^2 + c_3^2 = 1, \end{aligned}\right\} \qquad \text{(A')}$$

$$\left.\begin{aligned} b_1c_1 + b_2c_2 + b_3c_3 = 0, \\ c_1a_1 + c_2a_2 + c_3a_3 = 0, \\ a_1b_1 + a_2b_2 + a_3b_3 = 0. \end{aligned}\right\} \qquad \text{(B')}$$

by considering Ox', Oy', Oz' the primary system of axes, in which case the direction-cosines of Ox, Oy, Oz, will be (a_1, a_2, a_3), (b_1, b_2, b_3), (c_1, c_2, c_3). The equations (A') and (B') expressing the same facts as the equations (A) and (B), are of course deducible from those. Either system may be obtained from the identical equation $x^2 + y^2 + z^2 = x'^2 + y'^2 + z'^2$, substituting for x, y, z their equivalents given in equations (1), or similarly for x', y', z'.

118. The relations between these constants may also be expressed in the following convenient form.

From the equations

$$a_1a_2 + b_1b_2 + c_1c_2 = 0, \quad a_3a_1 + b_3b_1 + c_3c_1 = 0,$$

we obtain immediately

$$\frac{a_1}{b_2c_3 - b_3c_2} = \frac{b_1}{c_2a_3 - c_3a_2} = \frac{c_1}{a_2b_3 - a_3b_2};$$

each member of these equations is therefore equal to

$$\frac{(a_1^2+b_1^2+c_1^2)^{\frac{1}{2}}}{\{(b_2c_3-b_3c_2)^2+(c_2a_3-c_3a_2)^2+(a_2b_3-a_3b_2)^2\}^{\frac{1}{2}}}$$

$$=\frac{(a_1^2+b_1^2+c_1^2)^{\frac{1}{2}}}{\{(a_2^2+b_2^2+c_2^2)(a_3^2+b_3^2+c_3^2)-(a_2a_3+b_2b_3+c_2c_3)^2\}^{\frac{1}{2}}}=\pm 1,$$

by equations (A), (B).

In a similar manner, we obtain

$$\frac{a_2}{b_3c_1-b_1c_3}=\frac{b_2}{c_3a_1-c_1a_3}=\frac{c_2}{a_3b_1-a_1b_3}=\pm 1,$$

$$\frac{a_3}{b_1c_2-b_2c_1}=\frac{b_3}{c_1a_2-c_2a_1}=\frac{c_3}{a_1b_2-a_2b_1}=\pm 1.$$

By using the equations (B′) in a similar manner, we obtain

$$\frac{a_1}{b_2c_3-b_3c_2}=\frac{a_2}{b_3c_1-b_1c_3}=\frac{a_3}{b_1c_2-b_2c_1},$$

which shews that the ambiguities in the three systems of equations here obtained, must be taken all of the same sign.

Any two of these three systems of equations may be taken as completely expressing the relations between the nine constants: the third system being immediately deducible from the other two.

119. *Euler's formulæ for transforming from one system of rectangular co-ordinates to another having the same origin.*

There being in the formulæ already obtained for this purpose, nine constants connected by six invariable relations; it must be possible to obtain formulæ to effect this transformation which shall involve only three constants. The three chosen by Euler for this purpose are (1) the angle which the intersection of the planes of xy and $x'y'$ makes with the axis of x, (2) the angle made by the same straight line with the axis of x', (3) the angle between the planes of xy and $x'y'$.

Let Ox, Oy, Oz be the original, Ox', Oy', Oz' the transformed axes of co-ordinates; Ox_1 the intersection of the planes of xy,

$x'y'$: $xOx_1 = \theta$, $x'Ox_1 = \phi$, $zOz' = \psi$, which is the same as the angle between the planes of xy, $x'y'$.

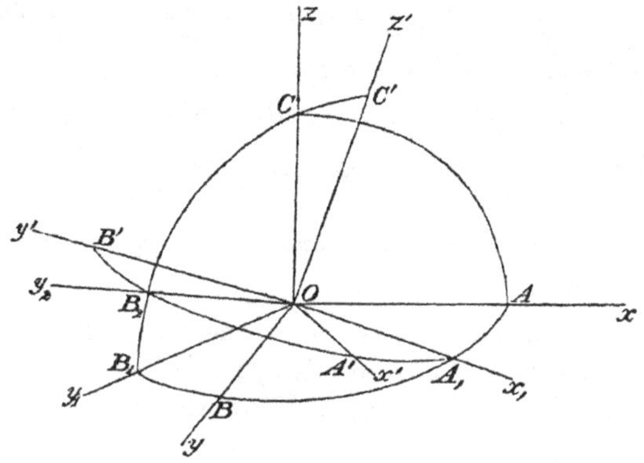

The transformations may be effected by successive transformations, each in one plane,

(1) through an angle θ, in the plane of xy, from Ox, Oy to Ox_1, Oy_1;

(2) through an angle ψ, in the plane of y_1z, from Oy_1, Oz to Oy_2, Oz';

(3) through an angle ϕ, in the plane of y_2x, from Ox_1, Oy_2 to Ox', Oy'.

The formulæ for these transformations are, using the same suffix for any one of the co-ordinates as for the corresponding axis,

$$\left.\begin{aligned} x &= x_1 \cos\theta - y_1 \sin\theta, \\ y &= x_1 \sin\theta + y_1 \cos\theta, \end{aligned}\right\}$$

$$\left.\begin{aligned} y_1 &= y_2 \cos\psi - z' \sin\psi, \\ z &= y_2 \sin\psi + z' \cos\psi, \end{aligned}\right\}$$

$$\left.\begin{aligned} x_1 &= x' \cos\phi - y' \sin\phi, \\ y_2 &= x' \sin\phi + y' \cos\phi. \end{aligned}\right\}$$

from which we obtain, by successive substitutions,

$$x = x'(\cos\theta\cos\phi - \sin\theta\sin\phi\cos\psi) - y'(\cos\theta\sin\phi + \sin\theta\cos\phi\cos\psi) + z'\sin\theta\sin\psi,$$

$$y = x'(\sin\theta\cos\phi + \cos\theta\sin\phi\cos\psi) - y'(\sin\theta\sin\phi - \cos\theta\cos\phi\sin\psi) - z'\cos\theta\sin\psi$$

$$z = x'\sin\phi\sin\psi + y'\cos\phi\sin\psi + z'\cos\psi.$$

120. *To obtain Euler's formulæ for transformation directly by Spherical Trigonometry.*

Describe a sphere with centre O, meeting the two systems of axes in XYZ, $X'Y'Z'$ respectively; and join these points by arcs of great circles, producing $X'Y'$ to meet XY in P.

Then

$$x = x'\cos X'X + y'\cos Y'X + z'\cos Z'X,$$

and similarly for y and z.

Also $\qquad XP = \theta,\ X'P = \phi,\ XPX' = \pi - \psi,$

and
$$\cos XX' = \cos XP\cos X'P + \sin XP\sin X'P\cos XPX'$$
$$= \cos\theta\cos\phi - \sin\theta\sin\phi\cos\psi,$$

$$\cos XY' = \cos XP\cos Y'P + \sin XP\sin Y'P\cos XPX'$$
$$= -\cos\theta\sin\phi - \sin\theta\cos\phi\cos\psi,$$

$$\cos XZ' = \sin XP\cos XPZ',\ PZ' \text{ being a quadrant,}$$
$$= \sin\theta\sin\psi,\ \angle Z'PY' \text{ being a right angle,}$$

$$\cos YX' = \cos YP\cos X'P + \sin YP\sin X'P\cos YPX'$$
$$= \sin\theta\cos\phi + \cos\theta\sin\phi\cos\psi,$$

$$\cos YY' = \cos YP\cos Y'P + \sin YP\sin Y'P\cos YPY'$$
$$= -\sin\theta\sin\phi + \cos\theta\cos\phi\cos\psi,$$

$$\cos YZ' = \sin YP\cos YPZ' = -\cos\theta\sin\psi,$$

$Y'PZ'$ being a right angle, and PZ' a quadrant.

$$\cos ZX' = \sin PX'\cos ZPX' = \sin\phi\sin\psi,$$

PZ being a quadrant, and ZPY a right angle.

$$\cos ZY' = \sin PY'\cos ZPY' = \cos\phi\sin\psi,$$

$$\cos ZZ' = \cos\psi.$$

121. These formulæ are too complicated and unsymmetrical to be generally employed. A modification of them is however sometimes useful in determining the nature of any plane section of a proposed surface. We may in that case, by using the first two transformations, make the plane of x_1y_2 coincide with the proposed plane section, and then, making $z'=0$, obtain the equation to the section in that plane. The results so obtained may be at once derived from our final equations by making $\phi=0$, $z'=0$, or directly by geometrical considerations, and we have the formulæ

$$x = x' \cos\theta - y' \sin\theta \cos\psi,$$

$$y = x' \sin\theta + y' \cos\theta \cos\psi,$$

$$z = y' \sin\psi,$$

by effecting which substitutions, we may obtain the equation to the curve which is the intersection of a given surface with a given plane.

Since the normal to this plane makes an angle ψ with the axis of z, and the plane containing the axis of z and the normal makes an angle $\theta - \frac{\pi}{2}$ with the plane of zx, the direction-cosines of the plane will be

$$\sin\psi \cos\left(\theta - \frac{\pi}{2}\right), \quad \sin\psi \sin\left(\theta - \frac{\pi}{2}\right), \quad \cos\psi;$$

$$\text{or } \sin\psi \sin\theta, \quad -\sin\psi\cos\theta, \quad \cos\psi.$$

Hence, if the equation of the plane be $lx + my + nz = 0$, the curve of intersection of which with the surface $f(x, y, z) = 0$ is required, we shall have

$$\cos\psi = \frac{n}{\sqrt{(l^2+m^2+n^2)}}, \text{ and } \tan\theta = -\frac{l}{m},$$

and (θ, ψ) being thus determined, the equation of the curve of intersection is

$$f(x\cos\theta - y\sin\theta\cos\psi,\ x\sin\theta + y\cos\theta\cos\psi,\ y\sin\psi) = 0.$$

122. As an example of the use of these formulæ, we will examine the nature of a section of the surface

$$\frac{x^2}{a^2}+\frac{y^2}{b^2}=\frac{z^2}{c^2},$$

which represents a cone, with a view to determine the circular sections, if any exist. The equation of the section made by a plane which passes through the point $(0, 0, \gamma)$, makes an angle ψ with the plane of xy, and whose trace on that plane makes an angle θ with the axis of x, is

$$\frac{(x'\cos\theta-y'\sin\theta\cos\psi)^2}{a^2}+\frac{(x'\sin\theta-y'\cos\theta\cos\psi)^2}{b^2}=\frac{(\gamma+y'\sin\psi)^2}{c^2},$$

the equation of a curve of the second order, which will be a circle, if

$$\frac{\cos^2\theta}{a^2}+\frac{\sin^2\theta}{b^2}=\frac{\cos^2\psi\sin^2\theta}{a^2}+\frac{\cos^2\psi\cos^2\theta}{b^2}-\frac{\sin^2\psi}{c^2},$$

$$\text{and}\quad \frac{\cos\psi\cos\theta\sin\theta}{a^2}-\frac{\cos\psi\cos\theta\sin\theta}{b^2}=0.$$

Hence, $\cos\psi$, $\cos\theta$, or $\sin\theta$ must be equal to zero, and we obtain the systems of solutions,

$$\text{(I)}\qquad \cos\psi=0,\quad \frac{\cos^2\theta}{a^2}+\frac{\sin^2\theta}{b^2}=-\frac{1}{c^2}:$$

$$\text{(II)}\qquad \cos\theta=0,\quad \frac{1}{b^2}=\frac{\cos^2\psi}{a^2}-\frac{\sin^2\psi}{c^2},$$

$$\text{or}\quad \cos\psi=\pm\sqrt{\left(\frac{\frac{1}{b^2}+\frac{1}{c^2}}{\frac{1}{a^2}+\frac{1}{c^2}}\right)}:$$

$$\text{(III)}\qquad \sin\theta=0,\quad \frac{1}{a^2}=\frac{\cos^2\psi}{b^2}-\frac{\sin^2\psi}{c^2}$$

$$\text{or}\quad \cos\psi=\pm\sqrt{\left(\frac{\frac{1}{a^2}+\frac{1}{c^2}}{\frac{1}{b^2}+\frac{1}{c^2}}\right)}.$$

Of these solutions, the first gives impossible values of θ, and the second or third impossible values of ψ, as $a >$ or $< b$.

Suppose $a > b$, the only possible circular sections are given by

$$\sin\theta = 0, \quad \cos\psi = \pm\sqrt{\left(\frac{\frac{1}{a^2}+\frac{1}{c^2}}{\frac{1}{b^2}+\frac{1}{c^2}}\right)},$$

or there are two systems of circular sections made by planes parallel to the axis of x, and equally inclined to it at an angle

$$\cos^{-1}\sqrt{\left(\frac{\frac{1}{a^2}+\frac{1}{c^2}}{\frac{1}{b^2}+\frac{1}{c^2}}\right)}.$$

123. *Transformation from a system of rectangular co-ordinates to a system of oblique co-ordinates, having the same origin.*

If $(a_1b_1c_1)$, $(a_2b_2c_2)$, $(a_3b_3c_3)$ be the direction-cosines of the second system, referred to the first, we shall have the equations, as in rectangular co-ordinates,

$$x = a_1x' + a_2y' + a_3z',$$
$$y = b_1x' + b_2y' + b_3z',$$
$$z = c_1x' + c_2y' + c_3z';$$

but the six equations of condition which in that case subsisted will now be reduced to the three

$$a_1^2 + b_1^2 + c_1^2 = 1,$$
$$a_2^2 + b_2^2 + c_2^2 = 1,$$
$$a_3^2 + b_3^2 + c_3^2 = 1,$$

and we have six disposable constants remaining.

124. *Transformation from one system of co-ordinates to another having the same origin, both systems being oblique.*

Let Ox, Oy, Oz; Ox', Oy', Oz' be the two systems; On, On', On'' the normals respectively to yz, zx, and xy, and let nx denote the angle nOx, and so for the others. Then the distance of a point whose co-ordinates in the two systems are respectively (x, y, z), (x', y', z') from the plane of yz, is $x\cos nx$, and is also

$$x'\cos nx' + y'\cos ny' + z'\cos nz'.$$

Whence $x \cos nx = x' \cos nx' + y' \cos ny' + z' \cos nz'$,

and similarly,

$$y \cos n'y = x' \cos n'x' + y' \cos n'y' + z' \cos n'z',$$

$$z \cos n''z = x' \cos n''x' + y' \cos n''y' + z' \cos n''z',$$

the required formulæ, involving in this form twelve constants, but as they may be written in the form

$$x = a_1x' + a_2y' + a_3z',$$
$$y = b_1x' + b_2y' + b_3z',$$
$$z = c_1x' + c_2y' + c_3z',$$

where $a_1 = \dfrac{\cos nx'}{\cos nx}$, and similarly for the others, we see that really only nine constants are involved, and these are connected by three equations on account of the angles between the original axes being fixed, so that there are still six disposable constants only.

125. *Transformation from any one system of axes to any other.*

If we wish in any of the above transformations of the directions of the axes also to remove the origin, we may first remove the origin to the point (α, β, γ), retaining the directions of the axes. This will give

$$x = x_1 + \alpha, \quad y = y_1 + \beta, \quad z = z_1 + \gamma,$$

x_1, y_1, z_1 being the co-ordinates of a point (x, y, z) referred to the system of axes through the new origin parallel to the primary system. Now changing the direction by transformations of the form

$$x_1 = a_1x' + a_2y' + a_3z', \text{ \&c.},$$

we see that the most general transformation possible is obtained by formulæ of the form

$$x = \alpha + a_1x' + a_2y' + a_3z',$$
$$y = \beta + b_1x' + b_2y' + b_3z',$$
$$z = \gamma + c_1x' + c_2y' + c_3z'.$$

126. *To shew that the degree of an equation cannot be changed by transformation of co-ordinates.*

We can now prove the important proposition, that the degree of an equation cannot be altered by any transformation of co-ordinates: the degree of an equation meaning the greatest number which can be obtained by adding the indices of the co-ordinates involved in any term. For let $Ax^py^qz^r$ be a term in an equation of the nth degree, such that $p+q+r=n$: this will be a type of all the terms of the nth degree involved in the equation, any one of which may be obtained by assigning to A, p, q, r suitable values. Now on any transformation this term becomes

$$A(\alpha+a_1x'+a_2y'+a_3z')^p(\beta+b_1x'+b_2y'+b_3z')^q(\gamma+c_1x'+c_2y'+c_3z')^r,$$

and no term in this product rises beyond the degree $p+q+r$ or n. Hence the degree of an equation cannot be raised by transformation of co-ordinates. Nor can it be depressed, for if by any transformation the degree be depressed, then on retransformation, the degree of the equation so depressed would be raised to its original value, which we have seen to be impossible.

127. *To transform from rectangular to polar co-ordinates.*

In the cases in which polar co-ordinates are required to be used, we may first transform the axes so that the axis of z is parallel to the line from which θ is measured, and the plane of zx parallel to the plane from which ϕ is measured. If when referred to these axes the co-ordinates of the pole be α, β, γ, the formula expressing the rectangular in terms of the polar co-ordinates will be

$$x=\alpha+r\sin\theta\cos\phi,\quad y=\beta+r\sin\theta\sin\phi,\quad z=\gamma+r\cos\theta.$$

128. *To transform from a four-plane to a three-plane co-ordinate system.*

This is immediately effected by the substitution

$$\alpha=p-lx-my-nz,$$

and by similar substitutions for β, γ, δ. (Art. 72.)

If the three planes terminating in D be taken for the three-plane system, and l, m, n be the sines of the angles which the edges DA, DB, DC make with the planes DBC, DCA, DAB respectively, we shall have

$$\alpha = lx, \quad \beta = my, \quad \gamma = nz;$$

and therefore $$\delta = p_4\left(1 - \frac{lx}{p_1} - \frac{my}{p_2} - \frac{nz}{p_3}\right)$$

for the formulæ of transformation.

129. *To transform from one four-point co-ordinate system to another.*

If the equations of the fundamental points of the second system, referred to the first, be

$$l\alpha + m\beta + n\gamma + r\delta = 0, \text{ \&c.},$$

and $\alpha, \beta, \gamma, \delta$; $\alpha', \beta', \gamma', \delta'$, be the co-ordinates of any plane in the two systems,

$$\alpha' = \frac{l\alpha + m\beta + n\gamma + r\delta}{l + m + n + r}, \text{ \&c., (Art. 95)}$$

from which equation the formulæ required for transformation can be deduced.

VIII.

1. If the expression $ax^2 + by^2 + cz^2 + 2a'yz + 2b'zx + 2c'xy$ become by transformation of co-ordinates,

$$\alpha x^2 + \beta y^2 + \gamma z^2 + 2\alpha' yz + 2\beta' zx + 2\gamma' zy,$$

shew that

$$a + b + c = \alpha + \beta + \gamma,$$

and

$$bc + ca + ab - a'^2 - b'^2 - c'^2 = \beta\gamma + \gamma\alpha + \alpha\beta - \alpha'^2 - \beta'^2 - \gamma'^2,$$

both systems of axes being rectangular.

2. If $(l_1m_1n_1)$, $(l_2m_2n_2)$, $(l_3m_3n_3)$ be the direction-cosines of a system of rectangular axes, and if $\frac{a}{l_1} + \frac{b}{m_1} + \frac{c}{n_1} = 0$, and $\frac{a}{l_2} + \frac{b}{m_2} + \frac{c}{n_2} = 0$, then will $\frac{a}{l_3} + \frac{b}{m_3} + \frac{c}{n_3} = 0$, and $a : b : c :: l_1l_2l_3 : m_1m_2m_3 : n_1n_2n_3$.

3. If $al_1^2 + bm_1^2 + cn_1^2 = 0 = al_2^2 + bm_2^2 + cn_2^2 = al_3^2 + bm_3^2 + cn_3^2$, shew that

$$l_1^2 - m_1^2 : l_2^2 - m_2^2 : l_3^2 - m_3^2 :: m_1^2 - n_1^2 : m_2^2 - n_2^2 : m_3^2 - n_3^2,$$

and that

$$l_1(m_2n_3 + m_3n_2) + l_2(m_3n_1 + m_1n_3) + l_3(m_1n_2 + m_2n_1) = 0.$$

4. Transform the equation $yz + zx + xy = a^2$, referred to rectangular axes, to an equation referred to another system, one of which makes equal angles with the original axes.

5. Shew that, by the same transformation as in the last problem, the equation

$$x^2 + y^2 + z^2 + yz + zx + xy = a^2$$

is reduced to the form

$$4x^2 + y^2 + z^2 = 2a^2.$$

6. The equations of the straight lines bisecting the angles between the straight lines given by the equations

$$lx + my + nz = 0, \quad ax^2 + 2bxy + cy^2 = 0,$$

are

$$lx + my + nz = 0, \quad x^2 \{anl - b(n^2 + l^2)\} + xy \{a(m^2 + n^2) - c(n^2 + l^2)\}$$
$$- y^2 \{cmn - b(m^2 + n^2)\} = 0.$$

7. The equations of the straight lines bisecting the angles between the straight lines given by the equations

$$lx + my + nz = 0, \quad ax^2 + by^2 + cz^2 = 0,$$

are

$$lx + my + nz = 0, \quad x^2(b - c) + y^2(c - a) + z^2(a - b)$$

$$= \frac{xy}{lm}\{l^2(b - c) + m^2(c - a) - n^2(a - b)\}$$

$$+ \frac{yz}{mn}\{-l^2(b - c) + m^2(c - a) + n^2(a - b)\}$$

$$+ \frac{zx}{nl}\{l^2(b - c) - m^2(c - a) + n^2(a - b)\}.$$

8. Shew, by transformation of four-point co-ordinates, that the center of gravity of a tetrahedron is also the center of gravity of the tetrahedron formed by joining the center of gravity of the faces.

9. Shew, by the same method, that the center of gravity of the surface of a tetrahedron is the center of the sphere inscribed in the tetrahedron formed by joining the centers of gravity of the faces.

CHAPTER VIII.

ON CERTAIN SURFACES OF THE SECOND ORDER.

130. BEFORE we proceed to discuss the general equation of the second order, we think it advisable for the student to render himself familiar with some of the properties of the surfaces which are represented by the general equation. We shall therefore introduce him to the equations of these surfaces in their simplest forms, in which the axes of co-ordinates being in the direction of lines symmetrically situated with regard to these surfaces, the nature and properties of the surfaces will be more easily deduced. The student will thus be enabled more clearly to understand the methods adopted in the general equations.

For this purpose we shall give geometrical definitions of the surfaces, and deduce equations from those definitions: and we shall shew *vice versâ* how from these equations the geometrical construction of those surfaces can be deduced.

The Sphere.

131. *To find the equation of a sphere.*

DEF. A sphere is the locus of a point, whose distance from a fixed point is constant. The fixed point is the center and the constant distance the radius of the sphere.

Let (a, b, c) be the center of the sphere, r the radius, (x, y, z) any point on the sphere;

$$\therefore\ (x-a)^2+(y-b)^2+(z-c)^2=r^2.$$

This equation may be written in the general form

$$x^2+y^2+z^2+Ax+By+Cz+D=0,$$

the equation required.

132. Since the general equation of the sphere contains four arbitrary constants, the sphere may be made to satisfy four specific conditions.

It may be seen from geometrical considerations that, when four conditions are given, there may be only one sphere, or a limited number, or an infinite number of spheres, which satisfy the equations; at the same time the four conditions must be consistent with the nature of a sphere, and if this be the case and the conditions be independent, there must be a limited number of spheres satisfying those conditions. For example, if four points be given through which a sphere is to pass, no three points can lie in one straight line, and if four points lie in one plane, they must also lie in a circle, otherwise no sphere could pass through them, and if such a condition is satisfied, an infinite number of spheres can be constructed each of which contains the circle in which the four points lie; if the four points do not lie in a plane so that the four conditions to be satisfied are independent, the sphere is completely determined.

Again, if four planes be given, each of which is to be touched by the sphere, no three of these must have one line of intersection, and the four cannot pass through one point, except under a condition, and in that case an infinite number of spheres can be drawn, touching the four planes. In other cases, eight spheres can be drawn satisfying the conditions.

Equation of a sphere under specific conditions.

133. *To find the equation of a sphere passing through a given point.*

Let (a, b, c) be the given point, and the equation of the sphere

$$x^2 + y^2 + z^2 + Ax + By + Cz + D = 0;$$

$$\therefore\ a^2 + b^2 + c^2 + Aa + Bb + Cc + D = 0;$$

$$\therefore\ x^2 + y^2 + z^2 + A(x - a) + B(y - b) + C(z - c) = a^2 + b^2 + c^2$$

is the equation required.

If the point be the origin, the equation becomes

$$x^2 + y^2 + z^2 + Ax + By + Cz = 0,$$

and the sphere may satisfy three more conditions.

134. *To find the equation of a sphere which passes through two given points in the axis of z.*

Let c_1, c_2 be the distances of the given points from O; when $x=0$, $y=0$, the equation must become $(z-c_1)(z-c_2)=0$; therefore the equation of the sphere is

$$x^2+y^2+(z-c_1)(z-c_2)+Ax+By=0.$$

If the sphere touches the axis of z, $c_1=c_2\equiv\gamma$,*

$$x^2+y^2+z^2+Ax+By-2\gamma z+\gamma^2=0.$$

135. *To find the equations of spheres which touch the three axes.*

Let the equation of the sphere be

$$x^2+y^2+z^2+Ax+By+Cz+D=0.$$

Since it touches the axis of x, let a be the distance from the origin; therefore when $y=0$, $z=0$,

$$x^2+Ax+D=0,$$

the roots of which are each equal to $\pm a$;

$$\therefore\ A=\pm 2a \text{ and } D=a^2.$$

Similarly, y^2+By+a^2 is a complete square;

$$\therefore\ B=\pm 2a \text{ and } C=\pm 2a,$$

and the equations of the spheres which satisfy the given conditions are

$$x^2+y^2+z^2\pm 2ax\pm 2ay\pm 2az+a^2=0,$$

which are eight in number for any given value of a, corresponding to the different compartments of the co-ordinate planes.

136. *To find the equation of a sphere touching the plane of xy in a given point.*

Since the sphere meets the plane of xy in the given point $(a, b, 0)$ only, when $z=0$, the equation must reduce to

$$(x-a)^2+(y-b)^2=0.$$

Therefore, the equation of the sphere is

$$(x-a)^2+(y-b)^2+z^2+Cz=0.$$

* The symbol $\equiv$, denoting identity, is employed by continental writers, because it is frequently convenient to distinguish between identity and equality.

137. *Interpretation of the expression*

$$(x-a)^2+(y-b)^2+(z-c)^2-r^2$$

in the equation of a sphere.

Let the equation of the sphere be

$$(x-a)^2+(y-b)^2+(z-c)^2-r^2=0,$$

and (x', y', z') be any point Q, C the center of the sphere, and let a straight line through Q intersect the sphere in the points P, P', and its equations be

$$\frac{x-x'}{l}=\frac{y-y'}{m}=\frac{z-z'}{n}=\rho;$$

therefore at the points P and P'

$$(l\rho+x'-a)^2+(m\rho+y'-b)^2+(n\rho+z'-c)^2-r^2=0,$$

if ρ_1, ρ_2 be the roots of this equation,

$$\rho_1\rho_2=(x'-a)^2+(y'-b)^2+(z'-c)^2-r^2;$$

therefore the left side of the equation for any point (x', y', z') is

$$QP.\,QP', \text{ or } -QP.\,QP',$$

according as Q is without or within the sphere.

If Q be without the sphere it is the square of a tangent drawn from Q to the sphere.

If Q be within, it is the square of the radius of the small circle on the sphere whose center is Q.

COR. All tangents drawn from an external point to the sphere are equal.

On the Relations of two or more Spheres.

138. *To find the equation of the radical plane of two spheres.*

DEF. The radical plane of two spheres is the locus of points, through which if lines be drawn intersecting the spheres, the product of the distances of the points of intersection from these points, measured in one direction, is the same for the two spheres.

Let the equations of the two spheres (A) and (B) be

$$(x-a)^2+(y-b)^2+(z-c)^2-r^2\equiv u=0,$$

and $$(x-a')^2+(y-b')^2+(z-c')^2-r'^2\equiv u'=0.$$

The equation of the radical plane is therefore

$$u - u' = 0.$$

139. *To shew that the six radical planes of four spheres intersect in one point.*

Let $u = 0, \ u' = 0, \ u'' = 0, \ u''' = 0$

be the equations, in this form, of the four spheres.

The equations of the six radical planes are given by

$$u = u' = u'' = u''',$$

which intersect in one point determined by these equations.

DEF. The point of intersection of the six radical planes is called the radical center of the four spheres.

Poles of Similitude.

140. DEF. If a point be found such that the tangents drawn to two spheres are proportional to the radii of the spheres, such points are called *Poles of Similitude.*

If the Pole of Similitude be in the line joining the centers of the spheres *produced*, such a pole is called the *External Pole of Similitude.*

If the pole of similitude be in the line joining the centers it is called the *Internal Pole of Similitude.*

If the spheres intersect the internal pole in a point through which if chords be drawn to each other, the rectangles under the segments are proportional to the squares of the radii.

Both poles of similitude are the vertices of the cones which envelope both spheres, if they do not intersect; if they intersect, this is true of the external pole alone.

Properties of the Poles of Similitude of four Spheres.

141. A sphere may be defined as the envelope of planes which are equidistant from the center; and if four-point co-ordinates be employed, the relation between the co-ordinate of such planes, which expresses this fact, is called the equation of the sphere.

The equations of the four spheres whose centers are the fundamental points A, B, C, D, and radii r_1, r_2, r_3, r_4, are

$$\alpha = r_1, \ \beta = r_2, \ \gamma = r_3, \ \delta = r_4.$$

The external and internal poles of similitude of the spheres (A) and (B) have equations

$$\frac{\alpha}{r_1} \mp \frac{\beta}{r_2} = 0, \text{ and similarly for the rest.}$$

(1) The external poles of (AB), (AC), and (AD) lie in a plane whose co-ordinates are connected by the equations

$$\frac{\alpha}{r_1} = \frac{\beta}{r_2} = \frac{\gamma}{r_3} = \frac{\delta}{r_4},$$

which evidently contains also the external poles of (BC), (CD) and (DA).

(2) The co-ordinates of the plane containing the external poles of (AB) and (AC) and the internal pole of AD satisfy the equations

$$\frac{\alpha}{r_1} = \frac{\beta}{r_2} = \frac{\gamma}{r_3} = -\frac{\delta}{r_4},$$

and the same plane evidently contains the external pole of (BC) and the internal poles of (BD) and (CD).

(3) The co-ordinates of the planes containing the external poles of (AB) and the internal of (AC) and (AD) satisfy the equations

$$\frac{\alpha}{r_1} = \frac{\beta}{r_2} = -\frac{\gamma}{r_3} = -\frac{\delta}{r_4},$$

and this plane evidently contains the external pole of (CD) and the internal poles of (BC) and (BD).

Hence one plane contains the six external poles, four planes contain each three external and three internal poles, and three contain each two external and four internal poles.

The poles of similitude lie in eight planes, each of which pass through six poles of similitude situated three and three in four straight lines.

The planes are called planes of similitude.

Thus for the six external poles

$$\frac{\alpha}{r_1} = \frac{\beta}{r_2} = \frac{\gamma}{r_3} = \frac{\delta}{r_4},$$

and $$\frac{\beta}{r_2} - \frac{\gamma}{r_3} = \frac{\alpha}{r_1} - \frac{\gamma}{r_3} - \left(\frac{\alpha}{r_1} - \frac{\beta}{r_2}\right) = 0;$$

therefore the pole of (BC) lies in the line joining those of (AB) and (AC).

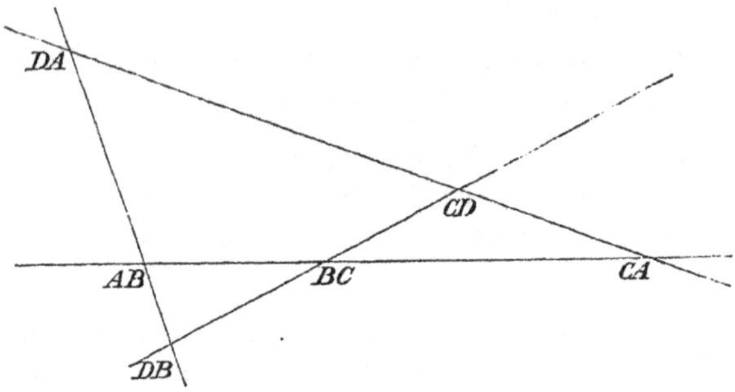

Similarly it lies in the lines joining those of (DB) and (DC).

Hence, the six external poles lie in the sides of a plane quadrilateral, as in the figure.

Cylindrical Surfaces.

142. It has been seen that the locus of an equation $F(x, y) = 0$, which involves only two of the co-ordinates, is a cylindrical surface, of which the generating lines are parallel to the axis of the co-ordinates omitted. We shall now shew how to obtain the equation of certain cylindrical surfaces in which the generating lines are in a general direction.

143. *To find the equation of the cylindrical surface, whose generating lines are in a given direction and guiding curve an ellipse traced on the plane of zx.*

Let the equations of the guiding ellipse be

$$\frac{x^2}{a^2} + \frac{y^2}{b^2} = 1, \text{ and } z = 0, \qquad (1),$$

and (l, m, n) the directions of the generating lines.

Let the equations of any generating line be

$$\left.\begin{aligned} nx &= lz + \alpha \\ ny &= mz + \beta \end{aligned}\right\} \qquad (2).$$

At the point of intersection of the generating line with the guiding curve, the values of x, y, z in (1) and (2) being the same, we obtain as a general equation, after eliminating x, y and z,

$$\frac{\alpha^2}{a^2} + \frac{\beta^2}{b^2} = n^2, \qquad (3),$$

and since this is true for all positions of the generating line, eliminating α, β between (2) and (3),

$$\frac{(nx - lz)^2}{a^2} + \frac{(ny - mz)^2}{b^2} = n^2,$$

is true for every point in the cylindrical surface, and is therefore its equation.

Conical Surfaces.

144. DEF. A conical surface is a surface generated by a straight line which constantly passes through a given point, called the vertex, and is subject to any other condition.

145. *To find the equations of a conical surface, whose vertex is the origin, and of which a guiding curve is an ellipse, whose center is in the axis of z, and plane parallel to the plane of xy.*

Let the equations of the guiding ellipse be

$$\frac{x^2}{a^2} + \frac{y^2}{b^2} = 1, \text{ and } z = c, \qquad (1),$$

those of a generating line in any position,

$$x = \alpha z, \quad y = \beta z. \qquad (2).$$

Eliminating x, y, z the point in which the generating line meets the guiding curve,

$$\frac{\alpha^2 c^2}{a^2} + \frac{\beta^2 c^2}{b^2} = 1. \qquad (3).$$

Since this equation is true for every position of the generating line, eliminating α, β from (2) and (3),

$$\frac{x^2}{a^2} + \frac{y^2}{b^2} = \frac{z^2}{c^2},$$

which is the required equation of the surface.

146. *To find the equation of an oblique circular cone.*

Let the equations of the guiding circle be

$$x^2 + y^2 = c^2, \quad z = 0, \qquad (1),$$

and the co-ordinates of the vertex be a, 0, b; and let the equations of the generating line in any position be

$$\left.\begin{aligned} x - a &= \alpha(z - b) \\ y &= \beta(z - b) \end{aligned}\right\} \qquad (2),$$

eliminating x, y, z the co-ordinates of the points in which the lines (1) and (2) intersect,

$$(a - \alpha b)^2 + \beta^2 b^2 = c^2, \qquad (3),$$

and this being true for every position of the generating line, we obtain from (2) and (3)

$$\{a(z-b) - b(x-a)\}^2 + b^2y^2 = c^2(z-b)^2;$$

$$\text{or,} \quad (az - bx)^2 + b^2y^2 = c^2(z-b)^2,$$

the equation of the oblique circular cone.

147. *To find the circular sections of the oblique circular cone.*

The equation of the cone

$$(az - bx)^2 + b^2y^2 = c^2(z-b)^2$$

may be written in the form

$$b^2(x^2 + y^2 + z^2 - c^2) = z\{2abx + (b^2 + c^2 - a^2)z - 2bc^2\}.$$

If a section be made by a plane whose equation is $z = k$, the points in the curve of intersection satisfy the equation

$$b^2(x^2 + y^2 + z^2 - c^2) = k\{2abx + (b^2 + c^2 - a^2)z - 2bc^2\},$$

which shews that the curve lies on a sphere (Art. 131), and is therefore a circular section.

If a section be made by a plane whose equation is

$$2abx + (b^2 + c^2 - a^2)z - 2bc^2 = p^3,$$

the points in the curve of intersection satisfy the equations

$$b^2(x^2 + y^2 + z^2 - c^2) = p^3 z,$$

which shews that this curve also lies on a sphere, and is therefore a circular section.

There are therefore two series of circular sections, made by two systems of parallel planes.

The trace of the cone on the plane of zx, putting $y = 0$, has for its equation,

$$(az - bx)^2 - c^2(z-b)^2 = 0,$$

and is therefore two of the generating lines; and the equation of two planes in opposite systems whose curves of intersections are circles,

$$(z - k)\{2abx + (b^2 + c^2 - a^2)z - 2bc^2 - p^3\} = 0,$$

by adding these equations, we obtain,

$$b^2 (x^2 + z^2 + Ax + Bz + C) = 0,$$

which shews that the four points, in which those generating lines meet the two planes, lie in a circle.

Hence, the first system makes the same angle with one generating line which the second does with the other.

The Spheroids.

148. Def. A spheroid may be generated by the revolution of an ellipse about either axis.

If the axis of revolution be the minor axis, the surface is called an *Oblate Spheroid*, and if the major axis, a *Prolate Spheroid.*

149. *To find the equation of a spheroid.*

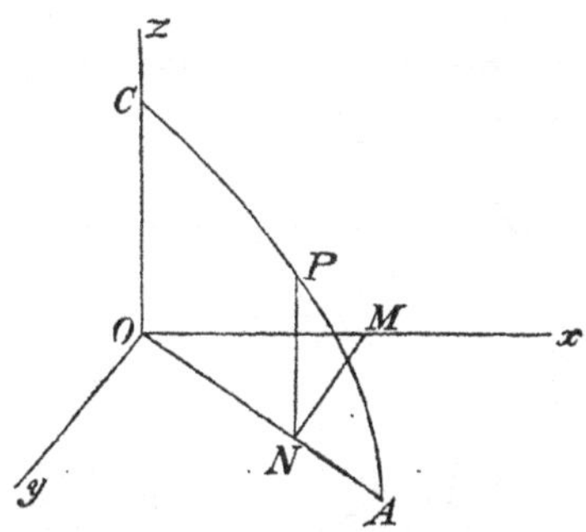

Let the center be taken as origin, the axis of revolution that of z, and let P be a point (x, y, z) in the ellipse CPA, which is the position of the revolving ellipse, when inclined at any angle to the plane of zx,

$$OM = x, \quad MN = y, \quad NP = z, \quad OA = a, \quad OC = b;$$

$$\therefore \frac{ON^2}{a^2} + \frac{NP^2}{c^2} = 1;$$

$$\therefore \frac{x^2 + y^2}{a^2} + \frac{z^2}{c^2} = 1.$$

This is the equation of an oblate or prolate spheroid according as c is less or greater than a.

The Ellipsoid.

150. *To find the equation of an ellipsoid.*

Def. An ellipsoid is the surface generated by the motion of a variable ellipse, which moves so that its plane is always parallel to a fixed plane, and which changes its form so that its vertices lie in two ellipses traced on perpendicular planes, to which its plane is perpendicular, and which have a common axis in the line of intersection of the planes.

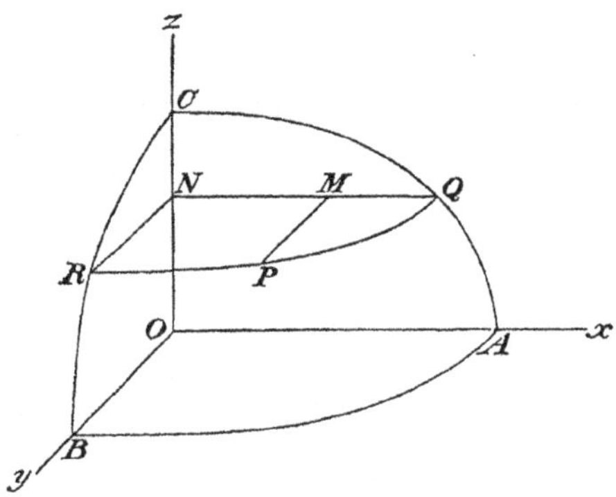

Let QRN be a variable ellipse in any position, Q, R being its vertices lying in two ellipses AC, BC, traced on perpendicular planes, taken for those of zx and yz; the plane of xy, to which the variable ellipse is parallel, being the plane containing the semi-axes OA, OB.

Let a, c, and b, c, be the semi-axes of AC and BC, and (x, y, z) any point P in QR, PM perpendicular to QN.

Then, $$\frac{y^2}{RN^2} + \frac{x^2}{QN^2} = 1,$$

and, since Q is a point in the ellipse AC,

$$\frac{QN^2}{a^2} = 1 - \frac{z^2}{c^2},$$

similarly,
$$\frac{RN^2}{b^2} = 1 - \frac{z^2}{c^2},$$

$$\therefore \frac{x^2}{a^2} + \frac{y^2}{b^2} + \frac{z^2}{c^2} = 1,$$

which is the equation required.

151. *To construct the surface whose equation is*

$$\frac{x^2}{a^2} + \frac{y^2}{b^2} + \frac{z^2}{c^2} = 1.$$

Let the surface be cut by a plane whose equation is $z = \gamma$. The projection of the curve of intersection on the plane of xy has the equation

$$\frac{x^2}{a^2} + \frac{y^2}{b^2} = 1 - \frac{\gamma^2}{c^2},$$

therefore, the curve is an ellipse whose semi-axes α, β are given by the equations

$$\frac{\alpha^2}{a^2} = 1 - \frac{\gamma^2}{c^2} = \frac{\beta^2}{b^2};$$

hence, the vertices lie in the two ellipses which are the traces of the surface on the planes of zx and yz.

Also, since $\frac{\alpha}{a} = \frac{\beta}{b}$, the variable ellipse remains always similar to a given ellipse, which is the trace on the plane of xy.

The surface may therefore be generated by the motion of a variable ellipse, whose plane, &c. (See Def.)

The Hyperboloid of one Sheet.

152. *To find the equation of the hyperboloid of one sheet.*

DEF. The hyperboloid of one sheet may be generated by the motion of a variable ellipse, which moves so that its plane is always parallel to a fixed plane, and which changes its form so that its vertices always lie in two hyperbolas traced on perpendicular planes, to which its plane is perpendicular, these hyperbolas having a common conjugate axis.

Let AQ, BR be the hyperbolas traced on the two perpendicular planes taken for the planes of zx, yz, OC their common semi-conjugate axis, being the direction of the axis of z.

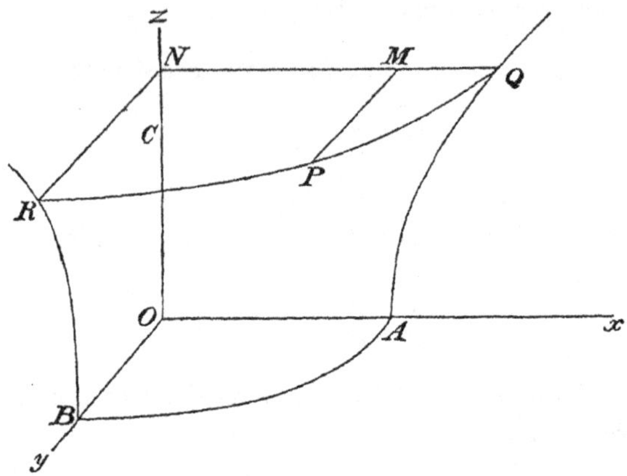

Let QPR be the variable ellipse in any position, P any point (x, y, z) in it, QN, RN its semi-axes.

Draw PM perpendicular to QN.

Then, $$MN = x, \quad PM = y, \quad ON = z,$$

and $$\frac{x^2}{QN^2} + \frac{y^2}{RN^2} = 1;$$

also, since Q, R are points in the hyperbolas,

if $$OA = a, \quad OB = b, \text{ and } OC = c,$$

$$\frac{QN^2}{a^2} = \frac{z^2}{c^2} + 1 = \frac{RN^2}{b^2},$$

$$\therefore \frac{x^2}{a^2} + \frac{y^2}{b^2} = \frac{z^2}{c^2} + 1,$$

$$\text{or, } \frac{x^2}{a^2} + \frac{y^2}{b^2} - \frac{z^2}{c^2} = 1,$$

which is the equation of the hyperboloid of one sheet.

153. *To construct the surface which is the locus of the equation,*

$$\frac{x^2}{a^2} + \frac{y^2}{b^2} - \frac{z^2}{c^2} = 1.$$

Let the surface be cut by a plane whose equation is $z=\gamma$, then the projection of the curve of intersection upon the plane of xy has for its equation,

$$\frac{x^2}{a^2}+\frac{y^2}{b^2}=1+\frac{\gamma^2}{c^2},$$

which is the equation of an ellipse, whose semi-axes α, β are given by the equations,

$$\frac{\alpha^2}{a^2}=1+\frac{\gamma^2}{c^2}=\frac{\beta^2}{b^2};$$

therefore, the vertices of the ellipse lie respectively on the hyperbolas, which are the traces of the surface on the planes of zx, yz.

Also since $\frac{\alpha}{a}=\frac{\beta}{b}$, this ellipse is always similar to the ellipse which is the trace of the surface on the plane of xy.

Hence, the locus may be generated by the motion of a variable ellipse which moves, &c. (See Def.)

154. If the surface be cut by a plane parallel to the plane of zx, whose equation is $x=\alpha$, the curve of intersection will be an hyperbola, the equation of whose projection on the plane of yz, will be

$$\frac{y^2}{b^2}-\frac{z^2}{c^2}=1-\frac{\alpha^2}{a^2}.$$

If $\alpha < a$, the semi-axes β, γ will satisfy the equation

$$\frac{\beta^2}{b^2}=\frac{\gamma^2}{c^2}=1-\frac{\alpha^2}{a^2}.$$

Hence, the extremities of the transverse axis 2β will lie on the ellipse which is the trace on the plane of xy.

If $\alpha > a$, we shall have $\frac{\beta^2}{b^2}=\frac{\gamma^2}{c^2}=\frac{\alpha^2}{a^2}-1.$

Hence, the extremities of the transverse axis 2γ will lie on the hyperbola, which is the trace on the plane of zx.

155. *To find the form of the surface at an infinite distance.*

If z be increased indefinitely,

$$\frac{x^2}{a^2}+\frac{y^2}{b^2}=\frac{z^2}{c^2}\left(1+\frac{c^2}{z^2}\right)=\frac{z^2}{c^2} \text{ ultimately.}$$

Let this surface and the hyperboloid be cut by a straight line drawn parallel to Oz through a point $(x', y', 0)$, and z_1, z_2 be the corresponding values of z,

$$\frac{x'^2}{a^2} + \frac{y'^2}{b^2} = \frac{z_1^2}{c^2},$$

$$\frac{x'^2}{a^2} + \frac{y'^2}{b^2} = \frac{z_2^2}{c^2} + 1;$$

$$\therefore \frac{z_1^2 - z_2^2}{c^2} = 1, \text{ and } z_1 - z_2 = \frac{c^2}{z_1 + z_2};$$

if x' or y' or both, and therefore z_1 and z_2, be indefinitely increased, $z_1 - z_2$ diminishes indefinitely, and ultimately vanishes;

$$\therefore \frac{x^2}{a^2} + \frac{y^2}{b^2} = \frac{z^2}{c^2}$$

is the equation of an asymptotic surface, which lies further from the plane of xy than the hyperboloid.

This asymptotic surface is a cone, for, if it be cut by any plane whose equation is $\frac{x}{a} = \frac{z}{c} \cos\theta$, all the points of intersection lie in the planes $\frac{y}{b} = \pm \frac{z}{c} \sin\theta$. The surface is therefore capable of being generated by a straight line which passes through the origin, and is guided by the ellipse whose equations are

$$\frac{x^2}{a^2} + \frac{y^2}{b^2} = 1, \text{ and } z = c.$$

The figure shews the position of the conical asymptote relative to the hyperboloid.

$ABab$ is the principal elliptic section, $A'B'a'b'$, $A''B''a''b''$ the sections of the hyperboloid and cone made by a plane parallel to that principal section, at a distance $OC = c$.

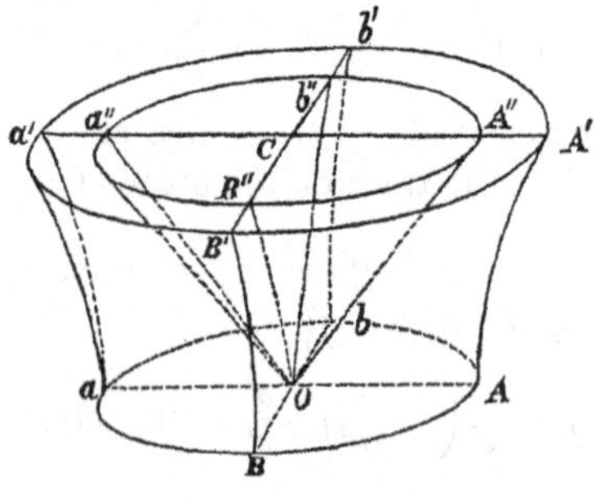

The Hyperboloid of two Sheets.

156. *To find the equation of the hyperboloid of two sheets.*

DEF. The hyperboloid of two sheets may be generated by the motion of a variable ellipse, which moves so that its plane is always parallel to a fixed plane, and which changes its form, so that its vertices lie always on two hyperbolas traced upon two perpendicular planes, having a common transverse axis, perpendicular to the fixed plane, to which the plane of the ellipse is parallel.

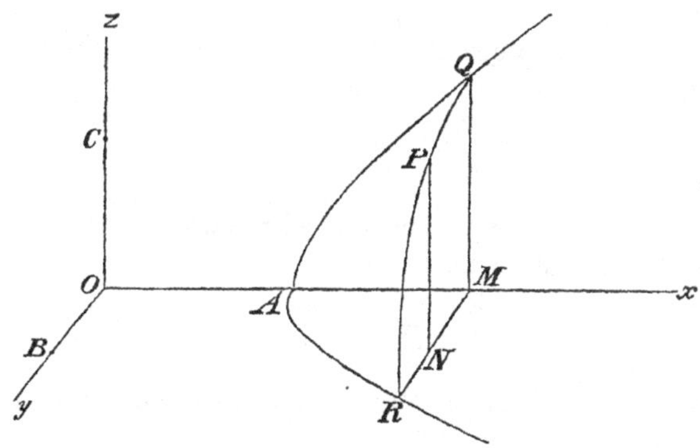

Let AQ, AR be the hyperbolas traced on two perpendicular planes, taken for the planes of zx, xy, and having the common semi-transverse axis OA, and let QPR be the variable ellipse in any position, whose axes are QM, RM, parallel to the plane of yz.

Take P any point (x, y, z) in the ellipse, and draw PN perpendicular to RM, then $OM = x$, $MN = y$, and $NP = z$; therefore, since P is a point in the ellipse,

$$\frac{y^2}{RM^2} + \frac{z^2}{QM^2} = 1;$$

and if a, c and a, b be the semi-axes of the two hyperbolas, AQ, AR,

$$\frac{RM^2}{b^2} = \frac{x^2}{a^2} - 1 = \frac{QM^2}{c^2};$$

$$\therefore \frac{y^2}{b^2} + \frac{z^2}{c^2} = \frac{x^2}{a^2} - 1;$$

$$\text{or,} \quad \frac{x^2}{a^2} - \frac{y^2}{b^2} - \frac{z^2}{c^2} = 1,$$

which is the equation of the hyperboloid of two sheets.

157. *To construct the locus of the surface whose equation is*

$$\frac{x^2}{a^2} - \frac{y^2}{b^2} - \frac{z^2}{c^2} = 1.$$

Let the surface be cut by a plane whose equation is $x = \alpha$; the equation of the projection on the plane of yz of the curve of intersection is

$$\frac{y^2}{b^2} + \frac{z^2}{c^2} = \frac{\alpha^2}{a^2} - 1,$$

which, if $\alpha > a$, is the equation of an ellipse whose semi-axes β, γ are given by the equations

$$\frac{\beta^2}{b^2} = \frac{\alpha^2}{a^2} - 1 = \frac{\gamma^2}{c^2},$$

therefore the vertices of the ellipse lie in two hyperbolas, whose equations are

$$\frac{x^2}{a^2} - \frac{y^2}{b^2} = 1, \text{ and } \frac{x^2}{a^2} - \frac{z^2}{c^2} = 1,$$

which are the traces of the surface on the planes of xy, zx, having a common transverse axis in the line Ox: and since $\frac{\beta}{b} = \frac{\gamma}{c}$, this ellipse is always similar to a given ellipse, axes $2b$, $2c$.

Hence, the locus may be constructed by the motion of a variable ellipse which, &c. (See Def.)

158. If the surface be cut by a plane parallel to the plane of xy, whose equation is $z = \gamma$, the curve of intersection will be an hyperbola, the equations of whose projection on the plane of xy, will be

$$\frac{x^2}{a^2} - \frac{y^2}{b^2} = 1 + \frac{\gamma^2}{c^2},$$

which may be written $\frac{x^2}{\alpha^2} - \frac{y^2}{\beta^2} = 1$, whose transverse and conjugate semi-axes will satisfy the equations

$$\frac{\alpha^2}{a^2} = 1 + \frac{\gamma^2}{c^2} = \frac{\beta^2}{b^2}.$$

Hence, the tranverse axis will have its extremities in the hyperbola, which is the trace on the plane of zx, and the hyperbolic section will be similar to the trace on the plane of xy.

159. *To find the form of the hyperboloid of two sheets at an infinite distance.*

If x be increased indefinitely, the equation $\frac{x^2}{a^2} = \frac{y^2}{b^2} + \frac{z^2}{c^2} + 1$ shews that y, or z, or both, are also increased indefinitely, and the equation becomes

$$\frac{x^2}{a^2} = \left(\frac{y^2}{b^2} + \frac{z^2}{c^2}\right)\left(1 + \frac{1}{\frac{y^2}{b^2} + \frac{z^2}{c^2}}\right)$$

$$= \frac{y^2}{b^2} + \frac{z^2}{c^2} \text{ ultimately.}$$

Let the hyperboloid, and the surface represented by this equation, be cut by a straight line parallel to the axis of x, drawn through the point $(0, y', z')$, x_1, x_2, the corresponding values of x, are given by the equations,

$$\frac{x_1^2}{a^2} = \frac{y'^2}{b^2} + \frac{z'^2}{c^2},$$

$$\text{and } \frac{x_2^2}{a^2} = \frac{y'^2}{b^2} + \frac{z'^2}{c^2} + 1;$$

$$\therefore \frac{x_2^2 - x_1^2}{a^2} = 1, \text{ and } x_2 - x_1 = \frac{a^2}{x_2 + x_1};$$

therefore, $x_2 - x_1$ diminishes indefinitely, and ultimately vanishes as y', or z', or both, increase indefinitely; hence the hyperboloid of two sheets continually approximates to the form of the surface whose equation is $\frac{x^2}{a^2} = \frac{y^2}{b^2} + \frac{z^2}{c^2}$, which is therefore called an asymptotic surface.

Also, if this surface be cut by a plane whose equation is $\frac{y}{b} = \frac{x}{a} \cos\theta$, all the points of intersection lie in the two planes $\frac{z}{c} = \pm \frac{x}{a} \sin\theta$; and the surface can therefore be generated by straight lines drawn through the origin, which intersect the ellipse, whose equations are $\frac{y^2}{b^2} + \frac{z^2}{c^2} = 1$, $x = a$.

This asymptotic surface is therefore a cone on an elliptic base, and lies nearer to the plane of yz than the hyperboloid, since $x_1^2 > x_2^2$.

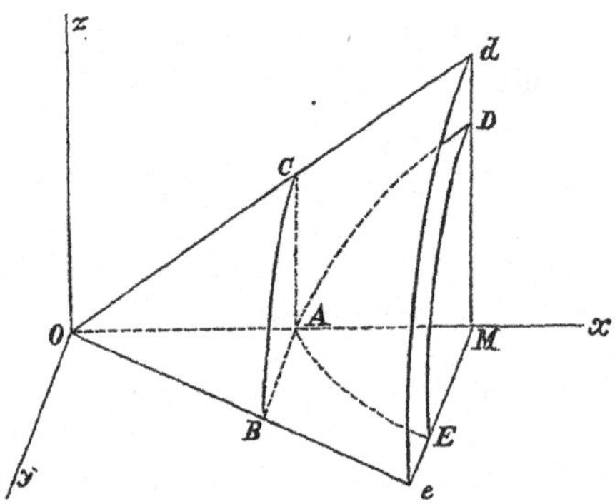

Its position relative to the hyperboloid is shewn in the figure in which BC is the section made by a plane parallel to yz through the extremity of the transverse axis, and DE, de are sections of the hyperboloid and conical asymptote, made by a plane parallel to yz.

The Elliptic Paraboloid.

160. *To find the equation of the elliptic paraboloid.*

Def. The elliptic paraboloid may be generated by the motion of a parabola, whose vertex lies in a parabola traced upon a fixed plane, to which its plane is always perpendicular, the axes of the two parabolas being parallel, and the concavities turned in the same direction.

Let xOy be the plane on which the fixed parabola OQ is traced, Ox the axis of OQ; QR the axis of the moveable parabola QP, P any point (x, y, z) in the parabola.

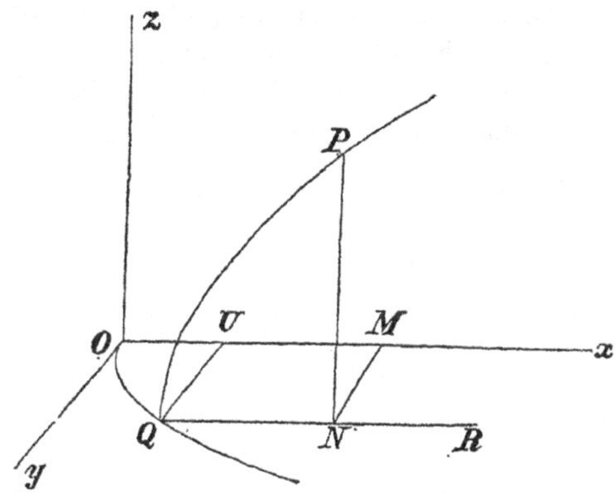

Draw PN perpendicular to QR, and QU, NM to Ox, then since P is a point in QP, if l, l' be the latera recta of OQ and QP,

$$PN^2 = l' . QN, \text{ and } QU^2 = l . OU;$$

$$\therefore \frac{y^2}{l} + \frac{z^2}{l'} = OU + QN = OM = x,$$

which is the equation of the elliptic paraboloid.

161. *To construct the locus of the equation,*

$$\frac{y^2}{l} + \frac{z^2}{l'} = x.$$

Let the locus be cut by a plane, whose equation is $y = \beta$, the projection of the curve of intersection upon the plane of zx has for its equation,

$$z^2 = l'\left(x - \frac{\beta^2}{l}\right),$$

which represents a parabola whose axis is parallel to the axis of x, the co-ordinates of whose vertex are $\frac{\beta^2}{l}$, β, 0; therefore the vertex of the parabolic section lies in the parabola whose equation is $y^2 = lx$, which is the trace on the plane of xy; therefore the locus may be constructed by the motion of a parabola, whose vertex, &c. (See Def.)

The Hyperbolic Paraboloid.

162. *To find the equation of the Hyperbolic Paraboloid.*

DEF. The Hyperbolic Paraboloid may be generated by the motion of a parabola, whose vertex lies in a parabola traced upon a fixed plane, to which its plane is perpendicular, the axes of the two parabolas being parallel, and the concavities turned in opposite directions.

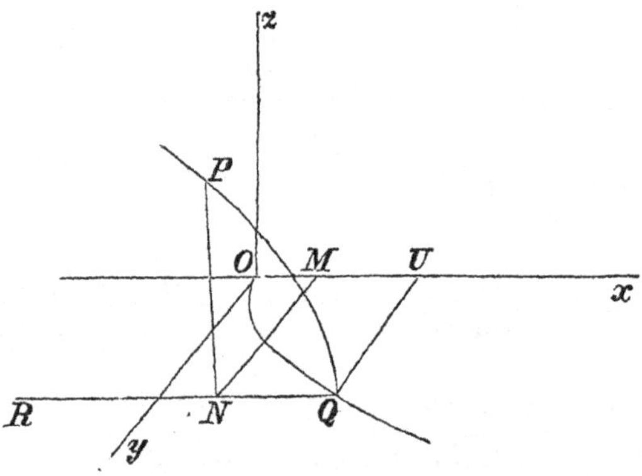

Let xOy be the fixed plane upon which the parabola is drawn, Ox the direction of the axis of the parabola: let QR be the axis of the moveable parabola QP, parallel to Ox, measured in the direction contrary to Ox.

Draw PN perpendicular to QR, and QU, NM to Ox; then, if P be any point (x, y, z) in QP, $OM=x$, $MN=y$, and $NP=z$.

Let l, l' be the latera recta of OQ, QP;

$$\text{therefore, } PN^2 = l' . QN, \text{ and } QU^2 = l . OU,$$

$$\text{and } \frac{QU^2}{l} - \frac{PN^2}{l'} = OU - QN = OM;$$

$$\therefore \frac{y^2}{l} - \frac{z^2}{l'} = x,$$

which is the equation of the hyperbolic paraboloid.

163. *To construct the locus of the equation*

$$\frac{y^2}{l} - \frac{z^2}{l'} = x.$$

Let the locus of the equation be cut by the plane, whose equation is $y = \beta$: the projection of the curve of intersection upon the plane of zx has for its equation,

$$z^2 = l'\left(\frac{\beta^2}{l} - x\right),$$

which represents a parabola, whose axis is measured in the direction contrary to Ox, and the algebraical distance of whose vertex from the plane yOz is $\frac{\beta^2}{l}$; therefore the section by the plane $y = \beta$ is a parabola, whose latus rectum is l' and the co-ordinates of whose vertex are, $\frac{\beta^2}{l}$, β, 0; or, the vertex lies in a parabola traced upon the plane of xy, whose equation is $y^2 = lx$.

Hence the locus may be generated by the motion of a parabola, whose vertex, &c. (See Def.)

164. The locus may also be generated by the motion of a hyperbola; for if it be cut by a plane parallel to that of yz on the positive side, whose equation is $x = \alpha$, the equation of the projection of the curve of intersection on the plane of yz will be $\frac{y^2}{l} - \frac{z^2}{l'} = \alpha$, whose transverse and conjugate semi-axes, β, γ, will satisfy the equations $\beta^2 = l\alpha$ and $\gamma^2 = l'\alpha$, the extremities of the transverse axis lie in the trace on the plane of xy, and the conjugate axis is equal to the double ordinate of the trace on the plane of zx corresponding to $x = -\alpha$.

If it be cut by a plane on the negative side of yz, the section will be an hyperbola whose transverse axis is in the direction of Oz.

165. *To find the form of the hyperbolic paraboloid at an infinite distance.*

If y and z be indefinitely increased while x remains finite,

$$\frac{y^2}{l} = \frac{z^2}{l'}\left(1 + \frac{l'x}{z^2}\right) = \frac{z^2}{l'} \text{ ultimately};$$

$$\therefore \frac{y}{\sqrt{l}} = \pm \frac{z}{\sqrt{l'}},$$

and if these planes and the hyperbolic paraboloid be cut by a straight line parallel to Oy, drawn through a point $(x', 0, z')$ y_1, y_2 the corresponding values of y are given by the equations,

$$\frac{y_1^2}{l} = \frac{z'^2}{l'}, \quad \text{and } \frac{y_2^2}{l} = \frac{z'^2}{l'} + x';$$

$$\therefore \frac{y_2^2 - y_1^2}{l} = x', \quad \text{or, } \frac{y_2 - y_1}{lx'} = \frac{lx'}{y_2 + y_1}.$$

Therefore, if x' remain finite or small compared with y_1 or y_2, $y_2 - y_1$ diminishes as z' increases and ultimately vanishes; and the two planes whose equations are $\frac{y}{\sqrt{l}} = \pm \frac{z}{\sqrt{l'}}$ give the form of the surface at an infinite distance for finite values of x, or for values of x which are small compared with y or z.

These planes will not form an asymptotic surface, except for points at which x vanishes compared with y or z, since $y_2 - y_1$ will not ultimately vanish in that case, and similarly for $z_2 - z_1$.

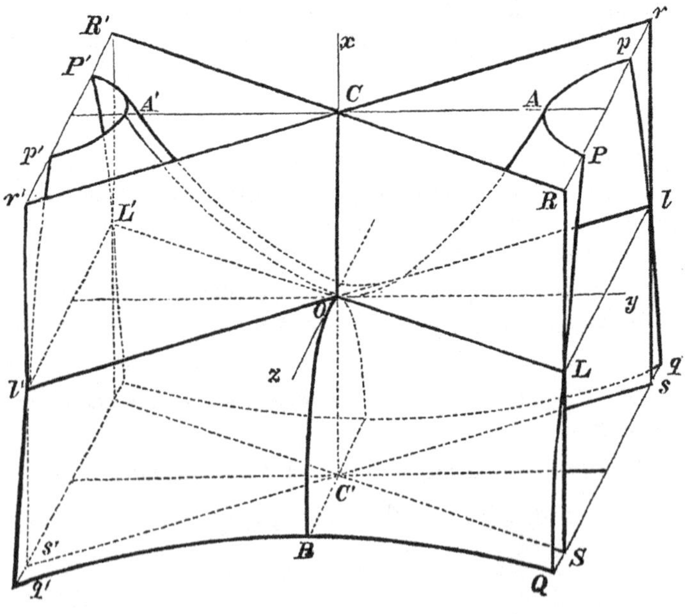

The figure is intended to shew the position of the asymptotic planes with reference to the hyperbolic paraboloid.

Ox is parallel to the axis of the generating parabola, of which OB is one position in the plane of zx.

PAp, $P'A'p'$ are opposite branches of a hyperbolic section perpendicular to Ox, the asymptotes of which RCR', rCr' are sections of the asymptotic surface, AA', the transverse axis, being parallel to Oy.

LL', ll' are the traces on the plane of yz of both the paraboloid and its asymptotic surface.

QBq' is a branch of a hyperbolic section on the negative side of Ox, SC', $sC's'$, the asymptotes are sections of the asymptotic surface, and the transverse axis BC' is parallel to Oz.

166. *To shew that the elliptic and hyperbolic paraboloid are particular cases of the ellipsoid, and the hyperboloid respectively.*

Let
$$\frac{x^2}{a^2} \pm \frac{y^2}{b^2} \pm \frac{z^2}{c^2} = 1$$

be the equation of an ellipsoid or hyperboloid, and remove the origin to the point $(-a, 0, 0)$. The transformed equation is

$$\frac{x^2}{a^2} \pm \frac{y^2}{b^2} \pm \frac{z^2}{c^2} = \frac{2x}{a}.$$

Let $\frac{b^2}{a}$, $\frac{c^2}{a}$ remain finite quantities, while a, b, c become infinite, and denote them by l and l'. The equation may then be written

$$\frac{x^2}{a} \pm \frac{y^2}{l} \pm \frac{z^2}{l'} = 2x,$$

which has for its limit, when a becomes infinite,

$$\pm \frac{y^2}{l} \pm \frac{z^2}{l'} = 2x,$$

which is the equation of an elliptic or hyperbolic paraboloid. The assumption that $\frac{b^2}{a}$ and $\frac{c^2}{a}$ remain finite is the same thing as assuming that the latera recta of the traces on the planes xy, zx, respectively, remain finite when the axes become infinite, and the corresponding ellipses or hyperbolas become parabolas.

It is obvious from the above that the elliptic paraboloid is a limiting case either of the ellipsoid or the hyperboloid of two sheets, and the hyperbolic paraboloid of the hyperboloid of one sheet.

167. The surfaces of the second order, which we have been discussing, are of the two forms,

$$Ax^2 + By^2 + Cz^2 = D, \tag{1}$$

and $$By^2 + Cz^2 = Ax; \tag{2}$$

and it will be shewn in a succeeding chapter that all surfaces of the second degree may by transformation of co-ordinates be reduced to one of these two forms.

The first form of equation includes all surfaces which have a center at a *finite* distance, and the second those which have a center at an *infinite* distance.

In the equation (1), if $-x$, $-y$, $-z$ be written respectively for x, y, z the equation is not altered, therefore, if (x, y, z) be a point in the surface, $(-x, -y, -z)$ is also a point in it, so that if POP' be any chord through the origin O, the chord is bisected in O, and O is a center of the surface.

Also, for any values of y and z, the values of x are equal and of opposite signs, therefore the plane of yz bisects the chords which are drawn perpendicular to it; and a plane which bisects the chords drawn perpendicular to it, is called a *principal plane* of the surface.

Hence, the planes xy, yz and zx are principal planes of the surface.

It is evident that the planes of yz, zx are principal planes of the surfaces whose equations are of the form (2).

The sections made by the principal planes are called *principal sections*.

That the surface represented by (2) has a center at an infinite distance, may be shewn by considering this equation as the limiting form of (1) when the origin is transferred to a point $(-\alpha, 0, 0)$, α being determined by the equation $A\alpha^2 = D$. The equation will then assume the form

$$Ax^2 + By^2 + Cz^2 = 2A\alpha x;$$

and this surface has a center on the axis of x, at distance α from the origin.

Now, if we suppose A to vanish, while $A\alpha$ remains finite, an equation of the form (2) is the result. But to satisfy these conditions α must be infinitely great; hence, a surface represented by (2) has a center at an infinite distance on the axis of x, and also a third principal section, parallel to the plane of yz, at an infinite distance.

IX.

(1) A straight line is drawn through a fixed point O, meeting a fixed plane in Q, and in this straight line is taken a point P such that $OP \,.\, OQ$ is equal to a given quantity; shew that P lies on a sphere passing through O, and whose center lies on the perpendicular from O upon the plane.

(2) Investigate the equation of a sphere conceived to be generated by the motion of a variable circle, whose diameter is one of a system of parallel chords of a given circle, to which the plane of the variable circle is perpendicular.

(3) Construct the sphere whose polar equation is

$$r = a \sin \theta \cos \phi.$$

(4) A straight line moves with three fixed points A, B, C in the three co-ordinate planes; shew that any other fixed point P of the straight line will lie on an ellipsoid whose semi-axes are PA, PB, and PC.

(5) Find the locus of a point whose distance from a given point bears a constant ratio to its distance, (1) from a fixed plane, (2) from a fixed straight line.

(6) Find the locus of a point which is equidistant from two fixed lines which do not intersect.

(7) The locus of a point, whose distance from a fixed plane is always equal to its distance from a fixed line, is a cone.

(8) The locus of a straight line, intersecting two fixed straight lines and remaining parallel to a fixed plane, is a hyperbolic paraboloid.

(9) The locus of the line of intersection of two planes at right angles to each other, each of which passes through one of two straight lines, inclined at an angle 2α and whose shortest distance is $2c$, is a hyperboloid of one sheet, one of whose axes is $2c$, and the others are as $\cos \alpha : \sin \alpha$.

(10) The surface generated by a straight line, revolving about a fixed straight line, with which it is supposed rigidly connected, will be a cone, or a hyperboloid, according as the straight lines do or do not intersect.

(11) Shew that the elliptic paraboloid may be generated by a variable ellipse, the extremities of whose axes lie on two parabolas having a common axis, and whose planes are at right angles to each other.

(12) Shew that an hyperboloid of one or of two sheets degenerates into a right elliptic cone, when its axes become indefinitely small, and preserve a finite ratio to each other.

(13) Three straight lines, mutually at right angles, are drawn from the origin to meet the ellipsoid $\frac{x^2}{a^2}+\frac{y^2}{b^2}+\frac{z^2}{c^2}=1$, shew that, if their lengths be r_1, r_2, r_3,

$$\frac{1}{r_1^2}+\frac{1}{r_2^2}+\frac{1}{r_3^2}=\frac{1}{a^2}+\frac{1}{b^2}+\frac{1}{c^2}.$$

(14) If A, B, C be the extremities of the axes of an ellipsoid, and AC, BC the sections containing the least axis, find the equations of the two cones whose vertices are A, B, and bases BC, AC respectively; shew that the cones have a common parabolic section, and if l be the latus-rectum of this parabola, and l_1, l_2 those of the sections AC, BC, then

$$\frac{1}{l^2}=\frac{1}{l_1^2}+\frac{1}{l_2^2}.$$

(15) The plane $lx+my+nz=0$ will cut the cone $ax^2+by^2+cz^2=0$ in two straight lines at right angles to each other, if

$$l^2(b+c)+m^2(c+a)+n^2(a+b)=0.$$

(16) Through a fixed point is drawn a straight line meeting an ellipsoid in two points, and on this straight line is taken a point such that its distance from the fixed point is, (1) an arithmetic, (2) a geometric, (3) a harmonic mean between the segments of the straight line made by the ellipsoid, find the locus of the point in the three cases.

(17) Find the locus of a point through which three straight lines can be drawn mutually at right angles, and passing through the perimeter of a given plane curve of the second order.

(18) The trace of an ellipsoid on the plane of xy is AB; shew that a cone which has AB for a guiding curve will intersect the ellipsoid in another plane curve, and that this plane intersects the plane of AB in the polar with respect to AB of the projection of the vertex on that plane.

CHAPTER IX.

ON GENERATION BY LINES AND CIRCLES.

168. In the preceding chapter we have shewn how certain surfaces of the second order may be generated by the motion of ellipses, hyperbolas and parabolas. In the case of the cylinder and cone we have investigated the equations by supposing a generation by the motion of a straight line subject to certain conditions: we have also shewn that the cone may be generated in two ways by the motion of a variable circle which moves parallel to a fixed plane, its magnitude being subject to fixed laws.

We shall in this chapter shew that the hyperboloid of one sheet, and the hyperbolic paraboloid, as well as the cone and cylinder are capable of being generated by the motion of a straight line, and that the ellipsoid, both hyperboloids, and the elliptic paraboloid can be generated by the motion of a variable circle, which moves parallel to either of two fixed planes.

Generating lines.

169. *To find the generating lines of an hyperboloid of one sheet.*

The equation of the hyperboloid of one sheet is

$$\frac{x^2}{a^2} + \frac{y^2}{b^2} - \frac{z^2}{c^2} = 1.$$

This equation may be written in the form

$$\frac{x^2}{a^2} + \frac{y^2}{b^2} = \left(\cos\theta \pm \frac{z}{c}\sin\theta\right)^2 + \left(\sin\theta \mp \frac{z}{c}\cos\theta\right)^2$$

for all values of θ;

$$\left.\begin{aligned} \therefore\ \frac{x}{a} &= \cos\theta \pm \frac{z}{c}\sin\theta \\ \text{and}\ \frac{y}{b} &= \sin\theta \mp \frac{z}{c}\cos\theta \end{aligned}\right\}, \qquad (1)$$

satisfy the equation, hence the two straight lines which, for a particular value of θ, have these for their equations, lie entirely in the surface.

By the variation of θ we obtain two systems of straight lines, which lie entirely in the surface, and either of these systems generates the hyperboloid. These equations may also be written in the form

$$\frac{x - a\cos\theta}{a\sin\theta} = \frac{y - b\sin\theta}{-b\cos\theta} = \pm\frac{z}{c},$$

from which equations it is manifest that parallel straight lines drawn through the centre will lie upon the asymptotic cone. Hence also, no three generators of the hyperboloid can be parallel to the same plane.

If $z = 0$, $x = a\cos\theta$, and $y = b\sin\theta$, therefore θ is the eccentric angle of the point of intersection of the two straight lines (1) with the trace of the hyperboloid on the plane of xy.

170. Any point of the hyperboloid may be represented by the co-ordinates

$$a\cos\theta\sec\phi,\quad b\sin\theta\sec\phi,\quad c\tan\phi,$$

since these satisfy the equation for all values of θ and ϕ. The equations of the generating lines through this point may readily be found to be

$$\frac{x - a\cos\theta\sec\phi}{a\sin(\theta\pm\phi)} = \frac{y - b\sin\theta\sec\phi}{-b\cos(\theta\pm\phi)} = \frac{z - c\tan\phi}{\pm c};$$

which shew that they meet the principal elliptic section in points whose eccentric angles are $\theta \pm \phi$.

171. *The projections of the generating lines upon the principal planes are tangents to the traces on those planes.*

The equation of the trace on the plane of zx is

$$\frac{x^2}{a^2} - \frac{z^2}{c^2} = 1,$$

and that of the projection of a generating line on the same plane

$$\frac{x}{a} = \cos\theta \pm \frac{z}{c}\sin\theta,$$

and the points of intersection are given by the equation

$$\frac{z^2}{c^2} + 1 - (\cos\theta \pm \frac{z}{c}\sin\theta)^2 = 0,$$

or $$\frac{z^2}{c^2}\cos^2\theta \mp \frac{2z}{c}\cos\theta\sin\theta + \sin^2\theta = 0,$$

which, giving coincident values of z, shews that the projection is a tangent to the trace upon the plane of zx.

Similarly, the projection on the plane of xy, and the trace on that plane, intersect in points given by the equations

$$\frac{x}{a}\cos\theta + \frac{y}{b}\sin\theta = 1, \qquad \frac{x^2}{a^2} + \frac{y^2}{b^2} = 1,$$

whence $$\left(\frac{x}{a}\sin\theta - \frac{y}{b}\cos\theta\right)^2 = 0.$$

Hence the points of intersection coincide, or the projection is a tangent to the trace on the plane of xy.

172. *To shew that two generating lines of the same system do not intersect.*

The equations of two generating lines of the same system are

$$\frac{x}{a} = \cos\theta \pm \frac{z}{c}\sin\theta, \qquad \frac{y}{b} = \sin\theta \mp \frac{z}{c}\cos\theta;$$

and $$\frac{x}{a} = \cos\theta' \pm \frac{z}{c}\sin\theta', \qquad \frac{y}{b} = \sin\theta' \mp \frac{z}{c}\cos\theta'.$$

If the two lines meet, we have at the points of intersection,

$$0 = \cos\theta - \cos\theta' \pm \frac{z}{c}(\sin\theta - \sin\theta'),$$

and $$0 = \sin\theta - \sin\theta' \mp \frac{z}{c}(\cos\theta - \cos\theta');$$

and the condition of intersection is

$$(\cos\theta - \cos\theta')^2 + (\sin\theta - \sin\theta')^2 = 0;$$

which cannot be satisfied unless $\theta = \theta'$.

Hence, generating lines of the same system do not intersect.

173. *To shew that generating lines of opposite systems intersect.*

The equations of two generating lines of opposite systems are

$$\frac{x}{a} = \cos\theta \pm \frac{z}{c}\sin\theta, \quad \frac{y}{b} = \sin\theta \mp \frac{z}{c}\cos\theta;$$

and
$$\frac{x}{a} = \cos\theta' \mp \frac{z}{c}\sin\theta', \quad \frac{y}{b} = \sin\theta' \pm \frac{z}{c}\cos\theta'.$$

If the two lines meet, we have on the points of intersection,

$$0 = \cos\theta - \cos\theta' \pm \frac{z}{c}(\sin\theta + \sin\theta'),$$

and
$$0 = \sin\theta - \sin\theta' \mp \frac{z}{c}(\cos\theta + \cos\theta'),$$

and the condition that they may intersect is

$$\cos^2\theta - \cos^2\theta' + \sin^2\theta - \sin^2\theta' = 0,$$

which being identically true, shews that any two generating lines of opposite systems intersect.

174. *No straight line lies on an hyperboloid which does not belong to one of the two systems of generating lines.*

For, if possible, let a straight line (C) lie entirely on the hyperboloid, then since each system generates the whole hyperboloid, (C) must meet an infinite number of straight lines of each system; let two of these (A) and (B) of opposite systems intersect (C) in two different points, in which case a plane can be drawn intersecting the surface in three straight lines; but the section of a surface of the second order by a plane must be a curve of the second degree, therefore no such line as (C) can exist.

175. *To shew that a hyperboloid may be generated by the motion of a straight line, intersecting three fixed straight lines, which do not intersect.*

Since any generating line intersects all the generating lines of the opposite system, let us take three fixed generating lines of the same system: these will therefore not intersect. If a straight line now be supposed to move in such a manner as always to intersect these three straight lines, it will trace out the hyperboloid of which they are generating lines.

For, the three points in which the moving line meets the three fixed lines are points of the hyperboloid, so that it meets the hyperboloid in three points, which is impossible, unless the straight line lies altogether upon the surface, since the equation determining the points of intersection of a straight line with a surface of the second order, being a quadratic equation, cannot be satisfied by more than two roots without being satisfied by an infinite number.

The straight line, therefore, in its different positions, will trace out the hyperboloid.

176. *To find the locus of the intersections of two generating lines of opposite systems, drawn through the points in the principal elliptic section, whose eccentric angles differ by a constant angle.*

Let $\theta+\alpha$, and $\theta-\alpha$, be the eccentric angles of the principal elliptic section, differing by a constant angle 2α.

The equations of the generating lines of opposite systems are

$$\frac{x}{a}=\cos(\theta+\alpha)\pm\frac{z}{c}\sin(\theta+\alpha),\quad \frac{y}{b}=\cos(\theta+\alpha)\mp\frac{z}{c}\cos(\theta+\alpha);$$

and

$$\frac{x}{a}=\cos(\theta-\alpha)\mp\frac{z}{c}\sin(\theta-\alpha),\quad \frac{y}{b}=\sin(\theta-\alpha)\pm\frac{z}{c}\cos(\theta-\alpha).$$

At the points of intersection,

$$0=\cos\theta\sin\alpha\mp\frac{z}{c}\sin\theta\cos\alpha,\qquad \therefore \frac{z}{c}=\pm\tan\alpha.$$

Also

$$\frac{x}{a}=\cos\theta\cos\alpha\pm\frac{z}{c}\cos\theta\sin\alpha=\cos\theta\sec\alpha,$$

$$\frac{y}{b}=\sin\theta\cos\alpha\pm\frac{z}{c}\sin\theta\cos\alpha=\sin\theta\sec\alpha;$$

$$\therefore\ \frac{x^2}{a^2}+\frac{y^2}{b^2}=\sec^2\alpha.$$

Therefore the locus of the intersections of the two pairs of opposite systems is the two elliptic sections, parallel to the plane of xy, which intersect the traces on the planes of zx, yz, at points whose eccentric angles are $\pm\alpha$, and θ is the eccentric angle of the ellipses at these points.

177. The geometrical construction of the locus may be made as follows; let PQ, PQ', and $P'Q$, $P'Q'$, be generating lines at points P and P' in the principal elliptic section; PT, $P'T$ their projections on that principal plane will be tangents to that ellipse; therefore (Art. 171) QTQ' will be the line of intersection of the planes containing the generating lines, and will be perpendicular to the principal plane.

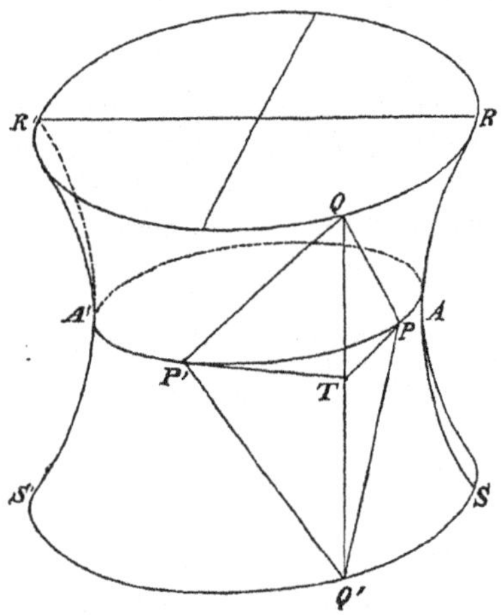

The co-ordinates of T are $a \cos\theta \sec\alpha$ and $b \sin\theta \sec\alpha$, as is easily shewn by means of the eccentric angle of the ellipse,

$$\text{hence, } \frac{QT^2}{c^2} = \cos^2\theta \sec^2\alpha + \sin^2\theta \sec^2\alpha - 1 = \tan^2\alpha,$$

$\therefore$ $QT = c\tan\alpha = Q'T$, or, the loci of Q, Q' are elliptic sections parallel to the principal elliptic section, at distances $c\tan\alpha$ from its plane.

178. The figure is meant to be a representation of the positions of sixteen generating lines of each system, correspond-

ing to eccentric angles differing by $\frac{\pi}{8}$. $ABab$ is the principal elliptic section, $A'B'a'b'$ and $A''B''a''b''$ are the parallel elliptic sections which intersect the conjugate axis of the hyperboloid at its extremities C', C'', the axes of which sections are in the ratio $\sqrt{2} : 1$ to the axes of the principal sections.

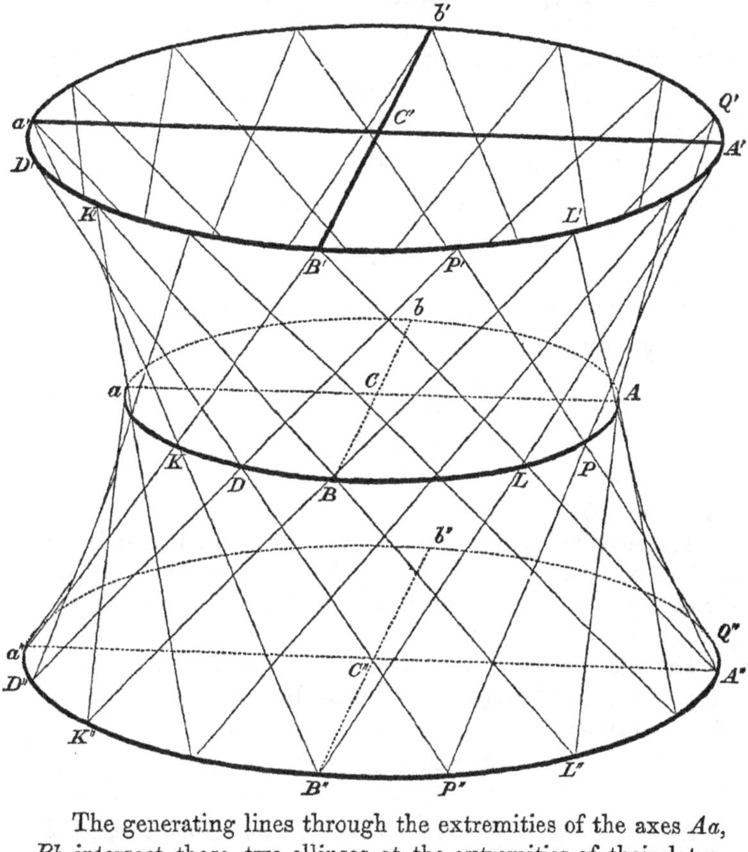

The generating lines through the extremities of the axes Aa, Bb intersect these two ellipses at the extremities of their latus rectums, as L', K', and L'', K'', and they are parallel to the asymptotes of the principal hyperbolic section through Bb: those through the extremities of the latus rectums, as L, K, pass through the extremities of the axes of the two ellipses.

The two ellipses $A'B'a'$ and $A''B''a''$ are the loci of the intersections of opposite systems of generating lines drawn through the extremities of conjugate diameters of the principal elliptic section.

The figure serves to represent that the intersection of generating lines of opposite systems drawn through points in the principal elliptic section, whose eccentric angles differ by a constant angle, lie in an ellipse parallel to the principal plane. As for example, such pairs of generating lines as LB', $P'D$, and BL', PP'.

179. *To find the generating lines of a hyperbolic paraboloid.*

The equation of the hyperbolic paraboloid,

$$\frac{y^2}{l} - \frac{z^2}{l'} = x,$$

is satisfied by the values of x, y, z for every point in the line whose equations are

$$\frac{y}{\sqrt{l}} \mp \frac{z}{\sqrt{l'}} = \frac{\alpha}{\sqrt{l'}}, \qquad (1)$$

$$\frac{y}{\sqrt{l}} \pm \frac{z}{\sqrt{l'}} = \frac{\sqrt{l'}}{\alpha} x, \qquad (2)$$

whatever be the value of α.

Therefore by giving α all values, we obtain two series of straight lines, all of which lie entirely in the surface; these are the two systems of lines which are rectilinear generators of the paraboloid.

The equation (1) shews that in the two systems all the generators are parallel respectively to the two asymptotic planes, whose equations are

$$\frac{y}{\sqrt{l}} \mp \frac{z}{\sqrt{l'}} = 0.$$

180. *To shew that generating lines of a hyperbolic paraboloid of the same system do not intersect; and that those of opposite systems do intersect.*

Let the equations of two generating lines of the same system be

$$\frac{y}{\sqrt{l}} \mp \frac{z}{\sqrt{l'}} = \frac{\alpha}{\sqrt{l'}}, \quad \frac{y}{\sqrt{l}} \pm \frac{z}{\sqrt{l'}} = \frac{\sqrt{l'}}{\alpha} x,$$

and $\dfrac{y}{\sqrt{l}} \mp \dfrac{z}{\sqrt{l'}} = \dfrac{\beta}{\sqrt{l'}}, \quad \dfrac{y}{\sqrt{l}} \pm \dfrac{z}{\sqrt{l'}} = \dfrac{\sqrt{l'}}{\beta} x.$

If the two lines intersect these equations are simultaneous, therefore $\dfrac{\alpha - \beta}{\sqrt{l'}} = 0$, which is impossible since α is not equal to β. Hence they do not intersect.

Changing the order of the signs in the ambiguities in the second set of equations, we have the equations of a line in the system opposite to that of the first.

If then the straight lines intersect,

$$\frac{\alpha + \beta}{\sqrt{l'}} = \frac{2y}{\sqrt{l}} = \sqrt{l'}\left(\frac{1}{\alpha} + \frac{1}{\beta}\right) x,$$

and
$$\frac{\alpha - \beta}{\sqrt{l'}} = \mp \frac{2z}{\sqrt{l'}} = \sqrt{l'}\left(\frac{1}{\beta} - \frac{1}{\alpha}\right) x,$$

and the consistency of these equations proves that two generating straight lines will always intersect if of opposite systems.

181. It may be shown by the reasoning employed in Art. 174, that no straight line can lie on the paraboloid which does not belong to one of these systems, and, as in Art. 175, that the paraboloid may be generated by the motion of a straight line which intersects two fixed straight lines, and is parallel to a fixed plane; also by a straight line which intersects three fixed straight lines which are themselves all parallel to the same plane.

It appears also from the latter construction, that if any straight line intersects three fixed straight lines, which are all parallel to the same plane, the intersecting straight line will in all its positions be parallel to another fixed plane.

182. *To shew that the projections of the generating lines on the principal planes, are tangents to the principal sections.*

Since the equations of the generating lines are

$$\frac{y}{\sqrt{l}} \mp \frac{z}{\sqrt{l'}} = \frac{\alpha}{\sqrt{l'}}, \quad \frac{y}{\sqrt{l}} \pm \frac{z}{\sqrt{l'}} = \frac{\sqrt{l'}}{\alpha} x.$$

The equation of their projections on the plane of zx, is

$$\pm \frac{2z}{\sqrt{l'}} = \frac{\sqrt{l'}}{\alpha} x - \frac{\alpha}{\sqrt{l'}},$$

which, being of the form $z = mx - \frac{l'}{4m}$ is the equation of a tangent to the parabola $z^2 = -l'x$, and similarly for the projection on the plane of xy.

183. The figure is intended to represent the manner in which the hyperbolic paraboloid is capable of being generated by straight lines.

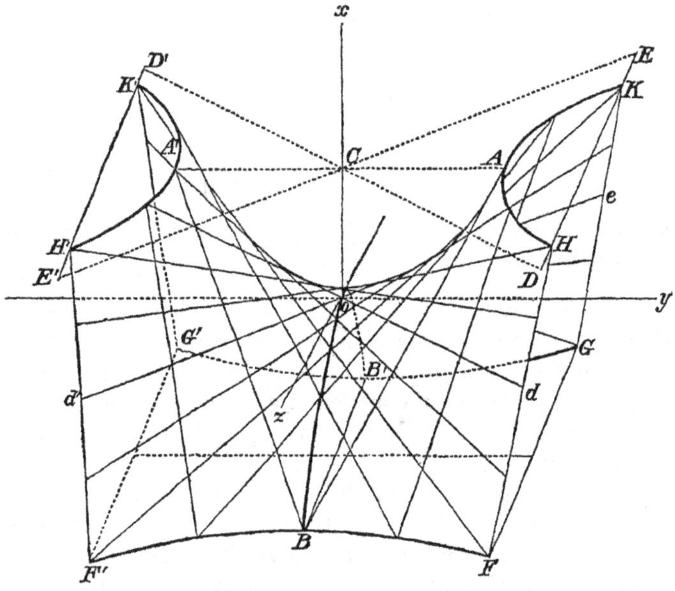

HAK, *H'A'K'* are portions of the branches of a hyperbolic section made by a plane parallel to that of yz, cutting Oz on the positive side; *ECE'*, *DCD'* are the asymptotes.

FBF', *GB'G'* are portions of the branches of the hyperbolic section parallel to yz on the negative side of Ox.

The two sections are so chosen that the generating lines through B, the extremity of the transverse axes of one section, pass through A, A', the extremities of the transverse axis of the other.

dO, $d'Oe$ are the traces of the paraboloid on the plane yz, where the hyperbolic sections degenerate into straight lines.

BOB' is the trace on the plane zx, and AOA' on the plane xy.

Generating Circles.

184. *To find the condition that a plane section of a central surface of the second order may be circular, and the locus of the centers of circular sections.*

Let $ax^2 + by^2 + cz^2 = 1$, be the equation of the surface,

and $lx + my + nz = p$, that of a cutting plane.

Any straight line drawn in this plane through the center of the section will be bisected by that point.

Let $\frac{x - x_0}{\lambda} = \frac{y - y_0}{\mu} = \frac{z - z_0}{\nu} = r$ be the equations of any diameter, (x_0, y_0, z_0) being the center, the values of r at the points in which this straight line meets the surface are given by the equation

$$a(x_0 + \lambda r)^2 + b(y_0 + \mu r)^2 + c(z_0 + \nu r)^2 = 1,$$

and since the values of r are equal and of opposite signs,

$$ax_0\lambda + by_0\mu + cz_0\nu = 0;$$

also
$$l\lambda + m\mu + n\nu = 0;$$

which equations being true for an infinite number of values of $\lambda : \mu : \nu$, we have

$$\frac{ax_0}{l} = \frac{by_0}{m} = \frac{cz_0}{n} = \frac{p}{\frac{l^2}{a} + \frac{m^2}{b} + \frac{n^2}{c}}.$$

The equations $\frac{ax}{l} = \frac{by}{m} = \frac{cz}{n}$ will therefore represent the locus of the centers of all sections made by planes whose direction-cosines are l, m, n.

Now, the values of r are given by the equation

$$ax_0^2 + by_0^2 + cz_0^2 - 1 + (a\lambda^2 + b\mu^2 + c\nu^2)\, r^2 = 0,$$

and since, for circular section, these values of r are equal for all values of λ, μ, ν consistent with the equation

$$l\lambda + m\mu + n\nu = 0;$$

$$\therefore\ a\lambda^2 + b\mu^2 + c\nu^2 = M(\lambda^2 + \mu^2 + \nu^2),$$

where M is independent of the values of λ, μ, ν; hence, eliminating λ, we have, for an infinite number of values of $\mu:\nu$,

$$(M-a)(m\mu+n\nu)^2+l^2\{(M-b)\mu^2+(M-c)\nu^2\}=0,$$

and therefore the coefficients of μ^2, $\mu\nu$, and ν^2 are each equal 0;

$$\therefore\ (M-a)\,mn=0.$$

If $M=a$, either $l=0$, or $M=b=c$, in which case the surface is spherical, and the equation is satisfied for any values of l, m, n, i. e. for any direction of the plane.

Also if $m=0$, the coefficient of $\mu^2=M-b=0$; similarly, if $n=0$, $M-c=0$.

Hence, if the surface be not spherical, we must have

$$l,\ m,\ \text{or}\ n=0.$$

Suppose $m=0$, then $M=b$,
and the coefficient of $\nu^2=(M-a)\,n^2+(M-c)\,l^2=0$;

$$\therefore\ \frac{l^2}{b-a}=\frac{n^2}{c-b}=\frac{1}{c-a}.$$

$$\text{Similarly, if}\quad l=0,\quad \frac{n^2}{a-c}=\frac{m^2}{b-a}=\frac{1}{b-c},$$

$$\text{and, if}\ n=0,\quad \frac{m^2}{c-b}=\frac{l^2}{a-c}=\frac{1}{a-b}.$$

One, and only one, of these three systems is possible, viz. the first, if a, b, c be in order of magnitude.

In this case, the co-ordinates of the center are given by the equations

$$\frac{ax_0}{\pm\sqrt{(b-a)}}=\frac{by_0}{0}=\frac{cz_0}{\pm\sqrt{(c-b)}}=\frac{pac}{\sqrt{(c-a)}}.$$

The equations of the locus of the centers are

$$\frac{ax}{\pm\sqrt{(b-a)}}=\frac{by}{0}=\frac{cz}{\pm\sqrt{(c-b)}}.$$

Therefore, any central surface may be generated by the motion of a variable circle parallel to either of two fixed planes, the center remaining on two corresponding lines which are diameters of the central surface.

I. The equation of the ellipsoid is

$$\frac{x^2}{a^2}+\frac{y^2}{b^2}+\frac{z^2}{c^2}=1.$$

If a, b, c be supposed to be in descending order of magnitude, the only possible directions of circular section are given by the equations

$$\frac{l^2}{\frac{1}{b^2}-\frac{1}{a^2}}=\frac{n^2}{\frac{1}{c^2}-\frac{1}{b^2}}=\frac{1}{\frac{1}{c^2}-\frac{1}{a^2}},\ \text{and } m=0,$$

the section being parallel to the mean axis.

II. The equation of the hyperboloid of one sheet is

$$\frac{x^2}{a^2}+\frac{y^2}{b^2}-\frac{z^2}{c^2}=1.$$

If $a > b$, the corresponding equations are

$$\frac{m^2}{\frac{1}{b^2}-\frac{1}{a^2}}=\frac{n^2}{\frac{1}{a^2}+\frac{1}{c^2}}=\frac{1}{\frac{1}{b^2}+\frac{1}{c^2}},\ \text{and } l=0,$$

the section being parallel to the greater real axis.

III. The equation of the hyperboloid of two sheets is

$$\frac{x^2}{a^2}-\frac{y^2}{b^2}-\frac{z^2}{c^2}=1.$$

If $b > c$, the corresponding equations are

$$\frac{l^2}{\frac{1}{b^2}+\frac{1}{a^2}}=\frac{n^2}{\frac{1}{c^2}-\frac{1}{b^2}}=\frac{1}{\frac{1}{a^2}+\frac{1}{c^2}},\ \text{and } m=0,$$

the section being parallel to the greater conjugate axis.

185. We may also determine the plane circular sections as follows.

The equation of the central surface being

$$ax^2+by^2+cz^2=1,$$

we may write it in the form

$$b(x^2+y^2+z^2)+(a-b)x^2-(b-c)z^2=1,$$

or $\{\sqrt{(a-b)}\,x+\sqrt{(b-c)}\,z\}\,\{\sqrt{(a-b)}\,x-\sqrt{(b-c)}\,z\}$
$$+\,b\,(x^2+y^2+z^2)=1.$$

If, therefore, a, b, c be in order of magnitude, the equation is satisfied by the points of intersection of the planes whose equations are

$$\sqrt{(a-b)}\,x \pm \sqrt{(b-c)}\,z = k,$$

and the sphere whose equation is

$$b\,(x^2+y^2+z^2)+k\,\{\sqrt{(a-b)}\,x \mp \sqrt{(b-c)}\,z\}-1=0.$$

Hence, plane circular sections are parallel to the mean axis in the ellipsoid, to the greater transverse axis in the hyperboloid of one sheet, and to the greater conjugate axis in the hyperboloid of two sheets, and there are two systems of such plane sections, all having the same inclination to the principal planes containing these axes, in opposite directions.

It is obvious that these are the only circular sections, since a plane not parallel to one of the axes, as that of y, being of the form

$$lx+my+nz=p,$$

could not reduce the expression $(a-b)\,x^2-(b-c)\,z^2$, to a linear form, for the points of intersection with the surface, which is requisite in order that they may lie upon a sphere.

186. *Any two circular sections of opposite systems lie in one sphere.*

The equations of the planes of two circular sections of opposite systems are

$$\{\sqrt{(a-b)}\,x-\sqrt{(b-c)}\,z-k\}\,\{\sqrt{(a-b)}\,x+\sqrt{(b-c)}\,z-k'\}=0\,;$$

or, $(a-b)\,x^2-(b-c)\,z^2-(k+k')\,\sqrt{(a-b)}\,x-(k-k')\,\sqrt{(b-c)}\,z+kk'=0.$

Hence, they intersect the surface in a sphere whose equation is

$$b\,(x^2+y^2+z^2)-1+(k+k')\,\sqrt{(a-b)}\,x+(k-k')\,\sqrt{(b-c)}\,z-kk'=0.$$

187. *To find the circular sections of the paraboloids.*

The equation of any paraboloid is

$$\frac{y^2}{b}+\frac{z^2}{c}=2x.$$

Let the equation of the cutting plane be

$$lx + my + nz = p,$$

and the equations of a diameter be

$$\frac{x - x_0}{\lambda} = \frac{y - y_0}{\mu} = \frac{z - z_0}{\nu} = r,$$

(x_0, y_0, z_0) being the center of the circular section;

$$\therefore\ l\lambda + m\mu + n\nu = 0,$$

$$\text{and } lx_0 + my_0 + nz_0 = p.$$

At the points in which the diameter meets the surface,

$$\frac{(y_0 + \mu r)^2}{b} + \frac{(z_0 + \nu r)^2}{c} = 2\,(x + \lambda\nu);$$

therefore the equations $\frac{y_0\mu}{b} + \frac{z_0\nu}{c} - 2\lambda = 0$,

$$m\mu + n\nu + l\lambda = 0,$$

are true for an infinite number of values of $\lambda : \mu : \nu$;

$$\therefore\ \frac{y_0}{bm} = \frac{z_0}{cn} = \frac{-2}{l} = \frac{p - lx_0}{bm^2 + cn^2} = \frac{-2lx_0}{2\,(bm^2 + cn^2) + lp}.$$

Also $\frac{\mu^2}{b} + \frac{\nu^2}{c}$ is constant $= M$, suppose;

$$\therefore\ M\lambda^2 + \left(M - \frac{1}{b}\right)\mu^2 + \left(M - \frac{1}{c}\right)\nu^2 = 0.$$

Hence, eliminating λ, we obtain the following equation which is true for all values of the ratio $\mu : \nu$,

$$M\,(m\mu + n\nu)^2 + l^2\left\{\left(M - \frac{1}{b}\right)\mu^2 + \left(M - \frac{1}{c}\right)\nu^2\right\} = 0;$$

$$\therefore\ Mmn = 0,$$

M cannot $= 0$, since b and c are not infinite; we must have, therefore, m or $n = 0$:

suppose $m = 0$, then $M = \frac{1}{b}$,

$$\text{and } Mn^2 + \left(M - \frac{1}{c}\right)l^2 = 0;$$

$\therefore\ \frac{1}{b} - \frac{l^2}{c} = 0$, and b and c must be of the same sign.

Hence, the hyperbolic paraboloid has no circular section; and, for the elliptic paraboloid,

$$\frac{l^2}{c} = \frac{1}{b} = \frac{n^2}{b-c}, \text{ and } m = 0.$$

Similarly, we have the system

$$\frac{l^2}{b} = \frac{1}{c} = \frac{m^2}{c-b}, \; n = 0.$$

One only of these is possible, namely, the first, if $b > c$. Hence, the circular section is parallel to the tangent at the vertex of the principal parabolic section which has the greater latus rectum.

The co-ordinates of the center of a circular section are given by the equations (1), which may be written,

$$\frac{y_0}{0} = \frac{z_0}{\pm\sqrt{(b-c)}} = 2\sqrt{c} = \frac{2cx_0}{2(b-c)\sqrt{c} \mp p\sqrt{b}}.$$

The equations of the locus of the centers of all circular sections are

$$z = \pm 2\sqrt{\{c(b-c)\}}, \text{ and } y = 0.$$

The elliptic paraboloid may therefore be generated by the motion of a variable circle parallel to either of two fixed planes, whose center remains on two corresponding lines, parallel to the axes of the principal parabolic sections.

188. We may also find the positions of the circular sections of the elliptic paraboloid as follows. Arrange the equation of the paraboloid in the form

$$\frac{x^2+y^2+z^2}{b} - \frac{x^2}{b} + z^2\left(\frac{1}{c} - \frac{1}{b}\right) - 2x = 0,$$

and let the surface be cut by a plane whose equation is of one of the forms

$$x = \pm z\sqrt{\left(\frac{b-c}{c}\right)} + k,$$

the points of intersection lie on one of the spheres whose equations are

$$\frac{x^2+y^2+z^2}{b} - \frac{k}{\sqrt{b}}\left\{\frac{x}{\sqrt{b}} \pm z\sqrt{\left(\frac{b-c}{bc}\right)}\right\} - 2x = 0;$$

that is, the sections by those planes are circular.

Also, these are the only circular sections, for no plane, except a plane parallel to one of the axes as that of y, could, by the combination of its equation with the equation of the paraboloid, reduce the expression $\frac{x^2}{b} - z^2\left(\frac{1}{c} - \frac{1}{b}\right)$ to a linear form for the points of intersection with the surface, which is necessary in order that they should lie on a sphere.

189. *Any two circular sections of an elliptic paraboloid, of opposite systems, lie on one sphere.*

For, the equations of the planes of two circular sections of opposite systems are

$$\left\{x - k + z\sqrt{\left(\frac{b-c}{c}\right)}\right\}\left\{x - k' - z\sqrt{\left(\frac{b-c}{c}\right)}\right\} = 0;$$

$$\text{or } (x-k)(x-k') - z^2\left(\frac{b}{c} - 1\right) + (k-k')z\sqrt{\left(\frac{b-c}{c}\right)} = 0.$$

Hence, they intersect the paraboloid in two circles lying on the sphere whose equation is

$$(x-k)(x-k') + y^2 + z^2 - 2bx + (k-k')z\sqrt{\left(\frac{b-c}{c}\right)} = 0.$$

X.

(1) The equations of the generating lines of the surface

$$yz + zx + xy + a^2 = 0,$$

drawn through the point $\left(0,\ am,\ -\frac{a}{m}\right)$, are

$$x(1 \pm m) = am - y = \mp(mz + a).$$

(2) At any point where the planes

$$x + y + z = \pm a$$

meet the surface

$$xy + yz + zx + a^2 = 0,$$

the two generating lines of the surface are at right angles to each other.

(3) If ϕ be the angle between the generating lines of the hyperboloid

$$\frac{x^2}{a^2} + \frac{y^2}{b^2} - \frac{z^2}{c^2} = 1,$$

which pass through a point at a distance r from the origin, and if p be the perpendicular from the origin upon the plane passing through them, shew that

$$2abc\cot\phi = p(r^2 - a^2 - b^2 + c^2).$$

(4) The tangent of the angle between the generating lines of the surface

$$\frac{x^2}{a} - \frac{y^2}{b} = z,$$

which pass through the point (x_0, y_0, z_0), is

$$\frac{\sqrt{\left(\frac{ab}{4} + \frac{bx_0^2}{a} + \frac{ay_0^2}{b}\right)}}{z_0 + \frac{a-b}{4}}.$$

(5) Generating lines of the hyperboloid

$$\frac{x^2}{a^2} + \frac{y^2}{b^2} - \frac{z^2}{c^2} = 1$$

are drawn through points in the plane of xy, whose eccentric angles are α, β : shew that their points are given by the equations

$$\frac{x}{a\cos\frac{\alpha+\beta}{2}} = \frac{y}{b\sin\frac{\alpha+\beta}{2}} = \frac{z}{\pm c\sin\frac{\alpha-\beta}{2}} = \frac{1}{\cos\frac{\alpha-\beta}{2}}.$$

Also, the shortest distance (δ) between two of the same system is given by the equation

$$\frac{4\sin^2\frac{\alpha-\beta}{2}}{\delta^2} = \frac{\sin^2\frac{\alpha+\beta}{2}}{a^2} + \frac{\cos^2\frac{\alpha+\beta}{2}}{b^2} + \frac{\cos^2\frac{\alpha-\beta}{2}}{c^2}.$$

(6) The eccentric angles of the points in which the principal hyperbolic sections are met by any generating line are complementary, and that of the point in which it meets the principal elliptic section is equal to one of these.

(7) In the hyperboloid

$$\frac{x^2}{a^2} + \frac{y^2}{b^2} - \frac{z^2}{c^2} = 1,$$

no three generating lines can be mutually at right angles, unless

$$\frac{1}{a^2} + \frac{1}{b^2} = \frac{1}{c^2},$$

and, if this condition be satisfied, an infinite number of such systems exist.

(8) The generating lines of the hyperboloid

$$\frac{x^2}{a^2} + \frac{y^2}{b^2} - \frac{z^2}{c^2} = 1$$

at any point where it is met by the cone

$$\frac{x^2}{a^6}+\frac{y^2}{b^6}-\frac{z^2}{c^6}=0,$$

are both perpendicular to some other generating line.

If the generating lines be themselves at right angles, the point lies also on the sphere $x^2+y^2+z^2=a^2+b^2-c^2$. Shew that the condition $\frac{1}{a^2}+\frac{1}{b^2}=\frac{1}{c^2}$ is necessary for the consistency of these equations.

(9) If three generating lines of the same system in a hyperboloid be mutually at right angles, the shortest distances between any two will also be generating lines.

(10) If three generating lines of the same system, mutually at right angles, be made the edges of a rectangular parallelopiped: shew that the angular points of the parallelopiped which are not on the hyperboloid, lie on the surface $x^2+y^2+z^2=a^2+b^2-c^2$, and on the surface whose equation is obtained by eliminating h between the equations

$$\frac{x^2}{h^2+a^2}+\frac{y^2}{h^2+b^2}+\frac{z^2}{h^2-c^2}+1=0,\quad \frac{a^2x^2}{(h^2+a^2)^2}+\frac{b^2y^2}{(h^2+b^2)^2}-\frac{c^2z^2}{(h^2-c^2)^2}=1.$$

(11) If two planes be drawn, passing respectively through two generating lines of the same system at the extremities of the major axis of the principal elliptic section, and intersecting in any third generating line, the traces of these planes on either of two fixed planes will be at right angles to each other.

(12) If a ray be reflected between two plane mirrors, inclined at any finite angle, shew that all the reflected rays lie on a hyperboloid of revolution; and find its position.

(13) The equations of the generating lines of the hyperboloid

$$\frac{x^2}{a^2}+\frac{y^2}{b^2}-\frac{z^2}{c^2}=1$$

which pass through the point (x_0, y_0, z_0), are

$$\frac{x-x_0}{\frac{x_0z_0}{c}\mp\frac{ay_0}{b}}=\frac{y-y_0}{\frac{y_0z_0}{c}\pm\frac{bx_0}{a}}=\frac{z-z_0}{c\left(1+\frac{z_0^2}{c^2}\right)}.$$

(14) The angle between two planes, each passing through the center, and through one of the generating lines at any point of an hyperboloid, is given by the equation

$$\frac{abc\cot\phi}{2r}=\frac{1}{p^2}-\frac{1}{a^2}-\frac{1}{b^2}+\frac{1}{c^2};$$

r being the distance from the center of the point and p that of the plane containing the generating lines.

(15) The perpendiculars from the origin upon the generating lines of the hyperboloid

$$\frac{x^2}{a^2}+\frac{y^2}{b^2}-\frac{z^2}{c^2}=1$$

lie upon the cone

$$\frac{a^2}{x^2}(b^2+c^2)^2+\frac{b^2}{y^2}(c^2+a^2)^2=\frac{c^2}{z^2}(a^2-b^2)^2.$$

(16) The perpendiculars from the origin on the generating lines of the paraboloid

$$\frac{x^2}{a^2}-\frac{y^2}{b^2}=\frac{z}{c}$$

lie upon the cone

$$\left(\frac{x}{a}\pm\frac{y}{b}\right)(ax\pm by)+2z^2=0.$$

(17) If θ be the angle between the perpendiculars from the center on the generating lines of an hyperboloid which pass through the point

$$(a\cos\alpha,\quad b\sin\alpha,\ 0),$$

$$\frac{c^2(a^2-b^2)^2}{\tan^2\frac{\theta}{2}}=\frac{a^2(b^2+c^2)^2}{\sin^2\alpha}+\frac{b^2(c^2+a^2)^2}{\cos^2\alpha}.$$

(18) A straight line moves so as to intersect the parabolas

$$y^2=ax,\quad z=0:\quad z^2=-bx,\quad y=0:$$

and remains parallel to one of the planes

$$\frac{y}{\sqrt{a}}=\pm\frac{z}{\sqrt{b}};$$

shew that its locus is the paraboloid

$$\frac{y^2}{a}-\frac{z^2}{b}=x.$$

(19) Find the locus of a variable circle which is perpendicular to the plane of an ellipse, and of which a diameter is one of a system of parallel chords of the ellipse. Shew that the same locus may be similarly generated by using another system of parallel chords.

(20) A series of parallel circles are described on parallel chords of a fixed circle as diameters; shew that their locus is an ellipsoid, the squares of whose axes are in arithmetical progression.

(21) If $\frac{my}{b}=\frac{nz}{c}$ be a central section of the hyperboloid

$$\frac{x^2}{a^2}+\frac{y^2}{b^2}-\frac{z^2}{c^2}=1,$$

and α, α' the eccentric angles of the points in which two generating lines of the same system at the extremities of a diameter of this central section meet the principal elliptic section, then

$$\tan\frac{\alpha}{2}\tan\frac{\alpha'}{2}=\frac{m+n}{m\sim n}.$$

CHAPTER X.

MODULAR AND UMBILICAL GENERATION OF SURFACES OF THE SECOND DEGREE.

Modular Generation.

190. In this chapter we shall give some account of the ingenious methods of generation of surfaces of the second degree which have been invented by Professor Mac Cullagh and Mr Salmon, and the student who wishes to examine these methods from a more extended point of view, is recommended to read some most valuable memoirs by Mr R. Townsend, published in the third volume of the *Cambridge and Dublin Mathematical Journal.*

The locus of a point whose distance from a fixed point is in a constant ratio to its distance from a fixed straight line, *measured parallel to a fixed plane*, is a surface of the second degree.

Since this locus contains *ten* disposable constants, viz. *three* dependent on the position of the fixed point, *four* on that of the fixed straight line, and *two* on the direction of the fixed plane, and *one* more, namely, the constant ratio, hence, the locus may, *in general*, be made to coincide with any surface which can be represented by an equation of the second degree in an infinite number of ways, since there will be nine equations only, connecting the ten disposable constants.

If it were possible to eliminate all but the three co-ordinates of the fixed point, there would result two final equations determining a curve locus of such points.

Similarly, if it were possible to eliminate all but the four constants which determine the position of the fixed line, we should obtain three final equations which, with the equations of the straight line, determine a ruled surface, which is the locus of the fixed line.

The fixed point is, from analogy with the focal generation of Conics, called a *modular focus*, the fixed line a *directrix*, the constant ratio the *modulus*, and the plane the *directing plane*.

191. *To find the locus of a point, whose distance from a focus is in a constant ratio to its distance from a directrix, measured parallel to a given directing plane.*

Let S the focus be taken for origin, Sz, Sy parallel to the directrix DN and the directing plane respectively, α, β the coordinates of D in xy, e the modulus, and ω the angle of inclination of the directing plane to xy.

Let PN be drawn from the point (x, y, z) to the directrix parallel to the directing plane, NM parallel to Sy, and PM

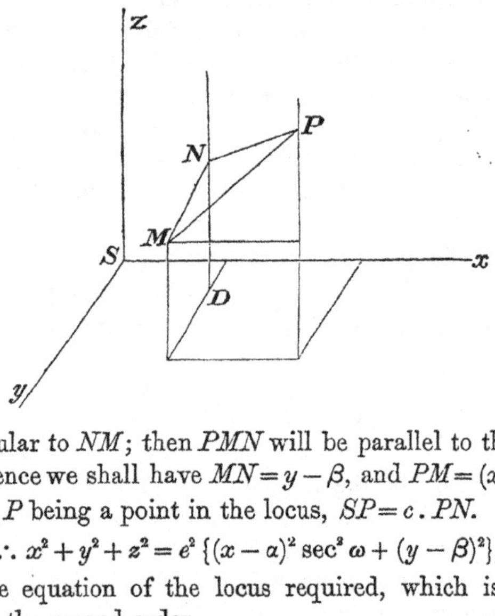

perpendicular to NM; then PMN will be parallel to the directing plane. Hence we shall have $MN = y - \beta$, and $PM = (x - \alpha) \sec \omega$.

Then, P being a point in the locus, $SP = e \, . \, PN$.

$$\therefore x^2 + y^2 + z^2 = e^2 \{(x - \alpha)^2 \sec^2 \omega + (y - \beta)^2\},$$

this is the equation of the locus required, which is always a surface of the second order.

Since $z = x \tan \omega + h$ is the equation of any section parallel to the directing plane, we have, at the points of intersection with the surface,

$$x^2 \sec^2 \omega + y^2 = x^2 + y^2 + (z - h)^2,$$

which, substituted in the equation of the locus, shews that the curve of intersection lies on a sphere, except when $e = 1$, in which case it lies on another plane.

Hence, all sections parallel to the directing planes are circular sections; or straight lines, when $e=1$.

Also, the form of the equation shews that the focus lies in the principal plane perpendicular to the directrix.

The equation may be reduced to the two forms

$$(1-e^2\sec^2\omega)\,x^2+(1-e^2)\,y^2+z^2+2e^2\alpha\sec^2\omega\,x+2e^2\beta y$$
$$=e^2\alpha\sec^2\omega^2+e^2\beta^2 \qquad (1),$$

and
$$(1-e^2\sec^2\omega)\left(x+\frac{\alpha e^2\sec^2\omega}{1-e^2\sec^2\omega}\right)^2+(1-e^2)\left(y+\frac{\beta e^2}{1-e^2}\right)^2+z^2$$
$$=\frac{e^2\sec^2\omega}{1-e^2\sec^2\omega}\,\alpha^2+\frac{e^2\beta^2}{1-e^2} \qquad (2).$$

If $e=\cos\omega$, (1) shews that the surface is in general an elliptic paraboloid; an elliptic cylinder if $\alpha=0$, and a parabolic cylinder if also $\omega=0$.

If $e=1$, it is a hyperbolic paraboloid, or cylinder if $\beta=0$.

If e have not those critical values, (2) shews that it is an ellipsoid or hyperboloid of one or two sheets, reducing to a cone, or a point, if $\alpha=0$, $\beta=0$.

The modulus is in every case the eccentricity of the principal section, real or imaginary, whose axes are parallel respectively to the directrix and to the directing plane.

Since in surfaces of revolution the circular sections are parallel to principal planes, therefore $\omega=0$, or 90°.

If $\omega=0$, we obtain the oblate spheroid or the hyperboloid of revolution of one sheet according as e is less or greater than unity.

If $\omega=90^\circ$, the equation (2) reduces to $(x-\alpha)^2=0$. Therefore the prolate spheroid and the hyperboloid of revolution of two sheets cannot be generated by the modular method.

192. *To find the modular focal and dirigent conics in the case of central surfaces.*

Let ξ,η be the co-ordinates of a focus in the principal plane xy of a central surface, Oz being parallel to the directrix, ξ',η' those of any point in the directrix.

Thus using the notation of equation (2)

$$\frac{\alpha e^2\sec^2\omega}{1-e^2\sec^2\omega}=\xi,\ \text{and}\ \frac{\beta e^2}{1-e^2}=\eta,$$

and the equation of the central surface is

$$(1-e^2\sec^2\omega)\,x^2+(1-e^2)\,y^2+z^2=\frac{1-e^2\sec^2\omega}{e^2\sec^2\omega}\,\xi^2+\frac{1-e^2}{e^2}\,\eta^2.$$

Comparing this with the equation

$$\frac{x^2}{a^2}+\frac{y^2}{b^2}+\frac{z^2}{c^2}=1,$$

we obtain $\quad a^2(1-e^2\sec^2\omega)=c^2=b^2(1-e^2);$

$$\therefore\ \frac{\xi^2}{a^2-c^2}+\frac{\eta^2}{b^2-c^2}=1,$$

which is the equation of the locus of the focus in that plane called a *modular focal conic.*

This conic passes through the foci of the principal sections containing the directrix, and is confocal with the principal section in which it lies.

This is true whether the foci be real or imaginary.

Similarly, if the directrix be taken parallel to Ox, the corresponding modular focal conic has for its equation

$$\frac{\zeta^2}{c^2-a^2}+\frac{\eta^2}{b^2-a^2}=1,\quad \text{and } e^2=1-\frac{a^2}{b^2}.$$

Again, $\quad \xi'=\xi+\alpha=\dfrac{\xi}{e^2\sec^2\omega}=\dfrac{a^2}{a^2-c^2}\,\xi,$

$$\eta'=\eta+\beta=\frac{\eta}{e^2}=\frac{b^2}{b^2-c^2}\,\eta,$$

$$\therefore\ \frac{a^2-c^2}{a^4}\,\xi'^2+\frac{b^2-c^2}{b^4}\,\eta'^2=1,$$

which is the equation of the locus of the directrix in that direction, or of its trace, called a *modular dirigent cylinder or conic;* that of the other is $\dfrac{c^2-a^2}{c^4}\,\zeta'^2+\dfrac{b^2-a^2}{a^4}\,\eta'^2=1.$

By changing the signs of a^2, b^2 and c^2 we obtain the focal and dirigent conics for all central surfaces, including cones and cylinders, if one or all of these quantities become infinite.

A third focal conic exists corresponding to imaginary circular sections, and is called indifferently a ***non-modular***, or *umbilical focal conic*, its equation is $\dfrac{x^2}{a^2-b^2}+\dfrac{z^2}{c^2-b^2}=1$, and it has received the latter name from the fact that, when real, it passes through the umbilici of the surface, or the points at which the circular section becomes a tangent plane.

The modular focal conics do not intersect the surfaces, but the non-modular focal conics do.

DEF. That principal axis which is parallel to the directing planes is called the *directive axis.*

193. *The focal and dirigent conics are reciprocals of each other with respect to the principal section in the plane of which they lie, and the line joining the foot of any directrix with the corresponding focus is a normal to the focal conic.*

The equation of a focal conic being

$$\frac{x^2}{a^2-c^2}+\frac{y^2}{b^2-c^2}=1,$$

the equation of the tangent at the point (ξ, η) is

$$\frac{x\xi}{a^2-c^2}+\frac{y\eta}{b^2-c^2}=1,$$

or

$$\frac{x\xi'}{a^2}+\frac{y\eta'}{b^2}=1,$$

whence it is the polar of (ξ', η'), the foot of the corresponding directrix with respect to the section in xy.

Also, since $a^2(\xi'-\xi)=c^2\xi'$, and $b^2(\eta'-\eta)=c^2\eta'$, the equation of the tangent may be written

$$x(\xi'-\xi)+y(\eta'-\eta)=c^2,$$

it is therefore perpendicular to the line joining (ξ, η) and (ξ', η'), whence the second part of the proposition.

194. *To find the focal conics for conical surfaces.*

If a, b, c be finite quantities proportional to the infinite principal axes of a conical surface, we obtain from the equation $\frac{x^2}{a^2}-\frac{y^2}{b^2}-\frac{z^2}{c^2}=0$, in which b corresponds to the directive axis, the two modular focal conics are

$$\frac{x^2}{a^2+c^2}-\frac{y^2}{b^2-c^2}=0, \quad \text{and} \quad \frac{z^2}{c^2+a^2}+\frac{y^2}{b^2+a^2}=0,$$

and the non-modular or umbilical focal conic

$$\frac{x^2}{a^2+b^2}+\frac{z^2}{b^2-c^2}=0.$$

The cone has therefore the property that all the three focal conics are real, having a common point in the vertex, two of them being evanescent ellipses in their transitions between real and imaginary existence, and the third the limit of an hyperbola consisting of two right lines intersecting in the vertex.

The vertex is therefore not only modular, but *doubly* modular, since it is a point in two modular focal curves, and it is also an umbilical focus, arising from the fact that it is the limit of the two species of hyperboloids, for both of which the real focal hyperbola is modular, and for one the real focal ellipse is modular, while for the other it is umbilical.

Of the two moduli in the modular generation of the cone, the less modulus belongs to the focal lines, and is called by Mac Cullagh the *linear modulus*, while the other, to which only a single focus corresponds, is called the *singular modulus*.

195. *Examination of the nature of the two moduli and modular conics in the principal central surfaces.*

I. In the ellipsoid, b is intermediate between a and c, the directrix which gives a real modulus and real focal and dirigent conics, is parallel to the least axis.

If the directrix be parallel to the greatest axis the modulus and modular conics are imaginary.

II. For the hyperboloid of two sheets, write $-b^2$ and $-c^2$ for b^2 and c^2 in the equation of the surface, b supposed $> c$.

If the directrix be parallel to an imaginary axis the modulus and modular conics are real.

If the directrix be parallel to the real axis the modulus is real, and the modular conics imaginary.

III. For the hyperboloid of one sheet, write $-c^2$ for c^2 in the equation of the ellipsoid, when $b > a$.

If the directrix is parallel to either axis the modulus and modular conics are real, the less modulus corresponding to the focal hyperbola.

196. *To find the relation between the moduli in central surfaces.*

Let e, e' be the moduli corresponding to the two directions of the directrix.

Then $$\frac{\cos^2\omega}{e^2}=\frac{a^2}{a^2-c^2}, \text{ and } \frac{\sin^2\omega}{e'^2}=\frac{c^2}{c^2-a^2};$$

$$\therefore \frac{\cos^2\omega}{e^2}+\frac{\sin^2\omega}{e'^2}=\frac{a^2-c^2}{a^2-c^2}=1,$$

which is the general relation between the moduli.

197. *To find the modular conics for non-central surfaces.*

I. For the elliptic paraboloid, in which $e=\cos\omega$.

The equation (1) reduces to

$$\sin^2\omega\,(y+\beta\cot^2\omega)^2+z^2=-2\alpha x+\alpha^2+\beta^2\cot^2\omega,$$

and comparing it with the equation $\frac{y^2}{b}+\frac{z^2}{c}=2x$, we shall have, ξ, η being the co-ordinates of a focus,

$$\xi=-\frac{\alpha^2+\beta^2\cot^2\omega}{2\alpha},\quad \eta=\beta\cot^2\omega,\quad \alpha=-c,$$

$$\frac{\sin^2\omega}{c}=\frac{1}{b}=\frac{\cos^2\omega}{b-c};$$

$$\therefore \eta^2=2c\cot^2\omega\left(\xi-\frac{c}{2}\right)=2(b-c)\left(\xi-\frac{c}{2}\right).$$

The focal conic is therefore a parabola having its vertex in the focus of the parabolic section parallel to the directrix, and confocal with the other parabolic section, since the abscissa of its focus is $\frac{c}{2}+\frac{b-c}{2}$, or $\frac{b}{2}$.

If ξ', η' be the co-ordinates of the corresponding directrix,

$$\xi'=\xi+\alpha=\xi-c,\ \text{and}\ \eta'=\eta+\beta=\frac{b}{b-c}\eta;$$

$$\therefore \eta'^2=\frac{2b^2}{b-c}\left(\xi'+\frac{c}{2}\right),$$

which is the reciprocal of the focal parabola with respect to the section $y^2=2bx$.

II. For the parabolic cylinder $\omega=0$ and $e=1$, and the equation (1) becomes

$$z^2+2\alpha x+2\beta y=\alpha^2+\beta^2,$$

and comparing with the equation $z^2=4ax$, we have $\beta=0$, and if ξ be the abscissa of the focus

$$\xi = \frac{-\alpha}{2} = a,$$

the focal conic becomes a straight line parallel to the generating lines, and containing the foci of all the parabolic principal sections,

$$\xi' = \xi + \alpha = -a,$$

or the dirigent conic is the locus of the feet of the directrices of these parabolic sections.

III. For the hyperbolic paraboloid, $e = 1$, and the equation (1) becomes

$$z^2 - \tan^2\omega\,(x - \alpha\operatorname{cosec}^2\omega)^2 = -2\beta y - \alpha^2\operatorname{cosec}^2\omega + \beta^2.$$

Comparing this with the equation $\frac{z^2}{c} - \frac{x^2}{a} = 2y$, we have, ξ, η being the co-ordinates of the focus,

$$\frac{\sin^2\omega}{c} = \frac{\cos^2\omega}{a} = \frac{1}{a+c}, \text{ and } \beta = -c,$$

$$\xi = -\alpha\operatorname{cosec}^2\omega = -\frac{a+c}{c}\,\alpha, \quad \eta = -\frac{\beta^2 - \alpha^2\operatorname{cosec}^2\omega}{2\beta};$$

$$\therefore \quad \xi^2 = -2\,(a+c)\left(\eta - \frac{c}{2}\right).$$

The focal conic is therefore a parabola having its vertex in the focus of the principal parabolic section parallel to the directrix and its focus in that of the other; since its abscissa is

$$\frac{c}{2} - \frac{a+c}{2} = -\frac{a}{2}.$$

If ξ', η' be the co-ordinates of the directrix

$$\xi' = \xi + \alpha = \frac{a}{a+c}\,\xi, \quad \eta' = \eta + \beta = \eta - c;$$

$$\therefore \quad \xi'^2 = -\frac{2a^2}{a+c}\left(\eta' + \frac{c}{2}\right),$$

which is the reciprocal of the focal conic with respect to the section $x^2 = -2ay$.

In both cases the line joining the focus and directrix is a normal to the focal curve.

Since tan ω may have any value, the plane containing a focal curve may be either of the principal planes containing the directive axis.

198. *To trace the changes of the surfaces and real focal conics corresponding to changes of the modulus from* 0 *to* ∞.

$e=0$,	Surface an infinitely small sphere. Focal conic an infinitely small ellipse.
$e<\cos\omega$,	Surface an ellipsoid. Focal conic an ellipse.
$e=\cos\omega$,	Surface an elliptic paraboloid. Focal conic a parabola.
$e>\cos\omega<1$,	Surface at first an hyperboloid of two sheets, passing through a cone, to an hyperboloid of one sheet, conjugate axis perpendicular to the directrix. Focal conic at first an hyperbola, transverse axis perpendicular to the directive axis, passing through the asymptotic limit, two straight lines to an hyperbola, transverse axis parallel to the directive axis.
$e=1$,	Surface an hyperbolic paraboloid. Focal conic a parabola.
$e>1$,	Suppose an hyperboloid of one sheet, conjugate axis parallel to the directrix, including an hyperboloid of revolution. Focal conic an ellipse, transverse axis parallel to the directive axis.

If $\omega=0$ the ellipsoid becomes an oblate spheroid. The prolate is inadmissible because the directrix cannot be parallel to the directing plane.

The hyperboloid of revolution of two sheets is lost between

$$e=1 \text{ and } e=\cos\omega.$$

Umbilical generation.

199. The locus of a point, the square of whose distance from a fixed point bears a constant ratio to the rectangle under

the perpendicular distances from two directing planes, is a surface of the second degree.

This locus contains ten disposable constants; *three* dependent on the position of the fixed point, *three* on the position of each of the directing planes, and *one* more, namely, the constant ratio.

The fixed point will, therefore, not generally be unique, but may be on any point of a curve locus.

The fixed point is called an *umbilical focus*, the intersection of the planes a *directrix*, and the constant ratio the *umbilical modulus*.

200. *To find the locus of a point, the square of whose distance from a focus is in a constant ratio to the rectangle under the distances from two fixed directing planes.*

Let the focus S be taken for the origin, the planes bisecting the angles between the directing planes being parallel to the planes of xy, yz.

Let also ω be the inclination of the directing planes to the plane of xy, α, γ the co-ordinates of any point in the directrix, and e the constant ratio.

From any point P, let PQ, PR be drawn perpendicular on the directing planes;

$$\therefore \; SP^2 = e\,PQ\,.\,PR,$$

the equations of the directing planes will be

$$(x-\alpha)\sin\omega \pm (z-\gamma)\cos\omega = 0;$$

therefore if x, y, z be the co-ordinates of P,

$$x^2+y^2+z^2 = e\{(x-\alpha)^2\sin^2\omega - (z-\gamma)^2\cos^2\omega\}$$

will be the equation of the locus, which is of the second degree.

If the surface be cut by a plane, parallel to either directing plane, whose equation is

$$(x-\alpha)\sin\omega \pm (z-\gamma)\cos\omega = p,$$

the curve of intersection obviously lies on a sphere, and is therefore a circle.

The equation may be written in the forms

$$(1-e\sin^2\omega)\,x^2+y^2+(1+e\cos^2\omega)\,z^2+2e\sin^2\omega\,.\,\alpha x-2e\cos^2\omega\,.\,\gamma z$$
$$=e\,(\alpha^2\sin^2\omega-\gamma^2\cos^2\omega)\;\ldots\ldots\ldots\ldots\ldots\;(1),$$

$$\text{or }(1-e\sin^2\omega)\left(x+\frac{e\sin^2\omega\,.\,\alpha}{1-e\sin^2\omega}\right)^2+y^2+(1+e\cos^2\omega)\left(z-\frac{e\cos^2\omega\,.\,\gamma}{1+e\cos^2\omega}\right)^2$$
$$=\frac{e\sin^2\omega}{1-e\sin^2\omega}\,\alpha^2-\frac{e\cos^2\omega}{1+e\cos^2\omega}\,\gamma^2\;\ldots\ldots\ldots\ldots\;(2).$$

Hence, the focus lies in the principal plane perpendicular to the circular sections.

The equation (1) shews that when $e=\operatorname{cosec}^2\omega$ or $-\sec^2\omega$, the surface is an elliptic paraboloid or evanescent cylinder.

And the equation (2) that, for other values of e, the surface is an ellipsoid or hyperboloid of two sheets or cone.

If $\omega=0$, e is negative,

and if $\omega=90^0$, e is positive.

In both cases the surface is a prolate surface of revolution, either a spheroid, hyperboloid, or paraboloid, including an evanescent circular cylinder as a limiting case.

201. *To find the umbilical focal and dirigent conics in the case of central surfaces.*

Let ξ, ζ be the co-ordinates of the umbilical focus referred to the principal planes of a central surface; ξ', ζ' those of any point in the corresponding directrix.

Then, using the notation of the last article,

$$\xi=\frac{\alpha e\sin^2\omega}{1-e\sin^2\omega}\quad\text{and}\quad\zeta=-\frac{\gamma e\cos^2\omega}{1+e\cos^2\omega},$$

and the equation of the surface, referred to its center, is

$$(1-e\sin^2\omega)\,x^2+(1+e\cos^2\omega)\,z^2+y^2=\frac{1-e\sin^2\omega}{e\sin^2\omega}\,\xi^2-\frac{1+e\cos^2\omega}{e\cos^2\omega}\,\zeta^2.$$

Comparing this with the equation $\dfrac{x^2}{a^2}+\dfrac{y^2}{b^2}+\dfrac{z^2}{c^2}=1$, we obtain

$$a^2(1-e\sin^2\omega)=b^2=c^2(1+e\cos^2\omega);$$

$$\therefore\; e=b^2\left(\frac{1}{c^2}-\frac{1}{a^2}\right),$$

$$\text{and} \quad \frac{\xi^2}{a^2 - b^2} - \frac{\zeta^2}{b^2 - c^2} = 1,$$

the equation of the umbilical focal conic which is confocal with the section by its plane, and passes through the foci of the other sections, and also the umbilici.

Again, $$\xi' = \xi + \alpha = \xi\left(1 + \frac{b^2}{a^2 - b^2}\right) = \frac{a^2}{a^2 - b^2}\,\xi,$$

$$\zeta' = \zeta + \gamma = \zeta\left(1 - \frac{b^2}{b^2 - c^2}\right) = -\frac{c^2}{b^2 - c^2}\,\zeta;$$

$$\therefore \quad \frac{a^2 - b^2}{a^2}\,\xi'^2 - \frac{b^2 - c^2}{c^2}\,\zeta'^2 = 1,$$

the equation of the umbilical dirigent conic, which is obviously the polar reciprocal of the focal conic with respect to the principal section in the same plane.

202. In the case of the cone, α and γ vanish, and the equation becomes

$$(1 - e\sin^2\omega)\,x^2 + y^2 + (1 + e\cos^2\omega)\,z^2 = 0,$$

and comparing this with the equation

$$\frac{x^2}{a^2} = \frac{y^2}{b^2} + \frac{z^2}{c^2},$$

$$e\sin^2\omega = 1 + \frac{b^2}{a^2}, \quad \text{and} \quad e\cos^2\omega = \frac{b^2}{c^2} - 1;$$

$$\therefore \quad e = b^2\left(\frac{1}{a^2} + \frac{1}{c^2}\right),$$

and the dirigent passes through the focus which is at the vertex.

203. *To find the umbilical focal and dirigent conics in the case of non-central surfaces.*

If $e\sin^2\omega = 1$, the equation (1) becomes

$$y^2 + ez^2 + 2\alpha x - 2e\cos^2\omega\,.\,\gamma z = \alpha^2 - e\gamma^2\cos^2\omega.$$

Comparing this equation with the equation

$$\frac{y^2}{b} + \frac{z^2}{c} = 2x,$$

if ξ, ζ be the co-ordinates of the focus,

$$\alpha = -b, \quad e = \frac{b}{c}, \quad \text{and } \sin^2\omega = \frac{c}{b}.$$

Also, $$\xi = \frac{e\gamma^2\cos^2\omega\sin^2\omega - \alpha^2}{2\alpha} = \frac{b^2 - \gamma^2\cos^2\omega}{2b}$$

and $\zeta = -\gamma\cos^2\omega$;

$$\therefore \zeta^2 = -\cos^2\omega \,.\, 2b\left(\xi - \frac{b}{2}\right)$$

$$= -2(b-c)\left(\xi - \frac{b}{2}\right),$$

the abscissa of whose focus is $\frac{b}{2} - \frac{b-c}{2} = \frac{c}{2}$, therefore the umbilical focal conic is a parabola, confocal with the parabolic section in whose plane it lies, and having its vertex in the focus of the other principal section.

If ξ', ζ' be the co-ordinates of the directrix,

$$\xi' = \xi + \alpha = \xi - b, \quad \zeta' = \zeta + \gamma = -\frac{c}{b-c}\zeta;$$

$$\therefore \zeta'^2 = -\frac{c^2}{b-c}\left(\xi' + \frac{b}{2}\right),$$

which is the reciprocal of the focal curve with respect to the section $z^2 = 2cx$.

204. *Surfaces capable of generation by the umbilical method.*

With a real focus and directrix the only surfaces which can be generated are the ellipsoid, the hyperboloid of two sheets, and the elliptic paraboloid, which is the limit of both; also the corresponding particular cases of these surfaces, viz. a cone, the limit of the hyperboloid of two sheets, a point or evanescent ellipsoid, and an infinitely slender cylinder or evanescent elliptic paraboloid.

The surfaces of revolution capable of being generated by this method are the prolate spheroid and the hyperboloid of revolution of two sheets.

DEF. A surface of the second degree shall in future be denominated a *Conicoid.*

Properties of conicoids deduced by the modular and umbilical methods of generation.

205. *If a section of a conicoid be made by a plane containing two directrices, the sum or difference of the distances of any point of the section from the corresponding foci is constant.*

Let P be any point of the section whose plane contains the directrices QD, $Q'D'$, and let F, F' be the corresponding foci.

Draw QPQ' perpendicular to the directrices and DPD' parallel to a directive plane.

Then, since the modulus is the same for both foci,

$$FP : PD :: F'P : PD';$$

$$\therefore\ FP : F'P :: PD : PD' :: PQ : PQ',$$

$$\text{and}\quad FP \pm F'P : PQ \pm PQ' :: FP : PQ.$$

Now, $PQ + PQ'$ or $PQ \sim PQ'$ is constant, according as P is or is not between the directrices, and $FP : PQ$ is constant, since $PQ : PD$ is so;

$$\therefore\ FP \pm F'P \text{ is constant.}$$

206. *If a straight line be drawn through any point in a directrix intersecting a conicoid in any two points, the line joining the corresponding focus with the point in the directrix bisects the angle between the focal distances of the points of intersection, or the supplement of that angle.*

Let the focus F correspond to the directrix DQ, and QPP' intersect the surface in P, P';

$$\text{then,}\qquad FP : PQ :: FP' : P'Q;$$

$$\therefore\ FP : FP' :: QP : QP',$$

which proves the proposition.

Cor. If QP be a tangent to the surface, QFP is a right angle.

207. *If a straight line touching a conicoid meet two parallel directrices, it makes equal angles with the lines drawn from the point of contact to the corresponding foci.*

For, if P be the point of contact, Q, Q' the points in which the tangent meets the directrices, F, F' the corresponding foci, since the modulus is the same for both foci,

$$FP : PQ :: F'P : PQ'.$$

Also the angles PFQ, $PF'Q'$ are right angles, therefore the triangles are similar, and the angles QPF, $Q'PF'$ are equal.

208. *If a cone, having its vertex in any directrix, envelope a conicoid, the plane of contact passes through the corresponding focus, and is perpendicular to the line joining it with the vertex.*

For if V be the vertex, and F the focus, VP any side of the cone touching the surfaces in P, PFV is a right angle.

Hence, the locus of P which is the plane of contact is a plane through F perpendicular to VF.

209. *If the vertex of a cone be any point in a focal curve of a conicoid, and the base be any plane section of the conicoid, the line joining the focus with the point in which the directrix meets the plane of section is an axis of the cone.*

Let the plane section cut the directrix in E, and EP be a tangent at P to the section, then FP is perpendicular to FE, which is therefore an axis.

Cor. 1. The second plane of section of the cone intersects the corresponding directrix in the same point as the first plane.

Cor. 2. If the first plane section passes through the directrix, the second will also pass through the directrix, and in this case, since there will be an infinite number of axes of the cone, it will be one of revolution.

210. *If the vertex of an enveloping cone of a conicoid be a point on a focal curve of the surface, the cone is one of revolution, and its internal axis is the tangent to the focal curve at the vertex.*

Let V be the vertex of the cone, VP, VP' the tangents to the trace of the surface on the plane of the focal curve, then PP' is a tangent to the dirigent conic at the foot of the corresponding directrix, (Art. 193); and since the plane of contact is perpendicular to the plane of the focal curve, it will contain the corresponding directrix, and therefore by Cor. 2 of the last article will be a cone of revolution.

Also, since the tangent at V is perpendicular to the directrix, and to the line joining V and the foot of the directrix (Art. 193), it is perpendicular to the plane of circular section, and is the internal axis of the cone.

Properties of Cones of the Second Degree.

211. *The sines of the angles, which any side of a cone makes with a focal line and the corresponding dirigent plane, are in a constant ratio.*

Let a plane pass through any directrix DQ and the corresponding focus F, and let P be any point in the section of the cone made by this plane: V the vertex of the cone.

Draw PR, PQ perpendicular to the dirigent plane and directrix.

Then $FP : PQ$ and $PR : PQ$, and therefore $FP : PR$ are constant ratios; and DF being perpendicular to VF, therefore VF is perpendicular to the plane of section, and PFV is a right angle.

Hence, the ratio of the sines proposed is $\frac{PF}{PV} : \frac{PR}{PV}$, and is therefore constant.

212. *The product of the sines of the angles which any side of a cone makes with the directing planes is constant.*

If V be the vertex of a cone, P any point on the cone, PL, PL' perpendicular on the directing planes through V, then by the umbilicar generation of the cone, (Art. 202), PV^2 is proportional to $PL . PL'$; or $\frac{PL}{PV} \cdot \frac{PL'}{PV}$ is constant, which is the property enunciated.

213. *The tangent plane of a cone makes equal angles with the planes through the side of contact and each of the focal lines.*

For, let the tangent QPQ' perpendicular to the side VP meet the dirigent planes in the points Q, Q', and take F, F' the foci corresponding to the directrices through Q, Q'; then FQ is perpendicular to VF, and also to PF, and therefore to the plane VPF; also VP is perpendicular to FQ and PQ, and therefore to FP; hence FPQ is the inclination of the planes VPF, VPQ, and being equal to $F'PQ'$, the proposition is proved.

COR. If a sphere be described having its center at the vertex, and meeting the focal lines in S, H, and the side of the cone in P, the great circle touching the curve of intersection in P makes equal angles with the arcs SP, HP.

214. *The sum of the angles which a side of the cone makes with the focal lines is constant.*

The statement of this proposition amounts to saying that the sum of the arcs SP, HP, in the corollary of the preceding article, is constant. This may be shewn immediately by limits, as in the case of the plane ellipse.

Reciprocal Cones.

215. *If a cone be constructed whose sides are perpendicular to the tangent planes of any given cone, the tangent planes will also be perpendicular to the sides of the given cone.*

For, let two tangent planes be drawn to a cone A, then two corresponding sides of the other cone B, perpendicular to those tangent planes, will be perpendicular to their line of intersection; the line of intersection of the tangent planes to A is, therefore, perpendicular to the plane containing the corresponding sides of B.

Proceeding to the limit, the line of intersection becomes a side of the cone A, and the plane containing the sides of B a tangent plane to B ultimately. Whence the truth of the proposition.

From this reciprocal property the cones are called *reciprocal cones.*

If $ax^2 + by^2 + cz^2 = 0$ be the equations of a cone,

$$\frac{x^2}{a} + \frac{y^2}{b} + \frac{z^2}{c} = 0$$

is that of the reciprocal cone.

216. *The directing planes of any cone are perpendicular to the focal lines of the reciprocal cone.*

The directing planes of the cone $ax^2 + by^2 + cz^2 = 0$ are inclined to the plane of zx at angles whose tangents are

$$\pm\sqrt{\left(\frac{a-b}{b-c}\right)},$$

and the equations of the focal lines of the reciprocal cone

$$\frac{x^2}{a}+\frac{y^2}{b}+\frac{z^2}{c}=0, \quad \text{are } \frac{x^2}{a-b}-\frac{z^2}{b-c}=0,$$

which, therefore, are perpendicular to the directing planes.

Hence, the focal lines of a cone correspond to the directing planes of the reciprocal cone.

217. The curves in which a sphere whose center is in the common vertex of reciprocal cones intersects the cones, are called *reciprocal spherical conics.*

The reciprocal property connecting the two may be stated thus: "Every point of a spherical conic is the pole of a great circle which touches the reciprocal spherical conic."

218. If two lines correspond respectively to two planes perpendicular to them, drawn through the common vertex of reciprocal cones, the plane which contains the two lines corresponds to the line of intersection of the corresponding planes.

Hence, theorems relating to any cone have reciprocal theorems in the reciprocal cone.

219. The following examples of theorems and their reciprocal theorems will be sufficient to illustrate the method of derivation.

Theorem. The intersections of a tangent plane to a cone with the directing planes make equal angles with the side of contact.

Reciprocal Theorem. The planes containing a side of a cone, and the two focal lines, make equal angles with the tangent plane along the side of the cone.

Theorem. The sum or difference of the angles, which a side of a cone makes with the focal lines, is constant.

Reciprocal Theorem. The sum or difference of the angles, which a tangent plane to a cone makes with the directing planes, is constant.

Theorem. The product of the sines of the angles, which any side of a cone makes with the directing planes, is constant.

Reciprocal Theorem. The product of the sines of the angles, which any tangent plane of a cone makes with the focal lines, is constant.

Such reciprocal theorems are easily translated into the corresponding theorems for reciprocal spherical conics.

XI.

1. If two conicoids have a common focus S, and a common directrix, and if a tangent to one of the surfaces at P meet the other surface in Q, Q' and the directrix in R, SP will bisect the angle QSQ'.

2. If the surfaces in (1) be condirective, the angle QSQ' will be constant.

3. A diameter of constant length revolving in a given central surface describes a cone, a tangent plane to which cuts the surface in a curve of which one axis is the line of contact; and the right lines in which the tangent plane cuts the directing planes make equal angles with the side of contact.

4. A spherical triangle has a given area, and two sides in a given direction, prove that its base touches a spherical conic, and is bisected by the point of contact.

5. If two planes be drawn through any point of a cone parallel to the directing planes, the side of the cone passing through the point makes equal angles with the tangents to their sections of the cone. State the reciprocal theorem.

6. A paraboloid is cut by any plane, prove that the cylinder passing through the curve of section, and having its sides parallel to the axis of the paraboloid, will be condirective with the surface.

7. A cone and a conicoid are concentric and condirective, prove that their curve of intersection lies in a sphere, and a cylinder containing this curve and having its axis parallel to an axis of the conicoid, will be condirective with both.

8. If a conicoid be intersected by a sphere whose center lies in a principal plane, the cylinder containing their curve of intersection, and whose axis is perpendicular to the given plane, will be condirective with the given surface.

9. Every sphere inscribed in a cone of revolution circumscribing an ellipsoid, will cut the ellipsoid in a plane curve.

10. When a series of ellipsoids are inscribed in a cone of revolution, so as to touch it along the same curve, $\dfrac{b^3}{\sqrt{\{(a^2-b^2)(b^2-c^2)\}}}$ will be constant for all.

11. The distance between a focus and the corresponding directrix of the section of an ellipsoid, made by the plane of contact with any enveloping cone of revolution, is constant.

12. If a sphere intersect an ellipsoid in two plane curves, the sphere and ellipsoid will have two common enveloping cones, whose vertices lie on opposite branches of the umbilical focal curve.

13. Any point in the plane of a focal curve of an ellipsoid will be the focus of two plane sections perpendicular to that, which will be real only when the point lies within the trace on that plane and without the focal curve.

Also, if the point lie on the focal curve, these planes will coincide and will contain the normal to the focal curve at that point.

14. The foci of a series of parallel sections of an ellipsoid perpendicular to the plane of a focal curve will lie on an ellipse which touches the trace on that plane, and the focal curve.

15. The cone described with its vertex on one focal curve, and its base on the other, with respect to any conicoid, will be a cone of revolution. This cone will intersect the conicoid in two planes which both pass through the directrix corresponding to the vertex of the cone, and each through a fixed point in the common axis of the focal curves.

16. If a section of an ellipsoid be taken passing through a focus P, and the corresponding directrix, and if P' be the point on the trace of the surface such that the eccentric angles of P, P' in the focal curve and the trace respectively are equal; D, D' the extremities of the diameters conjugate to these points, the eccentricity of the section is $\frac{\alpha\beta}{ab} \cdot \frac{OD'}{OD}$, O being the center, α, β, the semi-axes of the focal curve, and a, b, of the trace of the surface.

CHAPTER XI.

DISCUSSION OF THE GENERAL EQUATION OF THE SECOND DEGREE.

220. Our object in this chapter is to investigate the origin and axes of rectangular co-ordinates, to which, when referred, the equation of a surface represented by a proposed complete equation of the second degree, will assume its simplest form, and to investigate the relations among the coefficients which discriminate the various kinds of surfaces capable of being represented by this equation.

The extent of the simplification which may be effected by transformation of co-ordinates may be anticipated from the following considerations.

By a change of origin we introduce three arbitrary constants, which will enter only into the coefficients of the terms of one dimension in x, y, and z, (Art. 116), and by which, unless a certain relation between the coefficients subsists, these coefficients may be made to vanish.

By a change of the direction of the axes, we introduce nine constants, connected by six equations, or three independent arbitrary constants, which will enter into all the coefficients of the transformed equation. We may determine these constants by assuming any three relations among the coefficients, provided such relations furnish equations independent of and compatible with the six necessary relations connecting the constants of transformation. The three relations which we shall choose are, that the coefficients of yz, zx, and xy in the transformed equation shall severally vanish; and if these relations, together with the six connecting equations, furnish real values of the constants of transformation, the desired simplification will be effected.

The method which we shall adopt requires a preliminary investigation, which is effected in the two following articles.

221. *To determine the condition necessary in order that the equation*

$$U \equiv ax^2 + by^2 + cz^2 + 2a'yz + 2b'zx + 2c'xy = 0,$$

may represent two real or imaginary planes.

If a be finite, the proposed equation is equivalent to

$$(ax + c'y + b'z)^2 = (c'^2 - ab)\, y^2 + 2\,(b'c' - aa')\, yz + (b'^2 - ca)\, z^2 \equiv v,$$

$$\text{or } ax + c'y + b'z = \pm \sqrt{v}.$$

But, if the equation represent two planes, x must be capable of being expressed as a linear function of y and z in two ways, which can only happen when v is a complete square with respect to y and z. The condition for this is

$$(b'c' - aa')^2 = (c'^2 - ab)\,(b'^2 - ca),$$

$$\text{or } a\,(abc + 2a'b'c' - aa'^2 - bb'^2 - cc'^2) = 0,$$

$$\text{or } abc + 2a'b'c' - aa'^2 - bb'^2 - cc'^2 = 0;$$

since a was assumed to be finite.

The important function of the coefficients of U which forms the left-hand member of this equation, we shall, after Mr Salmon, denote by $H(U)$.

222. *When the equation $U = 0$ represents two planes, to determine their line of intersection.*

The line of intersection of the two planes will manifestly lie on the plane

$$ax + c'y + b'z = 0,$$

and by symmetry will also lie on the planes

$$c'x + by + a'z = 0,$$

$$b'x + a'y + cz = 0.$$

If we eliminate z from the first two of these equations, and x from the last two, we obtain the equations

$$x\,(b'c' - aa') = y\,(c'a' - bb') = z\,(a'b' - cc'),$$

which are the equations of the line of intersection.

It is obvious that $H(U)$ is the eliminant of these three equations.

223. *To determine the coefficients of the transformed equation, when $U = 0$ is referred to such a system of axes that the products of the variables vanish.*

Assume α, β, γ to be these coefficients; then since, by this transformation, U becomes

$$\alpha x^2 + \beta y^2 + \gamma z^2,$$

and since the expression $x^2 + y^2 + z^2$ is unaltered by any transformation, the expression

$$h(x^2 + y^2 + z^2) - U \equiv V$$

will become $\quad h(x^2 + y^2 + z^2) - \alpha x^2 - \beta y^2 - \gamma z^2 \equiv W,$
whatever be the value of h. Hence, if we give h such a value that V shall be separable into linear factors, W will also, for the same value of h, be separable into linear factors.

Therefore the equations

$$H(V) \equiv (h-a)(h-b)(h-c) - a'^2(h-a) - b'^2(h-b) - c'^2(h-c) - 2a'b'c' = 0,$$

and $\quad H(W) \equiv (h-\alpha)(h-\beta)(h-\gamma) = 0,$
have the same roots.

Hence, α, β, γ are the roots of the equation

$$(h-a)(h-b)(h-c) - a'^2(h-a) - b'^2(h-b) - c'^2(h-c) - 2a'b'c' = 0.$$

This equation is called the "Discriminating Cubic." We shall, for the present, assume what will hereafter be proved, (Art. 228), that its roots are always real; and proceed to investigate the position of the axes to which the equation is now referred, with respect to the original axes.

224. *To determine the equations of the axes, to which U must be referred, in order to assume the form*

$$\alpha x^2 + \beta y^2 + \gamma z^2.$$

When $h = \alpha$, W becomes $(\alpha - \beta) y^2 + (\alpha - \gamma) z^2$, and the two planes represented by $W = 0$ intersect in the new axis of x. Hence, the new axis of x is the line of intersection of the two planes represented by $V = 0$, with the same value of h.

Its equations are therefore, referred to the original axes,

$$x\{b'c' + a'(\alpha - a)\} = y\{c'a' + b'(\alpha - b)\} = z\{a'b' + c'(\alpha - c)\};$$

and similarly, the equations of the new axes of y and z are

$$x\{b'c' + a'(\beta - a)\} = y\{c'a' + b'(\beta - b)\} = z\{a'b' + c'(\beta - c)\},$$

$$x\{b'c' + a'(\gamma - a)\} = y\{c'a' + b'(\gamma - b)\} = z\{a'b' + c'(\gamma - c)\}.$$

If two of the roots of the discriminating cubic, as α, β, be equal, two of these systems of equations appear to become coincident. They are in reality indeterminate, as might be inferred from the circumstance that W, which in that case becomes

$$\alpha(x^2+y^2)+\gamma z^2,$$

will not be altered by any transformation of x and y in their own plane. This case, and the case of three equal roots, will be separately considered afterwards.

225. *To find the direction cosines of the new axes, referred to the old.*

It will be convenient here to denote

$$(\alpha-b)(\alpha-c)-a'^2,\quad (\alpha-c)(\alpha-a)-b'^2,\quad (\alpha-a)(\alpha-b)-c'^2,$$

by λ, μ, ν respectively, and

$$b'c'+a'(\alpha-a),\ c'a'+b'(\alpha-b),\ a'b'+c'(\alpha-c),$$

by λ', μ', ν' respectively. We shall then have, if $f(h)=0$ be the discriminating cubic, $f(\alpha)=0$, which may be written in any of the forms

$$\lambda\lambda'=\mu'\nu',\ \mu\mu'=\nu'\lambda',\ \nu\nu'=\lambda'\mu',\ \lambda'^2=\mu\nu,\ \mu'^2=\nu\lambda,\ \nu'^2=\lambda\mu.$$

$$\text{Also,}\qquad \lambda+\mu+\nu\equiv f'(\alpha).$$

Now, if l, m, n be the direction cosines of the new axis of x,

$$l\lambda'=m\mu'=n\nu';\quad \therefore\ l^2\mu\nu=m^2\nu\lambda=n^2\lambda\mu,$$

$$\text{or}\quad \frac{l^2}{\lambda}=\frac{m^2}{\mu}=\frac{n^2}{\nu}=\frac{l^2+m^2+n^2}{\lambda+\mu+\nu}=\frac{1}{f'(\alpha)}=\frac{1}{(\alpha-\beta)(\alpha-\gamma)}.$$

$$\text{Hence,}\ l^2=\frac{(\alpha-b)(\alpha-c)-a'^2}{(\alpha-\beta)(\alpha-\gamma)},\quad m^2=\frac{(\alpha-c)(\alpha-a)-a'^2}{(\alpha-\beta)(\alpha-\gamma)},$$

$$n^2=\frac{(\alpha-a)(\alpha-b)-c'^2}{(\alpha-\beta)(\alpha-\gamma)},$$

which may be written

$$l^2\frac{df(\alpha)}{d\alpha}+\frac{df(\alpha)}{da}=0,\ \&c.,\ \text{since}\ \lambda+\frac{df(\alpha)}{da}=0.$$

Similarly for the direction cosines of the axes of y and z.

226. *To find the conditions that the discriminating cubic may have two equal roots.*

If the equation $f(h)=0$ has two roots equal to α, we shall have

$$f(\alpha)=0\ \text{and}\ f'(\alpha)=0.$$

These give us, with the same notation as in the last article,

$$\lambda+\mu+\nu=0,\quad \lambda\lambda'=\mu'\nu',\quad \mu\mu'=\nu'\lambda',\quad \nu\nu'=\lambda'\mu'.$$

Hence, we have

$$\frac{\mu'\nu'}{\lambda'}+\frac{\nu'\lambda'}{\mu'}+\frac{\lambda'\mu'}{\nu'}=0 \text{ or } \mu'^2\nu'^2+\nu'^2\lambda'^2+\lambda'^2\mu'^2=0.$$

This equation can only be satisfied by the vanishing of two of the quantities λ', μ', ν'.

Assuming then $\lambda'=0$, $\mu'=0$, and substituting in $f'(\alpha)=0$ for $\alpha-a$ and $\alpha-b$, we obtain, if a' and b' be finite, $c'(\alpha-c)+a'b'=0$.

Hence if $f(h)=0$ have two roots equal to α, we have the three equations

$$a'(\alpha-a)+b'c'=b'(\alpha-b)+c'a'=c'(\alpha-c)+a'b'=0.$$

If c' also be finite, this supplies the two conditions

$$a-\frac{b'c'}{a'}=b-\frac{c'a'}{b'}=c-\frac{a'b'}{c'}.$$

The equations of the new axes of x and y become in this case indeterminate, and since

$$\gamma-a+\frac{b'c'}{a'}=\gamma-b+\frac{c'a'}{b'}=\gamma-c+\frac{a'b'}{c'},$$

by the conditions already fulfilled, the equations of the new axis of z will be

$$a'x=b'y=c'z. \qquad \text{(Art. 224).}$$

If a', b', or c' vanish, these conditions in this form become indeterminate; we will therefore obtain equivalent equations which shall remain finite in this case.

Since each of the equal quantities $a-\frac{b'c'}{a'}$, $b-\frac{c'a'}{b'}$, $c-\frac{a'b'}{c'}$, is equal to α, we have

$$\frac{b'c'}{a'}=a-\alpha,\qquad \frac{c'a'}{b'}=b-\alpha,\qquad \frac{a'b'}{c'}=c-\alpha,$$

and hence,

$$a'^2=(b-\alpha)(c-\alpha),\ b'^2=(c-\alpha)(a-\alpha),\ c'^2=(a-\alpha)(b-\alpha)\ \dots (1).$$

From these equations we obtain

$$\frac{b'^2-c'^2}{b-c}+a=\frac{c'^2-a'^2}{c-a}+b=\frac{a'^2-b'^2}{a-b}+c=\alpha.$$

From equation (1) we see that if $a'=0$, $\alpha=b$ or $\alpha=c$, and consequently c' or $b'=0$: if then a' and $b'=0$, we have

$$\alpha=c \text{ and } c'^2=(a-c)(b-c).$$

If a', b', and c' all vanish, we must have two of the three $a-\alpha$, $b-\alpha$, $c-\alpha$ zero, or two of the three a, b, c equal, which is otherwise obvious.

227. *To find the conditions that the discriminating cubic may have three equal roots.*

In this case, the transformed expression for U will be of the form

$$\alpha(x^2+y^2+z^2);$$

but this is incapable of being affected by transformation, and must therefore have been originally in the same form. We shall therefore have

$$a=b=c, \quad a'=b'=c'=0.$$

These conditions may of course be deduced from the cubic itself. We shall, in fact, have the equations

$$a+b+c=3\alpha,$$

$$bc-a'^2+ca-b'^2+ab-c'^2=3\alpha^2,$$

$$abc+2a'b'c'-aa'^2-bb'^2-cc'^2=\alpha^3.$$

Hence, $(a+b+c)^2=9\alpha^2=3(bc-a'^2+ca-b'^2+ab-c'^2)$,

or $(a-b)^2+(b-c)^2+(c-a)^2+6(a'^2+b'^2+c'^2)=0$,

which, for real coefficients, necessitates the before-mentioned conditions.

228. *To shew that the discriminating cubic has always real roots.*

We will first obtain a different form of the cubic, and one which is more convenient for its general discussion.

Let $h-a=p-\dfrac{b'c'}{a'}$, $\quad h-b=q-\dfrac{c'a'}{b'}$, $\quad h-c=r-\dfrac{a'b'}{c'}$.

On substitution, the equation becomes

$$u\equiv pqr-\frac{b'c'}{a'}qr-\frac{c'a'}{b'}rp-\frac{a'b'}{c'}pq=0 \quad \ldots\ldots\ (1),$$

$$\text{or,}\quad 1=\frac{\frac{b'c'}{a'}}{h-a+\frac{b'c'}{a'}}+\frac{\frac{c'a'}{b'}}{h-b+\frac{c'a'}{b'}}+\frac{\frac{a'b'}{c'}}{h-c+\frac{a'b'}{c'}};$$

$$\text{or,}\quad \frac{1}{a'b'c'}=\frac{1}{a'^2\left(h-a+\frac{b'c'}{a'}\right)}+\frac{1}{b'^2\left(h-b+\frac{c'a'}{b'}\right)}+\frac{1}{c'^2\left(h-c+\frac{a'b'}{c'}\right)}.$$

Now, assume that $a-\frac{b'c'}{a'}$, $b-\frac{a'c'}{b'}$, and $c-\frac{a'b'}{c'}$ are in descending order of magnitude, and that $a'b'c'$ is positive.

We may now trace the changes of sign of u for different values of h. They are exhibited in the following table:

h	p	q	r	u
∞	+	+	+	+
$a-\frac{b'c'}{a'}$	0	+	+	−
$b-\frac{c'a'}{b'}$	−	0	+	+
$c-\frac{a'b'}{c'}$	−	−	0	−

We thus see that, when $a'b'c'$ is positive, there is one value of h greater than the greatest of the three

$$a-\frac{b'c'}{a'},\quad b-\frac{c'a'}{b'},\quad \text{and } c-\frac{a'b'}{c'},$$

and one lying between each consecutive pair. Similarly, when $a'b'c'$ is negative, we may shew that there is one root less than the least of the three, and one between each pair. There are then always three real roots.

229. *To determine the surfaces which may be represented by the general equation of the second degree.*

We may now assume that any surface which can be represented by an equation of the second degree may also be represented by the particular form of this equation,

$$\alpha x^2 + \beta y^2 + \gamma z^2 + 2\alpha'' x + 2\beta'' y + 2\gamma'' z + f = 0,$$

and proceed to discuss the nature of these surfaces.

I. If α, β, γ be all finite, we may by a change of origin, destroy the terms involving the first powers of x, y, and z, and reduce the equation to the form

$$\alpha x^2 + \beta y^2 + \gamma z^2 + \delta = 0.$$

It has already been seen that this equation represents an ellipsoid if $\frac{\alpha}{\delta}$, $\frac{\beta}{\delta}$, $\frac{\gamma}{\delta}$ be all negative, an hyperboloid of one sheet if one of them be positive, and an hyperboloid of two sheets if two be positive. If all three be positive, the locus is impossible.

Also, if δ vanish, the locus is either a cone, or a point, according as α, β, γ have not, or have all the same sign.

II. If $\alpha = 0$, and α'' be finite, one of the co-ordinates of the center becomes infinite, and we cannot refer the surface to axes through its center. We may however determine the origin in this case, by making the constant term vanish, and the equation so transformed becomes

$$\beta y^2 + \gamma z^2 + 2\alpha'' x = 0,$$

which represents an elliptic, or hyperbolic, paraboloid, according as β, γ have or have not the same sign.

III. If $\alpha = 0$ and $\alpha'' = 0$, we can reduce the equation to the form

$$\beta y^2 + \gamma z^2 + \delta = 0,$$

which represents an elliptic cylinder, if $\frac{\alpha}{\delta}$, $\frac{\beta}{\delta}$ be both negative, an hyperbolic cylinder if one be negative, an impossible locus if both be positive.

If δ vanish, the locus is two planes, or a straight line, according as β, γ have not, or have the same sign.

One of the co-ordinates of the center is in this case indeterminate, or, there is a straight line every point of which is a center.

IV. If $\alpha=0$, $\beta=0$, and α'', β'' be finite, we may, by a change of origin, reduce the equation to the form

$$\gamma z^2 + 2\alpha''x + 2\beta''y = 0,$$

which, again, by a change of the direction of the axes of x and y in their own plane, may be further reduced to the form

$$\gamma z^2 + 2\delta x = 0,$$

and therefore represents a parabolic cylinder. In this case, there is a line of centers at an infinite distance.

V. If $\alpha=0$, $\beta=0$, $\alpha''=0$, $\beta''=0$, the equation, being a quadratic in z only, represents two parallel planes.

230. We have now shewn that the only real surfaces represented by any equations of the second degree are the ellipsoid, hyperboloid of one or two sheets, including cones as a limiting case; elliptic and hyperbolic paraboloids; elliptic, hyperbolic, and parabolic cylinders; and a pair of intersecting or parallel planes. We shall proceed to discuss the conditions to be satisfied by the coefficients of the general equation of the second degree, in these several cases. These may be most conveniently investigated by considering the properties of the center, or centers, of each surface. We will first consider the case of central surfaces.

231. *To investigate under what conditions the general equation of the second degree will represent any one of the central surfaces.*

The general equation of the second degree is

$$f(x, y, z) \equiv ax^2 + by^2 + cz^2 + 2a'yz + 2b'zx + 2c'xy$$
$$+ 2a''x + 2b''y + 2c''z + d = 0.$$

Let α, β, γ be its center; which is determined by the condition that it bisects all chords passing through it. Hence the equation

$$f(\alpha + lr, \beta + mr, \gamma + nr) = 0$$

must give equal and opposite values of r, for all values of the

M 2

ratios $l : m : n$; or $l\frac{df}{d\alpha}+m\frac{df}{d\beta}+n\frac{df}{d\gamma}=0$, for all values of $l : m : n$.

This leads to the equations $\frac{df}{d\alpha}=0,\quad \frac{df}{d\beta}=0,\quad \frac{df}{d\gamma}=0,$

or

$$\left.\begin{array}{l} a\alpha+c'\beta+b'\gamma+a''=0 \\ c'\alpha+b\beta+a'\gamma+b''=0 \\ b'\alpha+a'\beta+c\gamma+c''=0 \end{array}\right\} \qquad (A),$$

for determining the center. Since for central surfaces this must be at a finite distance, we have the condition

$$abc+2a'b'c'-aa'^2-bb'^2-cc'^2 > \text{ or } < 0$$

for all central surfaces. This might have been anticipated from the fact that the roots of the discriminating cubic are in this case all finite.

Now remove the origin to the center, and the equation becomes

$$ax^2+by^2+cz^2+2a'yz+2b'zx+2c'xy+f(\alpha,\beta,\gamma)=0,$$

$$\text{and } f(\alpha,\beta,\gamma)\equiv\frac{1}{2}\left(\alpha\frac{df}{d\alpha}+\beta\frac{df}{d\beta}+\gamma\frac{df}{d\gamma}\right)+a''\alpha+b''\beta+c''\gamma+d,$$

$$\text{or } f(\alpha,\beta,\gamma)=a''\alpha+b''\beta+c''\gamma+d\equiv\Delta.$$

If the values of α, β, γ be obtained from (A), and substituted, the value of Δ will be found to be

$$d+\frac{a''^2(bc-a'^2)+\ldots+2b''c''(b'c'-aa')+\ldots}{aa'^2+bb'^2+cc'^2-abc-2a'b'c'}.$$

If this value of Δ be zero, the equation of the surface becomes homogeneous in (x, y, z), and the surface will therefore be either a cone, or a point. If a point, the sections by the co-ordinate planes must also be points, and we must therefore have $bc-a'^2$, $ca-b'^2$, $ab-c'^2$ all positive. Also the roots of the discriminating cubic must be all of the same sign, or

$$(a+b+c)(abc+2a'b'c'-aa'^2-bb'^2-cc'^2)$$

must be positive.

If any of these conditions be not satisfied, the surface will be a cone.

If however Δ be finite, the equation of the surface becomes

$$Ax^2+By^2+Cz^2+2A'yz+2B'zx+2C'xy=1,$$

substituting A for $\frac{a}{-\Delta}$, and similarly for the other coefficients.

Referring this to its principal axes, we have

$$s_1x^2 + s_2y^2 + s_3z^2 = 1,$$

where s_1, s_2, s_3 are the roots of the equation

$$\frac{B'C'}{A'(S-A)+B'C'} + \frac{C'A'}{B'(S-B)+C'A'} + \frac{A'B'}{C'(S-C)+A'B'} = 0,$$

(Art. 228).

Now two of the roots of this equation lie between

$$A - \frac{B'C'}{A'}, \quad B - \frac{C'A'}{B'}, \quad \text{and } C - \frac{A'B'}{C'},$$

and the third is greater than the greatest, or less than the least of these, according as $A'B'C'$ is positive or negative (Art. 228).

Hence, remembering that

$$ABC + 2A'B'C' - AA'^2 - BB'^2 - CC'^2 \equiv H$$

is the product of the roots, we shall have the following cases, L, M, N denoting the above limits of the roots in descending order of magnitude.

(1) $H > 0$, $A'B'C' > 0$.

$L > 0$, $M > 0$, $N \gtrless 0$, three positive roots, and the surface is an ellipsoid.

$L \gtrless 0$, $M < 0$, $N < 0$, one positive, and two negative roots, and the surface is an hyperboloid of two sheets.

(2) $H < 0$, $A'B'C' > 0$.

$L > 0$, $M > 0$, $N > 0$, inconsistent with the other assumptions.

$L > 0$, $M \gtrless 0$, $N < 0$, one negative and two positive roots, and the surface is an hyperboloid of one sheet.

$L < 0$, $M < 0$, $N < 0$, three negative roots, and the locus is impossible.

(3) $H>0$, $A'B'C'<0$.

$L>0$, $M>0$, $N>0$, three positive roots; an ellipsoid.

$L>0$, $M\gtrless 0$, $N<0$, one positive and two negative roots; an hyperboloid of two sheets.

$L<0$, $M<0$, $N<0$, inconsistent with the other assumptions.

(4) $H<0$, $A'B'C'<0$.

$L>0$, $M>0$, $N\gtrless 0$, two positive and one negative root; an hyperboloid of one sheet.

$L\gtrless 0$, $M<0$, $N<0$, three negative roots; locus impossible.

Hence, the locus will be an ellipsoid, if

$$H>0,\ A'B'C'>0,\ L>0,\ M>0;$$
$$H>0,\ A'B'C'<0,\ L>0,\ M>0,\ N>0:$$

an hyperboloid of one sheet, if

$$H<0,\ A'B'C'>0,\ L>0,\ N<0;$$
$$H<0,\ A'B'C'<0,\ L>0,\ M>0:$$

an hyperboloid of two sheets, if

$$H>0,\ A'B'C'>0,\ M<0,\ N<0;$$
$$H>0,\ A'B'C'<0,\ L>0,\ N<0:$$

and impossible, if $\begin{cases} H<0,\ A'B'C'>0,\ L<0,\ M<0,\ N<0, \\ \text{or } H<0,\ A'B'C'<0,\ M<0,\ N<0. \end{cases}$

232. *To find the conditions that the general equation of the second degree shall be an elliptic or hyperbolic paraboloid.*

These surfaces have each one center at an infinite distance. The equations determining the center are

$$ax+c'y+b'z+a''=0,$$
$$c'x+by+a'z+b''=0,$$
$$b'x+a'y+cz+c''=0.$$

If these determine one point at infinity, we must have then

$$abc+2a'b'c'-aa'^2-bb'^2-cc'^2=0 \ \ldots\ldots\ldots\ldots\ (1),$$

and $a''(bc-a'^2)+b''(a'b'-cc')+c''(c'a'-bb')>\text{ or }<0,$

of which the latter may, by means of the former, be expressed in the form

$$\frac{a''}{b'c'-aa'}+\frac{b''}{c'a'-bb'}+\frac{c''}{a'b'-cc'} > \text{or} < 0 \text{.........} (2).$$

The paraboloid will be elliptic, or hyperbolic, according as the two finite roots of the discriminating cubic are of the same or opposite signs: that is, according as

$$bc+ca+ab-a'^2-b'^2-c'^2 > \text{or} < 0.$$

But the three quantities $bc-a'^2$, $ca-b'^2$, $ab-c'^2$ must be of the same sign, in order that the condition (1) may hold (Art. 221). Hence, for an elliptic paraboloid $bc-a'^2>0$, and for an hyperbolic paraboloid $bc-a'^2<0$, the conditions (1) and (2) being also necessary.

233. *To find the conditions that the general equation of the second degree shall be an elliptic or hyperbolic cylinder.*

In this case, there is a line of centers at a finite distance, or the three planes whose equations determine the center must intersect in one straight line at a finite distance. The conditions that this may be the case are that

$$abc+2a'b'c'-aa'^2-bb'^2-cc'^2=0 \text{ ....} (1),$$

$$\frac{a''}{b'c'-aa'}+\frac{b''}{c'a'-bb'}+\frac{c''}{a'b'-cc'}=0 \text{ ....} (2);$$

and the three quantities $b'c'-aa'$, $c'a'-bb'$, $a'b'-cc'$, must be finite; for (Art. 221) if one vanish, at the same time that (1) holds, the others will also vanish, and in that case the three planes determining the center will be parallel. The equations of the line of centers, when conditions (1) and (2) hold, may be found to be

$$x(b'c'-aa')-a'a''=y(c'a'-bb')-b'b''=z(a'b'-cc')-c'c''.$$

The cylinder will be elliptic, or hyperbolic, according as $bc-a'^2 \gtrless 0$, for the same reasons as in the last article.

234. *To find the condition that the general equation of the second degree shall be a parabolic cylinder.*

The parabolic cylinder has a line of centers at an infinite distance, and therefore the three planes determining the center must be parallel.

The conditions for this are

$$\frac{a}{c'} = \frac{c'}{b} = \frac{b'}{a'}, \quad \frac{c'}{b'} = \frac{b}{a'} = \frac{a'}{c},$$

four conditions which however are equivalent to the three

$$b'c' = aa', \quad c'a' = bb', \quad b'c' = cc' \quad \ldots\ldots\ldots\ldots\ldots\ldots \quad (1).$$

These are, as we have seen, the conditions for the vanishing of two roots of the discriminating cubic.

It is also necessary that the planes shall not become coincident, as in that case a plane of centers would exist, and the surface could only represent two parallel planes. Now the planes will become coincident, if in addition to the above conditions, we have

$$\frac{b'}{a'} = \frac{a''}{b''}, \quad \text{and } \frac{c'}{b'} = \frac{b''}{c''}, \text{ or if } a'a'' = b'b'' = c'c''.$$

Hence, for a parabolic cylinder, $a'a''$, $b'b''$, $c'c''$ must not be all equal, and the conditions (1) must hold.

235. *To find the conditions that the general equation of the second degree may represent two planes.*

The conditions for two parallel planes are, by the last article,

$$b'c' = aa', \quad c'a' = bb', \quad a'b' = cc'; \quad a'a'' = b'b'' = c'c''.$$

The conditions for two planes not parallel may be obtained from the consideration that such a surface is the limit of a hyperbolic cylinder, when its line of centers lies on the surface. Now the centers being given by the equations

$$ax + c'y + b'z + a'' = 0,$$
$$c'x + by + a'z + b'' = 0,$$
$$b'x + a'y + cz + c'' = 0;$$

if x, y, z, be a point on the surface, satisfying these equations, we shall have, multiplying by x, y, z, adding, and simplifying by using the equation of the surface,

$$a''x + b''y + c''z + d = 0.$$

Hence, the conditions will be

$$abc + 2a'b'c' - aa''^2 - bb''^2 - cc''^2 = 0,$$

$$\frac{a''}{b'c' - aa} + \frac{b''}{c'a' - bb'} + \frac{c''}{a'b' - cc'} = 0,$$

and
$$\frac{a'a''^2}{b'c' - aa'} + \frac{b'b''^2}{c'a' - bb'} + \frac{c'c''^2}{a'b' - cc'} + d = 0.$$

We must also have $bc - a'^2 < 0$. If $bc - a'^2 > 0$, the surface will be an evanescent elliptic cylinder.

XII.

1. FIND the nature of the surfaces represented by the following equations:

(1) $3z^2 - x^2 - y^2 + 4xy = a^2$,
(2) $x^2 - 2y^2 + 2z^2 + 3zx - xy - 2x + 7y - 5z - 3 = 0$,
(3) $x^2 + y^2 + 2(yz + zx + xy) = a^2$,
(4) .... $= az$,
(5) $x^2 + 2(yz + zx + xy) + 2(z - y - 1) = 0$,
(6) $x^2 + 6yz - 2z(x + y) = a^2$,
(7) $x^2 + (1 - m) yz + (1 + m) x(y + z) = ax$.

In (5) shew that the eccentricity of the principal elliptic section is $\sqrt{2 - \sqrt{2}}$, and in (7) examine the cases $m = -4$, $5m = 1$, and $m = 1$ respectively.

2. The equation $7x^2 + 8y^2 + 4z^2 - 7yz - 11zx - 7xy = a^2$ represents an hyperboloid of one sheet, whose greater real axis makes with the axis of z an angle $\tan^{-1}\sqrt{2}$.

3. The equation

$$x^2 + y^2 + 3z^2 + 3yz + zx + xy - 7x - 14y - 25z + d = 0$$

will represent an ellipsoid, a point, or an impossible locus, according as $d <=> 55$.

4. The equation

$$ax^2 + 4y^2 + 9z^2 + 12yz + 6zx + 4xy + 2a''x + 2b''y + 2c''z + d = 0$$

will in general represent an elliptic paraboloid, a parabolic cylinder, or a hyperbolic paraboloid, according as $a >=< 1$.

What surfaces will be represented in the following cases,

(1) $3b'' = 2c''$, $a >< 1$, (2) $6a'' = 3b'' = 2c''$, $a = 1$.

5. The equation $yz + zx + xy = a^2$ represents an hyperboloid of revolution of two sheets, the eccentricity of the generating hyperboloid being $\sqrt{\frac{3}{2}}$.

6. The equation $x^2+y^2+z^2+yz+zx+xy=a^2$ represents an oblate spheroid, whose polar axis is to its equatoreal as 1 : 2.

7. The equation $(y-z)^2+(z-x)^2+(x-y)^2=a^2$ represents a right circular cylinder, the equations of whose axis are $x=y=z$.

8. The equation $(cy-bz)^2+(az-cx)^2+(bx-ay)^2=1$ represents a right circular cylinder, the equations of whose axis are $\frac{x}{a}=\frac{y}{b}=\frac{z}{c}$.

9. The equation $a(y-z)^2+b(z-x)^2+c(x-y)^2=d^2$ represents a cylinder, which is hyperbolic when $bc+ca+ab$ is negative; and which, when $bc+ca+ab$ is positive, is elliptic or impossible according as $a+b+c$ is positive or negative.

If $a+b+c=0$, the principal section is a rectangular hyperbola.

10. The surface represented by the equation

$$a^2x^2+b^2y^2+c^2z^2-2bcyz-2cazx-2abxy=1,$$

is an hyperboloid of one sheet, and the sum of the squares on its real axes is equal to the square on its conjugate axis.

11. The equations

$$(ax+by+cz)^2+(a'x+b'y+c'z)^2+(a''x+b''y+c''z)^2=1$$

and

$$(ax+a'y+a''z)^2+(bx+b'y+b''z)^2+(cx+c'y+c''z)^2=1$$

represent similar and equal ellipsoids, which degenerate into elliptic cylinders, if

$$a(b'c''-b''c')+a'(b''c-bc'')+a''(bc'-b'c)=0.$$

12. If the general equation of the second degree

$$ax^2+by^2+cz^2+2ayz+2bzx+2cxy+2a''x+2b''y+2c''z+d=0$$

represent a paraboloid of revolution, shew that

$$\frac{a}{a'}+\frac{b'}{c'}+\frac{c'}{b'}=\frac{b}{b'}+\frac{c'}{a'}+\frac{a'}{c'}=\frac{c}{c'}+\frac{a'}{b'}+\frac{b'}{a'}=0;$$

and that $a'a''$, $b'b''$, $c'c''$ must be not all equal.

13. The equation $ax^2+by^2+cz^2+2a'yz+2b'zx+2c'xy=0$ will represent a right cone, whose vertical angle is θ, if

$$\frac{aa'-b'c'}{a'}=\frac{bb'-c'a'}{b'}=\frac{cc'-a'b'}{c'}=(a+b+c)\frac{1+\cos\theta}{1+3\cos\theta}.$$

14. The equation $ax^2+by^2+cz^2=1$ may, when referred to oblique axes, be transformed into the equation $2m(yz+zx+xy)=1$ in an infinite number of ways. If a', b', c' be the cosines of the angles between the axes, shew that

$$\frac{m}{a}+\frac{m}{b}+\frac{m}{c}=a'+b'+c'-\frac{3}{2},$$

$$m^2\left(\frac{1}{bc}+\frac{1}{ca}+\frac{1}{ab}\right)=b'c'+c'a'+a'b'-a'-b'-c',$$

and

$$\frac{2m^3}{abc}=1-a'^2-b'^2-c'^2+2a'b'c'.$$

If the oblique axes be mutually inclined at angles of 60°, shew that either $-2a=b=c$, $a=-2b=c$, or $a=b=-2c$.

15. If the equation $ax^2+by^2+cz^2=1$, when referred to oblique axes, inclined at angles whose cosines are a', b', c', assume the form $d(x^2+y^2-z^2)=1$, shew that

$$\frac{1}{d}=\frac{1}{a}+\frac{1}{b}+\frac{1}{c};\ \text{and that}\ \frac{1-a'^2-b'^2+c'^2}{a+b+c}=\frac{2c'(c'-a'b')}{a+b+c-d}=\frac{-abc}{(bc+ca+ab)^2}.$$

16. Shew that the hyperboloid, whose equation, referred to oblique axes inclined at angles $\cos^{-1}a'$, $\cos^{-1}b'$, $\cos^{-1}c'$, is

$$(1-a')\,yz+(1-b')\,zx+(1-c')\,xy=d,$$

is an hyperboloid of revolution, whose equation, referred to its principal axes, is

$$x^2+y^2-2\,\frac{(1-a')(1-b')(1-c')}{1-a'^2-b'^2-c'^2+2a'b'c'}\,z^2+2d=0.$$

17. If the equation of an hyperboloid, referred to oblique axes inclined at angles α, β, γ such that $\alpha+\beta+\gamma=\pi$, be

$$yz\cos\alpha+zx\cos\beta+xy\cos\gamma=d^2,$$

shew that the length of one of its axes is $4d$: and that the eccentricity (e) of its principal elliptic sections is given by the equation

$$\frac{4e^4}{1-e^2}=\frac{1-8\cos\alpha\cos\beta\cos\gamma}{\cos^2\alpha\cos^2\beta\cos^2\gamma}.$$

18. Discuss the different surfaces represented by the equation

$$x^2+(2m^2+1)(y^2+z^2)-2(yz+zx+xy)=2m^3-3m+1,$$

as m varies from $-\infty$ to $+\infty$: considering particularly the critical values -1, $\frac{1}{2}$, and 1.

CHAPTER XII.

DIAMETRAL SURFACES. DIAMETRAL PLANES. CONJUGATE DIAMETERS.

---

236. Def. A *diametral surface* is the locus of the middle points of a series of parallel chords of a given surface.

There are, therefore, for any given surface, an infinite number of corresponding diametral surfaces, varying in magnitude and position with the direction of the chords which they bisect. These surfaces are however all of the same degree, as is shewn by the following article.

237. *The degree of any surface which is diametral to a given surface of the nth degree, is $\frac{1}{2}n(n-1)$.*

Every straight line meets a surface of the nth degree in n real or imaginary points; these points, taken two and two, give $\frac{1}{2}n(n-1)$ chords of the surface, on each of which chords lies one point of the corresponding diametral surface: hence the whole number of real and imaginary points in which this straight line meets the diametral surface is $\frac{1}{2}n(n-1)$, which is therefore the *degree* of the diametral surface.

If we suppose the equation of the nth degree determining the points of intersection to have $2r$ imaginary roots, there will be r real roots on the diametral surface, corresponding each to a conjugate pair of these imaginary roots, and the whole number of real points on the diametral surface will be $r+\frac{1}{2}(n-2r)(n-2r-1)$, and therefore the number of imaginary points will be

$$\tfrac{1}{2}n(n-1)-\tfrac{1}{2}(n-2r)(n-2r-1)-r\equiv 2r(n-r).$$

In surfaces of the second degree, the diametral surfaces become planes, and meet the corresponding straight lines in real points, whether these straight lines meet the given surface in real or imaginary points. We shall discuss these *diametral planes* in the cases of central, and non-central surfaces, separately.

238. *To find the diametral plane, corresponding to a given series of parallel chords, in a given central surface of the second degree.*

Let the equation of the surface be

$$ax^2 + by^2 + cz^2 = 1,$$

l, m, n the direction-cosines of each of a series of parallel chords, and (x_0, y_0, z_0) the middle point of any one of them.

The equation of this chord will be

$$\frac{x - x_0}{l} = \frac{y - y_0}{m} = \frac{z - z_0}{n} = r,$$

and we shall have, for the points in which it meets the surface, the equation

$$a(x_0 + lr)^2 + b(y_0 + mr)^2 + c(z_0 + nr)^2 = 1.$$

But, since $(x_0 y_0 z_0)$ is the middle point of the chord, the values of r obtained from this equation will be equal, and of opposite signs, and we obtain therefore the equation

$$alx_0 + bmy_0 + cnz_0 = 0,$$

as the locus of the middle points of all such chords, or the diametral plane.

The form of this equation shews that it passes through the center, as it manifestly ought to do.

We shall have, conversely, that any central plane whose equation is $\lambda x + \mu y + \nu z = 0$, will bisect a series of chords parallel to the straight line $\frac{ax}{\lambda} = \frac{by}{\mu} = \frac{cz}{\nu}$, which is called the diameter *conjugate* to the plane. It appears from (Art. 184) that the locus of the centers of a series of sections of the surface parallel to a given central plane is the diameter conjugate to that plane.

If a surface of the second degree be referred to a diametral plane as that of xy, and the corresponding conjugate diameter as the axis of z, then since every straight line parallel to Oz is bisected by the plane of xy, the equation of the surface can only contain even powers of z. Hence if we can find three planes, such that the intersection of any two is conjugate to the third, the equation of the surface referred to these planes will be of the form $ax^2 + by^2 + cz^2 = 1$. We will investigate the conditions of the existence of such planes.

239. *To find the conditions that, of three central planes of a central surface of the second order, each may be diametral to the intersection of the other two.*

Let the direction-cosines of the planes be $(l_1m_1n_1)$, $(l_2m_2n_2)$, and $(l_3m_3n_3)$.

The equations of the diameters conjugate to the first are

$$\frac{ax}{l_1} = \frac{by}{m_1} = \frac{cz}{n_1},$$

and if this be parallel to the intersection of the other two, and therefore to both of them, we shall have

$$l_2\frac{l_1}{a} + m_2\frac{m_1}{b} + n_2\frac{n_1}{c} = 0, \text{ and } l_3\frac{l_1}{a} + m_3\frac{m_1}{b} + n_3\frac{n_1}{c} = 0.$$

Hence, if the three conditions

$$\frac{l_2l_3}{a} + \frac{m_2m_3}{b} + \frac{n_2n_3}{c} = \frac{l_3l_1}{a} + \frac{m_3m_1}{b} + \frac{n_3n_1}{c}$$
$$= \frac{l_1l_2}{a} + \frac{m_1m_2}{b} + \frac{n_1n_2}{c} = 0$$

be satisfied, the planes will be such as required.

These planes are called *conjugate planes*, and their intersections *conjugate diameters*.

Since we have only three relations between the six quantities which determine the planes, there will be an infinite number of such systems, and we can determine such a system satisfying any three other relations which we may choose, provided the resulting equations are not inconsistent with those already obtained. For example, we can, in general, determine a system of conjugate planes each of which shall pass through one of three given points.

240. *To find the relations between the co-ordinates of the extremities of a system of conjugate diameters of a central surface of the second degree.*

The equation of the surface being

$$ax^2 + by^2 + cz^2 = 1,$$

and the co-ordinates of the extremities of the diameters

$$(x_1y_1z_1),\ (x_2y_2z_2),\ (x_3y_3z_3),$$

we shall have, since the points lie on the surface,

$$ax_1^2 + by_1^2 + cz_1^2 = ax_2^2 + by_2^2 + cz_2^2 = ax_3^2 + by_3^2 + cz_3^2 = 1, \quad (1)$$

and, since the diameters through the points are conjugate,

$$ax_2x_3 + by_2y_3 + cz_2z_3 = ax_3x_1 + by_3y_1 + cz_3z_1$$
$$= ax_1x_2 + by_1y_2 + cz_1z_2 = 0. \quad (2)$$

The systems (1) and (2) shew that

$$x_1\sqrt{a},\ y_1\sqrt{b},\ z_1\sqrt{c};\ x_2\sqrt{a},\ y_2\sqrt{b},\ z_2\sqrt{c};\ x_3\sqrt{a},\ y_3\sqrt{b},\ z_3\sqrt{c};$$

are the direction-cosines of three straight lines at right angles to each other, and we know therefore (Art. 117) that they are equivalent to the systems

$$ax_1^2 + ax_2^2 + ax_3^2 = by_1^2 + by_2^2 + by_3^2 = cz_1^2 + cz_2^2 + cz_3^2 = 1,$$
$$y_1z_1 + y_2z_2 + y_3z_3 = z_1x_1 + z_2x_2 + z_3x_3 = x_1y_1 + x_2y_2 + x_3y_3 = 0.$$

Hence, in the ellipsoid,

$$\frac{x^2}{a^2} + \frac{y^2}{b^2} + \frac{z^2}{c^2} = 1,$$

we shall have

$$x_1^2 + x_2^2 + x_3^2 = a^2,\ y_1^2 + y_2^2 + y_3^2 = b^2,\ z_1^2 + z_2^2 + z_3^2 = c^2,$$

or the sum of the squares of the projections of three conjugate diameters on one of the axes is equal to the square of the principal diameter.

By adding these equations, we obtain, if r_1, r_2, r_3 be the lengths of these semi-conjugate diameters,

$$r_1^2 + r_2^2 + r_3^2 = a^2 + b^2 + c^2;$$

which with other relations will be investigated independently hereafter.

241. *To find the relations which exist between the lengths of a system of conjugate diameters of a central surface, and the angles between them.*

Let the equation of the surface referred to its principal axes be $ax^2 + by^2 + cz^2 = 1$, and its equation referred to a system of conjugate diameters, inclined at angles α, β, γ, be

$$a'x^2 + b'y^2 + c'z^2 = 1.$$

Then the expression

$$h(x^2 + y^2 + z^2) - ax^2 - by^2 - cz^2 \equiv U,$$

will, by transformation from the former to the latter system, become

$$h(x^2+y^2+z^2+2yz\cos\alpha+2zx\cos\beta+2xy\cos\gamma)$$
$$-a'x^2-b'y^2-c'z^2\equiv V,$$

h being any constant. Hence, the values of h which make $H(U)$ vanish, will also make $H(V)$ vanish; that is, the equations

$$(h-a)(h-b)(h-c)=0,$$

and

$$(h-a')(h-b')(h-c')-h^2\cos^2\alpha(h-a')-h^2\cos^2\beta(h-b')$$
$$-h^2\cos^2\gamma(h-c')+2h^3\cos\alpha\cos\beta\cos\gamma=0,$$

will have the same roots.

The conditions for this are

$$\frac{1}{a'}+\frac{1}{b'}+\frac{1}{c'}=\frac{1}{a}+\frac{1}{b}+\frac{1}{c},$$

$$\frac{\sin^2\alpha}{b'c'}+\frac{\sin^2\beta}{c'a'}+\frac{\sin^2\gamma}{a'b'}=\frac{1}{bc}+\frac{1}{ca}+\frac{1}{ab},$$

and

$$\frac{1-\cos^2\alpha-\cos^2\beta-\cos^2\gamma+2\cos\alpha\cos\beta\cos\gamma}{a'b'c'}=\frac{1}{abc}.$$

If r_1, r_2, r_3 be the lengths of the corresponding semi-diameters, a, b, c of the semi-axes, these equations become

$$r_1^2+r_2^2+r_3^2=a^2+b^2+c^2, \dots\dots\dots\dots\dots (1),$$

$$(r_2r_3\sin\alpha)^2+(r_3r_1\sin\beta)^2+(r_1r_2\sin\gamma)^2=b^2c^2+c^2a^2+a^2b^2 \dots (2),$$

and $$r_1^2r_2^2r_3^2(1-\cos^2\alpha-\cos^2\beta-\cos^2\gamma+2\cos\alpha\cos\beta\cos\gamma)$$
$$=a^2b^2c^2 \dots\dots (3).$$

If the surface be an ellipsoid, so that all these lengths are real, we see that (1) the sum of the squares of three conjugate radii is constant; (2) the sum of the squares of the faces of a parallelopiped having three conjugate radii as conterminous edges is constant; and (3) the volume of such a parallelopiped is constant.

In the hyperboloid of one sheet $a^2b^2c^2$ is negative, and consequently, since

$$1-\cos^2\alpha-\cos^2\beta-\cos^2\gamma+2\cos\alpha\cos\beta\cos\gamma$$

is positive, $r_1^2 r_2^2 r_3^2$ is negative, but r_1^2, r_2^2, r_3^2 cannot be all negative, hence one and only one is negative: that is, in a hyperboloid of one sheet, two of a system of conjugate diameters meet the surface in real points, and the third does not.

In the hyperboloid of two sheets, for b^2, c^2, substituting $-b^2, -c^2$, we see that $r_1^2\, r_2^2\, r_3^2$ is positive, hence two, or none, of the three r_1^2, r_2^2, r_3^2 are negative. If none be negative, we must have $a^2-b^2-c^2$ positive, and also $b^2c^2-a^2(b^2+c^2)$, which are easily shewn to be inconsistent. Hence two must be negative; or, in the hyperboloid of two sheets, one and only one of a system of conjugate diameters meets the surface in real points.

242. The relations (1) and (3) may be obtained geometrically, and (2) from the relations between the co-ordinates of the extremities obtained in the last article. We will give the proof in the case of the ellipsoid, and leave the other two cases as exercises for the student.

Let Ox, Oy, Oz be the directions of the axes of the surface, Ox', Oy', Oz' those of a system of conjugate diameters; A, B, C, A', B', C', the extremities of the diameters along those axes: a, b, c, a', b', c' their lengths. Also let the sections AB, $A'B'$ intersect in A_1, let OB_2 be the semi-diameter conjugate to OA_1 in the section $A'B'$, and let CB_2 meet AB in B_1: $OA_1=a_1$, $OB_1=b_1$, $OB_2=b_2$.

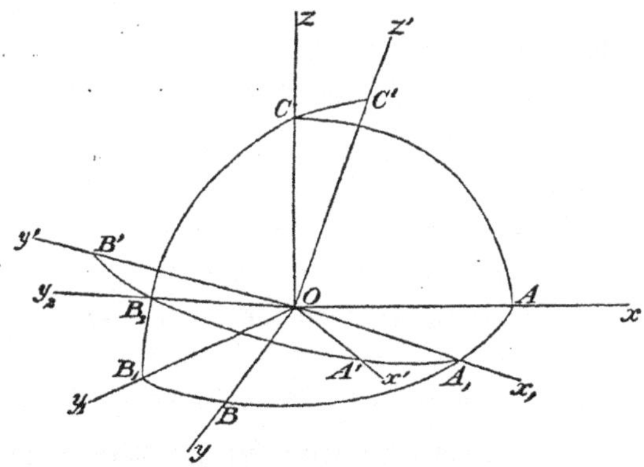

Then, the plane $A'B'$ being conjugate to OC', OA_1, OB_2, OC' will be a system of conjugate radii, or OA_1 will be conjugate to the plane $C'B_2$, and since OA_1 lies in a principal plane, $C'OB_2$ will be perpendicular to that plane, and will therefore contain OC; and OC, OB_1 being the principal semi-axes of the section B_1C', OA_1, OB_1, OC will be a system of conjugate diameters, and OA_1, OB_1 will be conjugate in the section AB.

Hence we have the equations

$$a_1^2+b_2^2=a'^2+b'^2,\ b_1^2+c^2=b_2^2+c'^2,\ a^2+b^2=a_1^2+b_1^2,$$

and from these we obtain the relation

$$a'^2+b'^2+c'^2=a^2+b^2+c^2.$$

Also, since the parallelogram of which OA', OB' are conterminous edges, is equal to that of which OA_1, OB_2 are conterminous edges, and similarly for the section B, C', AB, we have

$$\text{vol}\,(a', b', c')=\text{vol}\,(a_1, b_2, c')=\text{vol}\,(a_1, b_1, c)=\text{vol}\,(a, b, c);$$

denoting any parallelopiped by three conterminous edges. This is equivalent to relation (3) of the last article.

If now $(x_1y_1z_1)$, $(x_2y_2z_2)$, $(x_3y_3z_3)$ be the extremities of the three conjugate diameters, we shall have

$$(b'c'\sin\alpha)^2=(y_2z_3-y_3z_2)^2+(z_2x_3-z_3x_2)^2+(x_2y_3-x_3y_2)^2 \text{ (Art. 26)},$$

and if we put $x_1=al_1$, $y_1=bm_1$, $z_1=cm_1$, and similarly for the rest, we obtain, as in Art. 235, the direction-cosines of a system of rectangular co-ordinates, and therefore

$$(b'c'\sin\alpha)^2=b^2c^2(m_2n_3-m_3n_2)^2+c^2a^2(n_2l_3-n_3l_2)^2+a^2b^2(l_2m_3-l_3m_2)^2,$$
$$=b^2c^2l_1^2+c^2a^2m_1^2+a^2b^2n_1^2.\quad\text{(Art. 118)}.$$

Similarly,

$$(c'a'\sin\beta)^2=b^2c^2l_2^2+c^2a^2m_2^2+a^2b^2n_2^2,$$

and $$(a'b'\sin\gamma)^2=b^2c^2l_3^2+c^2a^2n_3^2+a^2b^2m_3^2.$$

Hence, adding these, we have the relation (2) of the last article,

$$(b'c'\sin\alpha)^2+(c'a'\sin\beta)^2+(a'b'\sin\gamma)^2=b^2c^2+c^2a^2+a^2b^2.$$

243. *To find the diametral plane bisecting a given system of parallel chords, in the case of non-central surfaces.*

Taking the equation of the surface to be

$$\frac{y^2}{b}+\frac{z^2}{c}=2x, \tag{1}$$

and (l, m, n) the direction of the chords, the equation of the diametral plane will be

$$m\frac{y}{b} + n\frac{z}{c} = l, \qquad (2)$$

shewing that all the diametral planes are parallel to the axis of the principal parabolic sections; a fact which might be anticipated from the consideration that these surfaces have their center on that axis at an infinite distance.

We cannot therefore, in these surfaces, have a system of three conjugate planes at a finite distance, but we can find an infinite number, such that, for two of them, each bisects the chords parallel to the other and to a third plane. By taking the origin where the intersection of these two meets the paraboloid, and referring to these three planes, the equation of the surface will assume the form

$$\frac{y^2}{b'} + \frac{z^2}{c'} = 2x.$$

Let the equations of the two diametral planes be

$$m_1 y + n_1 z = 1, \qquad (1)$$

$$\text{and } m_2 y + n_2 z = 1, \qquad (2)$$

and let the direction of the third plane be $(l_3 m_3 n_3)$. The direction-cosines of the chords bisected by (1) are in the ratios

$$1 : bm_1 : cn_1,$$

and if these be parallel to (2) and the third plane, we shall have

$$bm_1 m_2 + cn_1 n_2 = 0, \; l_3 + bm_3 m_1 + cn_3 n_1 = 0.$$

Similarly, in order that (2) may be conjugate to the intersection of the other two, we shall have

$$bm_1 m_2 + cn_1 n_2 = 0, \; l_3 + bm_2 m_3 + cm_2 m_3 = 0.$$

One of these is coincident with one of the former, and there being thus only three relations necessary, between the six quantities determining the planes, an infinite number of such systems can be determined.

244. *To find the equation of a diametral plane bisecting a given system of chords of any conicoid.*

Let the equation of the conicoid be

$$f(x, y, z) \equiv ax^2 + by^2 + cz^2 + 2a'yz + 2b'zx + 2c'xy$$
$$+ 2a''x + 2b''y + 2c''z + d = 0;$$

and let (l, m, n) be the direction of the chords to be bisected, (x_0, y_0, z_0) the middle point of any chord.

Then the equation

$$f(x_0 + lr,\ y_0 + mr,\ z_0 + nr) = 0$$

must have roots equal and of opposite signs, or the coefficient of r must vanish. This gives the condition

$$l\frac{df}{dx_0} + m\frac{df}{dy_0} + n\frac{df}{dz_0} = 0,$$

or $$l(ax_0 + c'y_0 + b'z_0 + a'') + m(c'x_0 + by_0 + a'z_0 + b'')$$
$$+ n(b'x_0 + a'y_0 + cz_0 + c'') = 0,$$

the equation of the diametral plane.

This plane passes through the point

$$\frac{df}{dx} = 0,\ \frac{df}{dy} = 0,\ \frac{df}{dz} = 0,\ \text{(Art. 227)},$$

whatever be the values of l, m, n, or every diametral plane passes through the center, as is otherwise obvious.

245. *To determine the principal planes of any conicoid.*

A *principal plane* is perpendicular to the system of chords which it bisects (Art. 167); hence we shall have, for the direction of a principal plane, the equations

$$al + c'm + b'n = sl,$$
$$c'l + bm + a'n = sm,$$
$$b'l + a'm + cn = sn;$$

s being some constant.

Eliminating the ratios $l : m : n$, we obtain the equation

$$(s-a)(s-b)(s-c) - a'^2(s-a) - b'^2(s-b) - c'^2(s-c) - 2a'b'c' = 0,$$

the *discriminating cubic* already discussed. This supplies three values of s, and to each corresponds one system of values of $l : m : n$; or there are in general three, and only three, principal planes.

The reduction of the general equation of the second degree may be effected in this manner, remembering that if a surface be referred to a *principal* plane as that of xy, the equation can only contain even powers of z.

It will be necessary however to supply the proposition that these three principal planes are mutually at right angles, as we shall now prove.

246. *To shew that the three principal planes of any conicoid are mutually at right angles.*

Let s_1, s_2, s_3 be the three roots of the discriminating cubic, and let the corresponding values of l, m, n be denoted by the same suffixes.

We shall then have

$$al_1 + c'm_1 + b'n_1 = s_1 l_1,$$
$$c'l_1 + bm_1 + a'n_1 = s_1 m_1,$$
$$b'l_1 + a'm_1 + cn_1 = s_1 n_1.$$

Multiplying by l_2, m_2, n_2 and adding, we obtain

$$l_1 (al_2 + c'm_2 + b'n_2) + m_1 (c'l_2 + bm_2 + a'n_2) + n_1 (b'l_2 + a'm_2 + cn_2)$$
$$= s_1 (l_1 l_2 + m_1 m_2 + n_1 n_2),$$

or $$l_1 . s_2 l_2 + m_1 . s_2 n_2 + n_1 . s_2 n_2 = s_1 (l_1 l_2 + m_1 m_2 + n_1 n_2),$$

whence $$(s_1 - s_2)(l_1 l_2 + m_1 m_2 + n_1 n_2) = 0.$$

Hence, if the roots of the discriminating cubic be unequal, the three principal planes are mutually at right angles; and the equation of the surface referred to them as co-ordinate planes, must assume the form

$$Ax^2 + By^2 + Cz^2 + d = 0.$$

It may readily be shewn also that A, B, C are the roots of the cubic; for the coefficient of x^2 in the transformed equation will be

$$al_1^2 + bm_1^2 + cn_1^2 + 2a'm_1 n_1 + 2b'n_1 l_1 + 2c'l_1 m_1,$$

which is equal to

$$l_1 (al_1 + c'm_1 + b'n_1) + m_1 (c'l_1 + bm_1 + a'n_1) + n_1 (b'l_1 + a'm_1 + cn_1);$$

that is, to $s_1 (l_1^2 + m_1^2 + n_1^2)$ or s_1: and similarly $B = s_2$, $C = s_3$.

The discussion of the cases of vanishing, or of equal roots, may now be proceeded with as in Chap. XII.

247. Another method of determining the principal planes, and the lengths of the axes, may be noticed, depending on the fact that these lengths are the maxima and minima values of

radii drawn from the center. Let the equation of a central conicoid be

$$ax^2 + by^2 + cz^2 + 2a'yz + 2b'zx + 2c'xy = d, \tag{1}$$

then, if xyz be the co-ordinates of the extremity of a radius, of length r, we shall have

$$x^2 + y^2 + z^2 = r^2, \tag{2}$$

and for the principal axes, r must be a maximum or minimum by variation of x, y, z, subject to (1). This gives the equations, employing an undetermined multiplier λ,

$$ax + c'y + b'z = \lambda x, \quad c'x + by + a'z = \lambda y, \quad b'x + a'y + cz = \lambda z. \tag{3}$$

Multiplying by x, y, z, and adding, we have by (1) and (2),

$$d = \lambda r^2,$$

and eliminating x, y, z from (3) we obtain the cubic equation

$$\left(\frac{d}{r^2} - a\right)\left(\frac{d}{r^2} - b\right)\left(\frac{d}{r^2} - c\right)$$

$$- a'^2\left(\frac{d}{r^2} - a\right) - b'^2\left(\frac{d}{r^2} - b\right) - c'^2\left(\frac{d}{r^2} - c\right) - 2a'b'c' = 0,$$

the roots of which are the lengths of the principal axes.

We shall be able to obtain from (3) the equations of any principal axes, in the form

$$a'x\left(\frac{d}{r^2} - a + \frac{b'c'}{a'}\right) = b'y\left(\frac{d}{r^2} - b + \frac{c'a'}{b'}\right) = c'z\left(\frac{d}{r^2} - c + \frac{a'b'}{c'}\right),$$

which, taking the three values of r^2, are the equations of the three axes. These coincide, as they ought to do, with the equations found in Art. 224.

Plane Sections of Conicoids.

248. *To determine the nature of the section of a conicoid made by any given plane.*

This may of course be done by the substitutions of Art. 121, but for surfaces of the second degree whose plane sections will be curves of the second degree, simpler methods may advantageously be employed. If it be required only to discover to what species of conics the section belongs, we may effect this immediately by taking any orthogonal projection of the curve of section, since an ellipse, parabola, or hyperbola, will be projected into a curve of the same species, though in general of different eccentricity. The only exception is when the plane of section is perpendicular to the plane of projection, but as no plane can be perpendicular to all the co-ordinate planes, there is at least one of the co-ordinate planes which may, in any proposed case, be taken as the plane of projection, and which will not be perpendicular to the plane of section.

As an example of this method, we may take the section of the paraboloid $ay^2 + bz^2 = x$ made by the plane $lx + my + nz = 0$. The equation of the projection of the curve of section on the plane of yz is $l(ay^2 + bz^2) + my + nz = 0$, which is always an ellipse, or always an hyperbola, according as a and b have like or unlike signs. If $l = 0$, the exceptional case above mentioned arises, and taking the projection on zx we have the equation

$$(n^2a + m^2b)\, z^2 = m^2x,$$

or the section is parabolic, unless $n^2a + m^2b = 0$, when it reduces to a straight line. Hence, in the paraboloids, all sections parallel to the axis of the principal sections are parabolas, and all other sections ellipses for the elliptic paraboloid, and hyperbolas for the hyperbolic paraboloid.

If, however, a more exact determination is required, we may conveniently use the angle between the asymptotes, real or impossible, as fixing the species of the curve.

Let the equation of the conicoid be

$$f(x, y, z) \equiv ax^2 + by^2 + cz^2 + 2a'yz + 2b'zx + 2c'xy$$
$$+ 2a''x + 2b''y + 2c''z + d = 0,$$

and that of the plane of section be

$$lx + my + nz = 0.$$

Then if (α, β, γ) be any point in the plane, and (λ, μ, ν) the directions of a line drawn in the plane through this point, we shall have the equation

$$f(\alpha + \lambda r, \beta + \mu r, \gamma + \nu r) = 0, \qquad (1)$$

to determine the points where the line meets the curve of section, λ, μ, ν being subject to the condition

$$l\lambda + m\mu + n\nu = 0. \qquad (2)$$

If now $(\lambda\mu\nu)$ be the direction of an asymptote, one value of r in (1) must be infinite, or the coefficient of r^2 must vanish. This gives for the directions of the asymptotes

$$a\lambda^2 + b\mu^2 + c\nu^2 + 2a'\mu\nu + 2b'\nu\lambda + 2c'\lambda\mu = 0. \qquad (3)$$

The equations (2) and (3) completely determine the directions of the asymptotes, and the section will be elliptic, parabolic, or hyperbolic, according as these directions are impossible, equal, or possible and unequal.

If $\lambda_1 : \mu_1 : \nu_1$, and $\lambda_2 : \mu_2 : \nu_2$ be the ratios obtained from (2) and (3), then, ω being the angle between the asymptotes,

$$\cos\omega = \frac{\lambda_1\lambda_2 + \mu_1\mu_2 + \nu_1\nu_2}{\sqrt{(\lambda_1^2 + \mu_1^2 + \nu_1^2)}\sqrt{(\lambda_2^2 + \mu_2^2 + \nu_2^2)}};$$

this, being a symmetric function of the roots, will always be real, and the sections will be elliptic, parabolic, or hyperbolic, according as it is $> = $ or < 1. The ratio of the squares of the axes of the section will be $\cos\omega + 1 : \cos\omega - 1$, and the species of the section will be completely determined.

Thus in the paraboloid $ay^2 + bz^2 = x$, we shall have the equations

$$a\mu^2 + b\nu^2 = 0, \quad l\lambda + m\mu + n\nu = 0,$$

and we may readily form the equations

$$\frac{\lambda_1\lambda_2}{an^2 + bm^2} = \frac{\mu_1\mu_2}{bl^2} = \frac{\nu_1\nu_2}{al^2} = \frac{\lambda_1\mu_2 + \mu_2\lambda_1}{-2blm} = \frac{\mu_1\nu_2 + \mu_2\nu_1}{0} = \frac{\nu_1\lambda_2 + \nu_2\lambda_1}{-2anl}$$

$$= \frac{\lambda_1\lambda_2 + \mu_1\mu_2 + \nu_1\nu_2}{a(n^2 + l^2) + b(l^2 + m^2)} = \frac{\sqrt{\{(\mu_1\nu_2 - \mu_2\nu_1)^2 \ldots\ldots\ldots\}}}{\sqrt{\{-4abl^2(l^2 + m^2 + n^2)\}}}.$$

Hence $$\cot\omega = \frac{a(n^2 + l^2) + b(l^2 + m^2)}{\sqrt{\{-4abl^2(l^2 + m^2 + n^2)\}}}.$$

(1) For hyperbolic sections, ab must be negative, and l finite.

(2) For parabolic sections, $\omega = 0$, and therefore $l = 0$.

(3) For elliptic sections, ab must be positive, and l finite.

(4) For rectangular hyperbolic sections,

$$a(n^2 + l^2) + b(l^2 + m^2) = 0.$$

(5) For circular sections, $\cot\omega = \sqrt{(-1)}$, or

$$a^2(n^2 + l^2)^2 + b^2(l^2 + m^2)^2 + 2ab\{m^2n^2 - l^2(l^2 + m^2 + n^2)\} = 0,$$

and this may be obtained in the form

$$\{(l^2+n^2)\,a-(l^2+m^2)\,b\}^2+4abm^2n^2=0,$$

and since by (3) ab is positive, we have the systems

$$m=0,\ \ \frac{l^2}{a}=\frac{n^2}{b-a};\quad n=0,\ \ \frac{l^2}{b}=\frac{m^2}{a-b}.$$

Of these only one is possible, and the results coincide with those of Art. 187.

It appears from (1), (2), (3) that the elliptic paraboloid has no hyperbolic sections, and the hyperbolic paraboloid no elliptic sections, and that all sections parallel to the axis in either surface are parabolas. To the latter there is one case of exception, namely, when $l=0$, and $an^2+bm^2=0$, in which case $\cot\omega$ assumes an indeterminate form. We have however for all sections satisfying the condition $an^2+bm^2=0$,

$$\cot\omega=\frac{(a+b)\,l}{\sqrt{(-4ab)}},$$

which gives for $l=0$, $\omega=\frac{\pi}{2}$, or a rectangular hyperbola. The section is however in this case really two straight lines, one at a finite distance, and one at infinity. For we may write the equation of the plane of section

$$y\sqrt{a}\pm z\sqrt{(-b)}=k,$$

and when this meets the surface $ay^2+bz^2=x$, we shall have

$$k\,\{y\sqrt{a}\mp z\sqrt{(-b)}\}=x,$$

or the only points at a finite distance in which the plane meets the surface lie on the straight line

$$\frac{x}{k}=2y\sqrt{a}-k=k\mp 2z\sqrt{(-b)}.$$

These sections are parallel to the asymptotic planes of Art. 165, and are themselves asymptotic in the same sense.

In the central surface $ax^2+by^2+cz^2=k$, the equations for the directions of the asymptotes are

$$a\lambda^2+b\mu^2+c\nu^2=0,\quad l\lambda+m\mu+n\nu=0.$$

The first of these shews that every asymptote is parallel to a generating line of the asymptotic cone; hence for parabolic sections, in which the directions of the asymptotes coincide, the cutting plane must be parallel to a tangent plane of the asymptotic cone. The section of an hyperboloid made by such a tangent plane will be two parallel straight lines, since the center of the hyperboloid must be a center of the section, and the section parabolic.

We shall, in general, have the equations

$$\frac{\lambda_1\lambda_2}{bn^2+cm^2}=\frac{\mu_1\mu_2}{cl^2+an^2}=\frac{\nu_1\nu_2}{am^2+bl^2}$$

$$=\frac{\mu_1\nu_2+\mu_2\nu_1}{-amn}=\frac{\nu_1\lambda_2+\nu_2\lambda_1}{-bnl}=\frac{\lambda_1\mu_2+\lambda_2\mu_1}{-clm}$$

$$=\frac{\lambda_1\lambda_2+\mu_1\mu_2+\nu_1\nu_2}{a(m^2+n^2)+b(n^2+l^2)+c(l^2+m^2)}$$

$$=\frac{\sqrt{\{(\mu_1\nu_2-\mu_2\nu_1)^2+\ldots\ldots\ldots\}}}{\sqrt{\left\{-4abc\,(l^2+m^2+n^2)\left(\frac{l^2}{a}+\frac{m^2}{b}+\frac{n^2}{c}\right)\right\}}}.$$

Hence $\cot\omega=\dfrac{a(m^2+n^2)+b(n^2+l^2)+c(l^2+m^2)}{2\sqrt{\left\{-abc\,(l^2+m^2+n^2)\left(\frac{l^2}{a}+\frac{m^2}{b}+\frac{n^2}{c}\right)\right\}}}.$

(1) The section will be hyperbolic, parabolic, or elliptic, according as $l^2bc+m^2ca+n^2ab$ is negative, zero, or positive. This of course coincides with the conditions that the parallel plane through the center shall meet the asymptotic cone in real and different, coincident, or impossible straight lines.

(2) For a rectangular, hyperbolic section,

$$a(m^2+n^2)+b(n^2+l^2)+c(l^2+m^2)=0,$$

unless at the same time $l^2bc+m^2ca+n^2ab=0$, but it is easily shewn that these conditions are inconsistent for any real finite values of a, b, c.

(3) For a circular section,

$$\{a(m^2+n^2)+b(n^2+l^2)+c(l^2+m^2)\}^2$$
$$=4(l^2+m^2+n^2)(l^2bc+m^2ca+n^2ab),$$

or $$l^4(b-c)^2+\ldots+2m^2n^2(a-b)(a-c)+\ldots\ldots=0.$$

If a, b, c be in order of magnitude, this may be written

$$\{l^2(b-c)+m^2(c-a)-n^2(a-b)\}^2+4l^2m^2(a-c)(b-c)=0,$$

whence we have two systems of circular sections,

$$l = 0, \quad \frac{m^2}{a-b} = \frac{n^2}{c-a}; \quad m = 0, \quad \frac{l^2}{a-c} = \frac{n^2}{b-c};$$

of which only one is possible. Compare the results of Art. 184.

It appears from (1) that all sections of an ellipsoid are ellipses, but that for the hyperboloids we may have all three species of conics.

249. *The sections, made by parallel planes, of similar and similarly situated conicoids, are similar.*

This appears from the fact that the equations determining the direction of asymptotes involve only the coefficients of the terms of two dimensions in x, y, and z, so that the asymptotes of a section of the surface

$$a(x-\alpha)^2 + b(y-\beta)^2 + c(z-\gamma)^2 = d,$$

made by any plane, are determined in direction by equations depending only on a, b, c and on the direction of the plane. Hence the sections by parallel planes of all the surfaces represented by this equation for different values of α, β, γ, and d will be similar: and a like proof holds for non-central surfaces.

In the similarity here determined, we consider an hyperbola and its conjugate as similar curves. Indeed, in the sections of the same surface by a series of parallel planes which cut it in hyperbolas, though the asymptotes of all the sections are parallel, the curve of section lies, for one portion of the surface, in one pair of angles made by the asymptotes, and in the other portion in the second pair, the two series being separated by the plane section for which the curve degenerates into two straight lines.

250. *To determine the area of a section of a central conicoid made by a given plane.*

Let the equation of the conicoid be $ax^2 + by^2 + cz^2 = 1$, and of the given plane $lx + my + nz = p$.

Let (x_0, y_0, z_0) be the center of the section, and r the length of any central radius whose direction-cosines are λ, μ, ν.

We shall then have the equations

$$\frac{ax_0}{l}=\frac{by_0}{m}=\frac{cz_0}{n}=\frac{p}{\frac{l^2}{a}+\frac{m^2}{b}+\frac{n^2}{c}}, \quad \text{(Art. 184)},$$

and $r^2=\dfrac{1-ax_0^2-by_0^2-cz_0^2}{a\lambda^2+b\mu^2+c\nu^2}\equiv\rho^2\left(1-ax_0^2-by_0^2-cz_0^2\right).$

Then, for the lengths of the principal axes, we shall have

$$\frac{1}{\rho^2}\equiv a\lambda^2+b\mu^2+c\nu^2$$

a maximum or minimum by variation of λ, μ, ν, subject to the conditions

$$1=\lambda^2+\mu^2+\nu^2,$$
$$0=l\lambda+m\mu+n\nu.$$

This gives us the equations

$$a\lambda d\lambda+l\mu d\mu+c\nu d\nu=0,$$
$$\lambda d\lambda+\mu d\mu+cd\nu=0,$$
$$ld\lambda+md\mu+nd\nu=0;$$

or, using two undetermined multipliers k, k',

$$(a+k)\lambda+lk'=0, \quad (b+k)\mu+mk'=0, \quad (c+k)\nu+nk'=0.$$

Multiplying by λ, μ, ν and adding we obtain, $\dfrac{1}{\rho^2}+k=0$, and therefore

$$\frac{\lambda}{k'}=\frac{l}{\frac{1}{\rho^2}-a}, \quad \frac{\mu}{k'}=\frac{m}{\frac{1}{\rho^2}-b}, \quad \frac{\nu}{k'}=\frac{n}{\frac{1}{\rho^2}-c},$$

whence the equation

$$0=\frac{l^2}{\frac{1}{\rho^2}-a}+\frac{m^2}{\frac{1}{\rho^2}-b}+\frac{n^2}{\frac{1}{\rho^2}-c},$$

whose roots are the lengths of the semi-axes.

The rectangle under these lengths is then

$$\sqrt{\left(\frac{l^2+m^2+n^2}{l^2bc+m^2ca+n^2ab}\right)},$$

and the area of the section is

$$\pi\sqrt{\left(\frac{l^2+m^2+n^2}{l^2bc+m^2ca+n^2ab}\right)}\left(1-ax_0^2-by_0^2-cz_0^2\right);$$

or, substituting for x_0, y_0, z_0,

$$\pi\sqrt{\left(\frac{l^2+m^2+n^2}{l^2bc+m^2ca+n^2ab}\right)}\left(1-\frac{p^2abc}{l^2bc+m^2ca+n^2ab}\right).$$

(1) In the case of the ellipsoid, this may be put into a convenient form as follows. Let the plane move parallel to itself till the elliptic section vanishes, and let ϖ be the perpendicular upon it from the center when in that position. The center of the section being then a point on the surface we shall have

$$\varpi^2=\frac{l^2}{a}+\frac{m^2}{b}+\frac{n^2}{c},$$

and if A, A' be the areas of a central section, and of a parallel section at a distance p, we have

$$A'=A\left(1-\frac{p^2}{\varpi^2}\right).$$

(2) Similarly, in the case of the hyperboloid of two sheets,

$$A'=A\left(\frac{p^2}{\varpi^2}-1\right).$$

(3) In the hyperboloid of one sheet, the section can never vanish, but if we take ϖ the perpendicular on the parallel plane for which the section in the conjugate hyperboloid of two sheets vanishes, we shall have

$$A'=A\left(1+\frac{p^2}{\varpi^2}\right).$$

(4) If we take two conjugate hyperboloids $ax^2+by^2+cz^2=\pm 1$, and the asymptotic cone to both, $ax^2+by^2+cz^2=0$, the area of the section of the latter may be found, from those of the former, by making a, b, c infinitely large, preserving their ratios. Hence if A_1, A_2, A_3 be the sections of the three surfaces made by any plane cutting them all in ellipses, and A the area of the parallel central section of the hyperboloid of one sheet, we shall have

$$A_1=A\left(1+\frac{p^2}{\varpi^2}\right),\quad A_2=A\left(\frac{p^2}{\varpi^2}-1\right),\quad A_3=A\,\frac{p^2}{\varpi^2},$$

whence $A_1+A_2=2A_3$, or the section of the cone is an arithmetic mean between the sections of the two hyperboloids.

Also, if V be the volume of the cone cut off by a plane touching the hyperboloid of two sheets, we shall have

$$V=\frac{1}{3}A_3\varpi=\frac{1}{3}A\varpi.$$

Now $A=\dfrac{\pi}{\sqrt{(l^2bc+m^2ca+n^2ab)}}$, and $\varpi^2=\dfrac{l^2}{a}+\dfrac{m^2}{b}+\dfrac{n^2}{c}$;

$$\therefore\ V=\frac{1}{3}\frac{\pi}{\sqrt{(abc)}},\text{ and is constant.}$$

251. *To find the area of a section of an elliptic paraboloid made by a given plane.*

Let the equation of the paraboloid be $\dfrac{y^2}{a}+\dfrac{z^2}{b}=2x$, and of the given plane $lx+my+nz=p$.

Let (x_0, y_0, z_0) be the center of the section, and r the length of any central radius of the section whose direction-cosines are λ, μ, ν.

We shall then have the equations

$$\frac{lx_0}{m^2a+n^2b+lp}=\frac{y_0}{-ma}=\frac{z_0}{-nb}=\frac{1}{l}.\quad\text{(Art. 187).}$$

Also $$r^2\left(\frac{\mu^2}{a}+\frac{\nu^2}{b}\right)=2x_0-\frac{y_0^2}{a}-\frac{z_0^2}{b}=\frac{m^2a+n^2b+2lp}{l^2}.$$

Hence, for the semi-axes, we shall have $\dfrac{\mu^2}{a}+\dfrac{\nu^2}{b}\equiv u$ a maximum or minimum by variation of μ, ν subject to the conditions

$$\lambda^2+\mu^2+\nu^2=1,\quad l\lambda+m\mu+n\nu=0.$$

This leads, as in the last article, to the equation

$$l^2+\frac{m^2}{1-au}+\frac{n^2}{1-bu}=0.$$

If u_1, u_2 be the roots of this equation, $u_1u_2=\dfrac{l^2+m^2+n^2}{l^2ab}=\dfrac{1}{l^2ab}$,

and the area of the section is

$$\pi\frac{m^2a+n^2b+2lp}{l^2\sqrt{(u_1u_2)}},\quad\text{or }\frac{\pi\sqrt{(ab)}}{l}(m^2a+n^2b+2lp).$$

If the plane move parallel to itself till the section vanishes, and ϖ be the perpendicular upon it in that position, we shall have $0 = m^2a + n^2b + 2l\varpi$, and the area of the section may be written in the form $2\pi\sqrt{(ab)}\,(p - \varpi)$.

Hence the areas of all sections are equal, which are made by planes at equal distances from the tangent planes parallel to them respectively.

XIII.

(1) If α, β, γ be the angles between three equal conjugate radii of an ellipsoid, shew that $\cos^2\alpha + \cos^2\beta + \cos^2\gamma$, and $\cos\alpha\cos\beta\cos\gamma$, are constant; and that $\alpha + \beta + \gamma$ will always be equal to π, if a certain relation holds between the lengths of the axes.

(2) In the hyperboloid $\dfrac{x^2}{a^2} + \dfrac{y^2}{b^2} - \dfrac{z^2}{c^2} = 1$, shew that if a^2 and b^2 be each $> c^2$, the equation of the surface may be put into the form $y^2 + z^2 - x^2 = d^2$, and if α, β, γ be the angles between (yz), (zx), (xy) in this case, $\cos\alpha\,(\cos\beta\cos\gamma - \cos\alpha) = \dfrac{(a^2+b^2)\,(a^2-c^2)\,(b^2-c^2)}{2\,(a^2+b^2-c^2)^3}$.

(3) If $(x_1y_1z_1)$, $(x_2y_2z_2)$, $(x_3y_3z_3)$ be the extremities of three conjugate diameters of the ellipsoid $\dfrac{x^2}{a^2} + \dfrac{y^2}{b^2} + \dfrac{z^2}{c^2} = 1$, the equation of the plane passing through these points is

$$\frac{x}{a^2}(x_1 + x_2 + x_3) + \frac{y}{b^2}(y_1 + y_2 + y_3) + \frac{z}{c^2}(z_1 + z_2 + z_3) = 1.$$

(4) Also the locus of the center of gravity of the triangle formed by joining their extremities is

$$\frac{x^2}{a^2} + \frac{y^2}{b^2} + \frac{z^2}{c^2} = \frac{1}{3},$$

and the locus of the intersection of planes drawn through their extremities parallel to the conjugate planes respectively, is

$$\frac{x^2}{a^2} + \frac{y^2}{b^2} + \frac{z^2}{c^2} = 3.$$

(5) The section of a hyperbolic paraboloid by a plane which makes angles α, β with the planes of the principal parabolic sections, whose latera-recta are a, b, will be a rectangular hyperbola if

$$a\sin^2\alpha = b\sin^2\beta.$$

(6) The co-ordinates of the vertex of the section of the paraboloid

$$\frac{y^2}{a} + \frac{z^2}{b} = x,$$

made by the plane $y \cos \alpha + z \sin \alpha = p$, are given by the equations

$$\frac{x}{p} = \frac{y}{a \cos \alpha} = \frac{z}{b \sin \alpha} = \frac{p}{a \cos^2 \alpha + b \sin^2 \alpha};$$

and the latus-rectum of the section is $\dfrac{ab}{a \cos^2 \alpha + b \sin^2 \alpha}$.

(7) In a paraboloid of revolution, the eccentricity of any section is the cosine of the inclination of the plane to the axis of the surface, and the foci of the section are the points of contact with spheres inscribed in the surface.

(8) The section of the surface $yz + zx + xy + a^2 = 0$, by the plane $lx + my + nz = p$, will be an ellipse, parabola, or hyperbola, according as $l^2 + m^2 + n^2 < = \text{ or } > 2(mn + nl + lm)$. What will be the condition for a rectangular hyperbola?

(9) If e be the eccentricity of the section of the cone

$$\sqrt{x} + \sqrt{y} + \sqrt{z} = 0$$

made by the plane $lx + my + nz = p$, shew that

$$\frac{e^4}{1-e^2} = \frac{(mn - l^2 + nl - m^2 + lm - n^2)^2}{(l^2 + m^2 + n^2)(mn + nl + lm)}.$$

(10) Shew that, if any section of the surface given in (9) make angles α, β, γ with the co-ordinate planes, the eccentricity of the section will be constant if $\cos \alpha + \cos \beta + \cos \gamma$ be constant.

Shew also that the section will be hyperbolic if $\cos \alpha + \cos \beta + \cos \gamma$ lie between $+1$ and -1, parabolic at these limits, and elliptic for all other possible values.

(11) The section of the conicoid

$$ax^2 + by^2 + cz^2 + 2a'yz + 2b'zx + 2c'xy + 2a''x + 2b''y + 2c''z + d = 0,$$

by the plane $lx + my + nz = p$, will be an ellipse, parabola, or hyperbola, according as

$$\alpha \equiv (bc - a'^2)\, l^2 + \ldots\ldots\ldots \quad - 2(aa' - b'c')\, mn - \ldots\ldots\ldots$$

is positive, zero, or negative, and a rectangular hyperbola, if

$$\beta \equiv l^2 (b + c) + \ldots\ldots\ldots \quad - 2a'mn - \ldots\ldots\ldots = 0.$$

Shew that the angle between the asymptotes is $\cot^{-1} \dfrac{\beta}{2\sqrt{-\alpha}}$.

(12) The foci of the section of the ellipsoid $\dfrac{x^2}{a^2} + \dfrac{y^2}{b^2} + \dfrac{z^2}{c^2} = 1$, made by the plane $lx + my + nz = 0$, are given by the equations

$$\frac{x(mz - ny)}{a^2(b^2 - c^2)} = \frac{y(nx - lz)}{b^2(c^2 - a^2)} = \frac{lmn}{l^2a^2 + m^2b^2 + n^2c^2}; \quad lx + my + nz = 0.$$

(13) A plane drawn through the origin perpendicular to any generating line of the cone

$$x^2(a^2-d^2)+y^2(b^2-d^2)+z^2(c^2-d^2)=0$$

will intersect the ellipsoid $\frac{x^2}{a^2}+\frac{y^2}{b^2}+\frac{z^2}{c^2}=1$ in a section of constant area.

(14) A plane touching the asymptotic cone of an hyperboloid will intersect a confocal ellipsoid in a section of constant area.

(15) If sections of an ellipsoid $\frac{x^2}{a^2}+\frac{y^2}{b^2}+\frac{z^2}{c^2}=1$ be made by planes passing through the center, and through another given point $(x'y'z')$, the sections of greatest and least area will be at right angles to each other, and the areas will be $\frac{\pi abc}{r_1}$, $\frac{\pi abc}{r_2}$, r_1, r_2 being the semi-axes of the section made by the plane $\frac{xx'}{a}+\frac{yy'}{b}+\frac{zz'}{c}=0$.

Shew that the product of the areas will be constant, if the point lie on the curve of intersection of the ellipsoid and a concentric sphere.

CHAPTER XIII.

DEGREES OF SURFACES AND CURVES. INTERSECTIONS OF SURFACES. NUMBER OF CONDITIONS TO WHICH SURFACES CAN BE SUBJECTED. CLASSES OF SURFACES.

252. HAVING already fully investigated the nature of the surfaces represented by the general equation of the second degree, we will proceed to the loci of equations of higher degrees, which we may consider as equations either in three plane or four plane co-ordinates: in the latter case, we may consider the equations homogeneous, without loss of generality.

253. Surfaces which are represented by rational and integral algebraical equations are arranged according to the degrees of these equations.

A surface is of the nth degree, if the equation of which it is the locus is of the nth degree in the co-ordinates of any point of the locus; or, if any arbitrary straight line meets it in n points, real or imaginary, which is the geometrical equivalent.

Curves are also arranged according to degrees, a curve of the nth degree being one which intersects any arbitrary plane in n points, real or imaginary.

254. Among the methods of treating of curves which have been proposed, one is to consider them as the intersection of surfaces whose equations are given. In this method the difficulty arises to which allusion has been made (Art. 18), viz. that extraneous curves may be introduced, which are not the subjects of investigation.

If any curve be supposed to be given in space, it is impossible generally to determine two surfaces, which shall contain no other points but points which lie on the proposed curve, but among all the surfaces which might be drawn through a curve, it is desirable to obtain, where it is possible, the simplest forms of surfaces of which the curve shall be the partial intersection.

255. The number of points in which three surfaces intersect, which are of the mth, n^{th}, and p^{th} degrees respectively, is mnp, unless they intersect in a common curve, in which case it is infinite.

For the proof of this proposition the student is referred to Salmon's treatise on *Higher Algebra* (Art. 39), on the number of solutions of three equations in three unknown quantities.

The student may be able to satisfy himself of the truth of the proposition, by considering that the number of points, in which the surfaces intersect, will by the law of continuity be unaltered, if we substitute particular instead of the general forms of the surfaces. If the surfaces be equations of m, n, p arbitrary planes it is obvious that their common points of intersection are in number mnp, each point being the intersection of three planes, taken one from each system.

256. *The complete intersection of two surfaces of the mth and n^{th} degrees respectively, is a curve of the mn^{th} degree.*

Let a plane intersect the surfaces, the number of points of intersection of the plane with the surfaces is mn, and this is therefore the number of points in which the plane cuts the curve, or the curve is of the mn^{th} degree.

257. *To find the number of conditions which a surface of the nth degree may be made to satisfy.*

The number of constants in the general equation of the nth degree is evidently the number of homogeneous products of four things of n dimensions, and is therefore

$$= \frac{4.5 \ldots (4+n-1)}{1.2 \ldots n} \equiv \frac{(n+1).(n+2)(n+3)}{1.2.3};$$

but in estimating the number of constants with reference to the number of conditions which the locus can be made to satisfy, we must diminish this number by one, since the equation is unaltered if we divide by any one of the constants.

The number of disposable constants so obtained is

$$\frac{(n+1)(n+2)(n+3)}{1.2.3} - 1 \equiv \frac{n(n^2+6n+11)}{6} \equiv \phi(n).$$

Thus $\phi(2) = 9$, $\phi(3) = 19$, $\phi(4) = 34$,
$\phi(5) = 55$, $\phi(6) = 83$, and so on.

Since when a point is given we may substitute its co-ordinates in the general equation of a given degree, and thus obtain a linear equation of condition between the constants; a surface of the third degree may be made to pass through 19 arbitrarily chosen points, and one of the fourth through 34, &c., and $\phi(n)$ arbitrarily chosen points will completely determine the position and dimensions of a surface of the nth degree.

A surface of the nth degree is also determined by $\phi(n)$ independent linear equations of any kind between its coefficients.

258. *All surfaces of the nth degree which pass through $\phi(n)-1$ given points have a common curve of intersection.*

Let $u=0$, $v=0$ be the equation of two surfaces passing through the given points, $\lambda u + \mu v = 0$ is the equation of another surface of the nth degree which passes through the $\phi(n)-1$ given points; and since, by giving proper values to the ratio $\lambda : \mu$, this surface may be made to pass through any additional point which is not common to the two surfaces $u=0$, $v=0$, this equation will be the general equation of all surfaces which contain the $\phi(n)-1$ given points, if $\lambda : \mu$ receive all values from $-\infty$ to $+\infty$. But this equation is also satisfied by the co-ordinates of all points which lie on the curve of intersection of $u=0$ and $v=0$, which is therefore a common curve of intersection of all surfaces containing the $\phi(n)-1$ points, and is of the degree n^2.

259. By reasoning similar to the above it can be seen that, if a surface be of such a nature that m points or m linear equations of condition completely determine it, we may assert, that if $m-1$ such conditions be given, all surfaces of this kind which satisfy these conditions will have a common curve of intersection.

260. Conversely, if $\phi(n)-1$ points be given, we may eliminate from the general equation of the surface of the nth degree all the constants but one, which will enter into the resulting equation in the first power only. This equation will then be of the form $u + \lambda v = 0$, where u, v are of the nth degree, and λ an undetermined constant. All surfaces represented by this equation will

pass through the curve given by the equations

$$u = 0, \ v = 0;$$

which curve is therefore completely determined. For example, eight points determine a curve which is the complete intersection of two conicoids.

In the case of complete intersection of surfaces the nature of the curve is not given when the degree is given, except in the case of prime numbers, in which it is a plane curve.

For example, a curve of the twelfth degree might be the complete intersection of pairs of surfaces of the degrees (1, 12), (2, 6), (3, 4), and these different species, belonging to the same degree, would require a different number of given points completely to determine the surfaces.

The following proposition serves to obtain the number of given points sufficient to determine a surface of the nth degree which, by its complete intersection with a surface of a lower degree, gives a curve of the nq^{th} degree: this is given by Plucker, but may also be proved directly by a theorem given by Cayley, *Nouvelles Annales*, XII. p. 396.

261. *All surfaces of the nth degree which pass through*

$$\phi(n) - \phi(n-q) - 1$$

given points of a surface of the qth degree cut this last surface in one and the same curve.

Of $\phi(n) - 1$ given points, $\phi(p)$ lie on a surface of the pth degree whose equation is $u_p = 0$, and if the rest, viz.

$$\phi(n) - \phi(p) - 1,$$

lie on a surface of the qth degree, where $n = p + q$, whose equation is $u_q = 0$, then $u_p u_q = 0$ is one of the surfaces which contain the $\phi(n) - 1$ points, and may be obtained by giving a certain value to the ratio $\lambda : \mu$ in the equation $\lambda u + \mu v = 0$, so that

$$\lambda u + \mu v \equiv u_p u_q.$$

The curve of intersection of all the surfaces of the nth degree containing these points lies on the surfaces $u_p = 0$ and $u_q = 0$.

Hence if $\phi(n) - \phi(n-q) - 1$ points be taken on any fixed surface $u_q = 0$, all surfaces of the nth degree, which pass through these points, intersect the surface of the qth degree in the same curve.

Thus, if $q = 1$, the proposition is reduced to the following:

All surfaces of the nth degree which pass through

$$\frac{(n+1)(n+2)}{1.2}-1$$

given points in a plane determine a fixed curve of the nth degree.

If $q=2$, the proposition becomes,

All surfaces of the nth degree which pass through $n(n+2)$ points on a conicoid, intersect the conicoid in the same curve.

A curve of the sixth degree, which is the complete intersection of surfaces of the second and third degrees, is determined by 15 given points; if the surfaces be of the first and sixth, 27 points are required.

One of the tenth degree requires 65 or 35, according as it is a plane curve or lies on surfaces of the second and fifth degrees.

262. When it is said that a curve is determined by a certain number of points, these points must be supposed arbitrarily taken, for it is possible to select the same number of points, which would not be sufficient. Thus, a plane cubic is generally determined by 9 points, but, if those be the nine points of intersection of two of such curves, an infinite number may be drawn through them. A curve of the fourth degree of one species can be determined completely by 8 arbitrary points, but if these given points are the intersections of three conicoids which have not a common curve of intersection, taking these surfaces two and two, we may obtain three curves of that species passing through the same eight points.

263. If a curve of the pth degree be the complete intersection of two surfaces of the mth and n^{th} degrees, so that $p=mn$, and if $m>n$, then the number of points which determine the curve is

$$\phi(m)-\phi(m-n)-1,$$

or $\quad \dfrac{mn}{2}(m-n+4)+\dfrac{1}{6}(n-1)(n-2)(n-3)\equiv\rho \quad$ (Art. 261),

then, if $\quad \dfrac{mn}{2}\left(\dfrac{mn}{n+1}-n+3\right)+\dfrac{1}{6}n(n-1)(n-2)\equiv\rho',$

$$\rho-\rho'=\frac{mn}{2}\left(\frac{mn}{n(n+1)}+1\right)-\frac{1}{2}(n-1)(n-2),$$

and since $mn>(n-1)(n-2)$, $\rho>\rho'$. Hence, as n increases, while mn remains constant, ρ increases.

If therefore mn can be divided in a different manner as $m'n'$, the nearer n is to m the smaller is the number of points requisite to determine a curve of the mn^{th} degree.

Thus $$mn = 60,$$

60.1	requires	1890	points,
30.2	....	960	....
30.3	....	630	....
15.4	....	451	....
12.5	....	334	....
10.6	....	250	....

Nouvelles Annales, XI. p. 361.

264. An extension of the theorem of Art. 261, is also given by Plucker, as follows.

If of the $\phi(n) - 1$ given points, $\phi(p) + m$ points lie on a surface of the pth degree, and the remainder $\phi(n) - \phi(p) - m - 1$ exceed the number which is sufficient for determining a surface of the $(n-p)^{\text{th}}$ degree, so that

$$\phi(n) - \phi(p) - m - 1 \overline{\gtrless} \phi(n-p),$$

and therefore,

$$m < \phi(n) - \phi(p) - \phi(n-p) < \frac{p(n-p)(n+4)}{2},$$

it follows that, if of $\phi(n) - 1$ points through which surfaces of the nth degree pass $\phi(p) + m$ lie on a surface of the pth degree, and

$$m < \frac{p(n-p)(n+4)}{2},$$

the curves of intersection of all the surfaces lie on two surfaces respectively of the degree p and $n-p$.

265. The theory of *partial* intersections of surfaces is discussed by Salmon in Vol. V. of the *Quarterly Journal*. Without an examination of such partial intersections it is not possible to analyze different species of curves of the same degree. If we considered only *complete* intersections of surfaces, curves of the third degree could only be considered as plane curves, whereas it will be seen that they may also be *partial* intersections of conicoids.

266. In order to find the surfaces which may contain a curve of the mth degree, it is observed that through $\phi(k)$ points a

surface ot the kth degree can be made to pass. Now, the total number of points which are common to a proper curve of the mth degree with such a surface, supposed not to contain the curve entirely, are mk, since this is the number of points in which k planes intersect the curve; and the law of continuity makes the statement general.

If $\phi(k) = mk + 1$, one such surface can be drawn containing the curve, if $\phi(k) > mk + 1$, two surfaces of the kth degree can be drawn, and therefore an infinite number. Thus, for a curve of the third degree, if $k = 2$, $\phi(k) = 9 > 3.2 + 1$, hence an infinite number of conicoids may be drawn containing any curve of the third degree.

Again, if $\phi(k) = mk + 1$, one surface of the kth degree contains the curve and the simplest surface of a higher degree can be found by trial; which will be of the $(k+1)^{th}$ degree, whenever $(k+2)(k+3) > 2m$.

Hence, all curves of the fourth degree can be obtained by the intersections of proper surfaces of the second and third degrees: and all curves of the sixth degree from surfaces of the third and fourth degrees.

Modifications are required if the surfaces are not proper surfaces. Salmon gives as examples of this modification, a plane curve of the third degree through which it is possible to describe an infinite number of conicoids, but since each conicoid must necessarily consist of the plane of the curve and an arbitrary plane, the intersection of the plane and conicoid will not determine the curve: again, if a curve of the fifth degree, which, according to the above laws, ought necessarily to be determined by surfaces of the third degree, lie entirely on a conicoid, all the surfaces of the third degree which contain the curve may be a compound of the conicoid and a plane, and we must recede to surfaces of the fourth degree to determine the curve.

If a curve be given of the mth degree, and k, l be the lowest degrees of surfaces upon which it can lie, any surface of the kth degree constructed to pass through $mk + 1$ points will contain the curve, and similarly for the other surface.

If $ml + 1$ points known to lie on the curve be given, and $l > k$, all the rest can be found.

267. The number of *arbitrary* points through which a curve of the mth degree can be drawn cannot exceed a certain superior limit which is easily determined, for suppose k arbitrary points be given and a cone be constructed containing the curve, and having its vertex in one of the assumed points, the degree of this cone will be $m-1$, and the number of its generating lines sufficient for its complete determination is the same as that of the number of points necessary to determine a plane curve of the $\overline{m-1}]^{\text{th}}$ degree, viz. $\frac{m(m+1)}{2}-1$.

The greatest value of k for which such a cone can be constructed is $\frac{m(m+1)}{2}$; this is therefore a superior limit, although other lower limits to the number k may be obtained in general from other considerations.

Thus, a curve of the third degree cannot be made to pass through more than six arbitrarily chosen points.

268. *If $\phi(n)-2$ points be given, all surfaces of the nth degree which can be drawn through these points, will pass through $n^3-\phi(n)+2$ more fixed points.*

Let $u=0$, $v=0$, $w=0$ be the equations of three surfaces of the nth degree which pass through $\phi(n)-2$ points, and which have not a common curve of intersection, they will pass through n^3 common points, and $\lambda u+\mu v+\nu w=0$ is the equation of another surface of the nth degree, which passes through the same points, and by giving different values to $\lambda:\mu:\nu$ we can obtain *all* surfaces which pass through these points. Any surface will be particularized when two points are given, which do not lie on all three of the surfaces, or both on the same two: and all such surfaces will contain $n^3-\phi(n)-2$ common points besides the given points.

Thus, all conicoids which pass through seven points will pass through a fixed eighth, as is easily seen if each conicoid be two parallel planes, the seven points being angular points of a parallelopiped.

A surface of the third degree, drawn through 17 points, passes through 10 others.

269. *All surfaces of the nth degree which pass through a certain number of given points in the curve of intersection of two surfaces of inferior degrees, q and s, that number being*

$$\phi(n) - \phi(n-q) - \phi(n-s) - 2, \text{ if } n < q + s,$$

and $\phi(n) - \phi(n-q) - \phi(n-s) + \phi(n-q-s) - 1$, *if* $n \overset{=}{>} q+s$,

will intersect that curve in another number of fixed points which is the excess of nqs above the number of given points.

Let $u = 0$ be a surface of the nth degree passing through the given points on the curve of intersection of $v_q = 0$ and $w_s = 0$.

I. Let $n < q + s$.

Take on $u = 0$, and $v_q = 0$, $\phi(n-s)$ points, and on $u = 0$, $w_s = 0$, $\phi(n-q)$ points, which do not lie on the curve of intersection of $v_q = 0$, $w_s = 0$.

These two sets of points completely determine surfaces

$$w_{n-s} = 0 \text{ and } v_{n-q} = 0;$$

$$\therefore\ v_{n-q}v_q = 0, \text{ and } w_{n-s}w_s = 0,$$

are composite surfaces which each contain $\phi(n) - 2$ fixed points.

Hence $\lambda u + \mu v_{n-q}v_q + \nu w_{n-s}w_s = 0$ is the equation of all surfaces of the nth degree which pass through these $\phi(n) - 2$ points, which surfaces will all intersect in n^3 points distributed over the four groups of surfaces, the numbers being

$$nqs + n(n-q)(n-s) + ns(n-q) + nq(n-s);$$

therefore the number of points in which the surfaces meet the curve of intersection of $v_q = 0$, $w_s = 0$, besides the given points, is

$$nqs - \phi(n) + \phi(n-q) + \phi(n-s) + 2.$$

II. Let $n \overset{=}{>} q + s$. In this case, since

$$\mu v_{n-q}v_q + \nu w_{n-s}w_s \equiv \mu(v_{n-q} + \rho\nu w_s \Omega_{n-q-s})\, v_q$$
$$+ \nu(w_{n-s} - \rho\mu\Omega_{n-q-s})\, w_s,$$

in which $\rho\Omega_{n-q-s} = 0$ contains $\phi(n-q-s) + 1$ constants, and will diminish the number to be determined in $v_{n-q} = 0$, and $w_{n-s} = 0$, by so many.

Hence, $\phi(n-q) + \phi(n-s) - \phi(n-q-s) + 1$ points determine the most general surfaces of the degrees $n-q$ and $n-s$, which are required to obtain the equation of the general surfaces of the nth degree through the $\phi(n) - 2$ points, to be distributed as before. Therefore all surfaces of the nth degree which pass through the given number of points also pass through other fixed points in number

$$nqs - \phi(n) + \phi(n-q) + \phi(n-s) - \phi(n-q-s) + 1.$$

270. The form in which the number appears in the two cases will be the same for all values of n which make

$$\phi(n-q-s) + 1 \equiv \frac{(n-q-s+1)(n-q-s+2)(n-q-s+3)}{6} = 0,$$

or $$n = q+s-1,\ q+s-2, \text{ or } q+s-3,$$

hence, whenever $n > q+s-4$, it is easily shewn that the number of new points is $\dfrac{qs(q+s-4)}{2} + 1$, which is independent of the degree of the surface $u = 0$.

If $s = 1$ and $q < n$ the proposition may be stated as follows: if any plane curve of the nth degree passes through

$$nq - \frac{q(q-3)}{2} - 1$$

fixed points lying on a curve of the qth degree, $q < n$, the number of the remaining points of intersection will be fixed, viz.

$$\frac{q(q-3)}{2} + 1 \equiv \frac{(q-1)(q-2)}{2};$$

this is the same result as in Salmon's *Higher Curves* (Art. 26).

271. The following propositions, connected with this part of the subject, are of importance in some investigations in which it is required to determine the number of points of intersections of three surfaces: the surfaces under consideration in particular cases may have common lines in any degree of multiplicity, and it becomes necessary to determine to how many points of intersection these lines are equivalent.

272. *These surfaces of the mth, n^{th}, and p^{th} degrees, contain a multiple straight line in the degrees of multiplicity μ, ν and ϖ respectively; to find the number of points of intersection which correspond to these multiple lines.*

The number of points of intersection of any surfaces will be unaltered, if we suppose the surfaces to degenerate into proper surfaces of inferior degrees, so long as the sum of the degrees be the degree of any of the surfaces so broken up.

We will, therefore, suppose the surface of the mth degree to consist of μ planes, and of a proper surface of the $(m-\mu)^{\text{th}}$ degree; and similarly for the others.

The whole number mnp of points of intersection will then be made up of intersections (1) of the three proper surfaces, (2) of proper surfaces from the two systems with planes from the remaining systems, (3) of a proper surface of one system with planes from the two remaining systems, and (4) of planes from the three systems: the number of these intersections are

$$(1)\ (m-\mu)(n-\nu)(p-\varpi),\qquad (2)\ (n-\nu)(p-\varpi)\mu+\ldots$$
$$(3)\ (m-\mu)\nu\varpi+\ldots,\qquad (4)\ \mu\nu\varpi.$$

If now we suppose the planes all to pass through the same straight line, we have the case of surfaces with multiple lines; and those of the mnp points, which lie on the multiple line, will be clearly taken from the groups (3) and (4).

The multiple line therefore corresponds to the number of points
$$(m-\mu)\nu\varpi+(n-\nu)\varpi\mu+(p-\varpi)\mu\nu$$
$$\equiv m\nu\varpi+n\varpi\mu+p\mu\nu-2\mu\nu\varpi,$$
which coincides with the particular case given by Salmon in the *Cambridge and Dublin Mathematical Journal*, Vol. II. page 71.

273. *Three surfaces of the mth, n^{th}, and p^{th} degrees have a common curve line, of the μth, ν^{th}, ϖ^{th} degree of multiplicity respectively, the curve being the intersection of two surfaces of the degrees k and l: to find the number of points which correspond to this multiple line.*

Let the surfaces be broken up into proper surfaces, and the multiple lines be thrown out of gear.

The first shall be composed of μ surfaces of the degree k and one of the degree $m-\mu k$, the second of ν surfaces of the degree k, ν of degree l and one of the degree $n-\nu l$, the third of ϖ of the degree l, and one of the degree $p-\varpi l$.

The number of points which lie on the intersection of surfaces of the degrees k and l will be

$$(m-\mu k)\,\nu k\,.\,\varpi l+(n-\nu l-\nu k)\,\mu k\,.\,\varpi l+(p-\varpi l)\,\mu k\,.\,\nu l$$
$$+\varpi l\,.\,\mu k\,.\,(\nu k+\nu l)\equiv lk\,\{m\nu\varpi+n\varpi\mu+p\mu\nu-\mu\nu\varpi\,(l+k)\},$$

which is the number of points required, coinciding with the result of the preceding proposition when $l=k=1$.

Application to the Four-point System.

274. If $\alpha, \beta, \gamma, \delta$ be the co-ordinates of a plane referred to a four-point system, corresponding to each system of values which satisfy a given equation $F(\alpha, \beta, \gamma, \delta)=0$, a plane is determined, and the surface to which each plane of the system is a tangent plane is therefore determined by the equation, in the same sense as in a plane co-ordinate system the equation determines a surface which is the locus of all the points whose co-ordinates satisfy the equation.

275. Lemma. *To find the co-ordinates of a plane which passes through the line of intersection of two planes whose co-ordinates are given.*

Let $(\alpha', \beta', \gamma', \delta')$ and $(\alpha'', \beta'', \gamma'', \delta'')$ be the given planes P, P', $(\alpha, \beta, \gamma, \delta)$ the required plane Q passing through their line of intersection. The perpendiculars from the fundamental point A on the three planes all lie in a plane, and the relation between them may be found from the trilinear co-ordinates corresponding to an evanescent fundamental triangle, whose angles are the angles between the planes or the supplements of these angles; hence

$$\alpha \sin (P, P')=\alpha' \sin (P', Q)+\alpha'' \sin (P, Q),$$
$$\therefore\ \alpha=\lambda\alpha'+\mu\alpha'', \text{ where } \lambda^2 \pm 2\lambda\mu \cos (P, P')+\mu^2=1.$$

276. Cor. *To find the angle between two given planes.*

For all values of the co-ordinates of a plane,

$$\frac{\alpha^2}{p_1^2}+\frac{\beta^2}{p_2^2}+\ldots\ldots-\frac{2\alpha\beta}{p_1p_2}\cos CD-\ldots\ldots=1, \ldots \text{Art. (106)}.$$

Hence writing

$$\lambda\alpha'+\mu\alpha'',\ \lambda\beta'+\mu\beta'',\ \ldots \text{ for } \alpha, \beta, \ldots$$

we obtain

$$\lambda^2+\mu^2+2\left(\lambda'\frac{\alpha''}{p_1}+\mu'\frac{\beta''}{p_2}+\ldots\right)\lambda\mu=1,$$

where

$$\lambda' = \frac{\alpha'}{p_1} - \frac{\beta'}{p_2}\cos CD - \frac{\gamma'}{p_3}\cos DB - \frac{\delta'}{p_4}\cos BC, \; \mu' = \&c.$$

$$\cos(P, P') = \lambda'\frac{\alpha''}{p_1} + \mu'\frac{\beta''}{p_2} + \nu'\frac{\gamma''}{p_3} + \rho'\frac{\delta''}{p_4}$$

$$= \frac{\alpha'\alpha''}{p_1^2} + \frac{\beta'\beta''}{p_2^2} + \ldots - \frac{\alpha'\beta'' + \alpha''\beta'}{p_1 p_2}\cos CD - \ldots$$

which gives the required angle.

277. The class of a surface is the number the planes of which can be drawn through a given straight line so as to touch the surface.

If $(\alpha, \beta, \gamma, \delta)$ and $(\alpha', \beta', \gamma', \delta')$ be two planes, the co-ordinates of any plane passing through their line of intersection will be $\lambda\alpha + \mu\alpha'$, $\lambda\beta + \mu\beta'$, ...; $\lambda : \mu$ being any arbitrary ratio, and the particular planes which touch a surface, whose equation is $F(\alpha, \beta, \gamma, \delta) = 0$, supposed a homogeneous algebraical equation of the nth degree, will be determined by the values of $\lambda : \mu$ which satisfy the equation

$$F(\lambda\alpha + \mu\alpha', \ldots\ldots) = 0;$$

the number of values of the ratio is n, and this is therefore the class of the surface, and corresponds to the degree of the surface in the plane co-ordinate system.

278. It is easy to express in the language of four-point co-ordinates the results of this chapter.

Thus, a surface of the nth class is determined if $\phi(n)$ tangent planes be given.

If surfaces of the nth class be drawn touching $\phi(n) - 1$ tangent planes, they will all be touched by a common developable surface.

If three surfaces of the nth class touch $\phi(n) - 2$ planes they will have n^3 common tangent planes.

Similarly for other theorems.

XIV.

1. PROVE that two cones of the second degree, having a common generating line, intersect in a curve of the third degree.

2. Every cone containing a curve of the third degree, in which the vertex lies, is of the second degree.

3. Prove that an infinite number of curves of the third degree can be drawn through five points arbitrarily chosen in space, but that six determine the curve: what limitations are necessary that such a curve shall pass through the points?

4. When a curve of the third degree is traced on an hyperboloid of one sheet, it meets in two points all generating lines of one system, and in one point those of the opposite system.

5. Through a curve of the third degree, and a straight line meeting the curve in one point only, a hyperboloid can be drawn, of which the generating lines, which intersect the given line, meet the curve each in two points.

6. Through any point in space a straight line can be drawn which meets a curve of the third degree, not a plane curve, in two points.

7. The projection of a curve of the third degree, not plane, upon any plane by straight lines drawn from a given point, is a curve of the third degree having a double point.

8. When a straight line joins two points of a curve of double curvature of the third degree, an hyperboloid can be drawn through the curve, the straight line, and a given point in space.

9. If two hyperboloids contain the same curve of the third degree, they must have a common generating line, and the curve meets all generating lines of the same system in two points, and of the opposite in one.

10. When two curves of the third degree lie on the same hyperboloid, and meet each in two points the same generating line, the curves meet in four points. If one meets it in two, and the other in only one point, the curves meet in five points.

11. Three hyperboloids which have a common generating line meet only in four points besides the generating line.

12. Through five points of an hyperboloid, we can draw two curves of the third degree lying entirely in the hyperboloid.

13. A curve of the fourth degree being traced on a conicoid, an infinite number of surfaces of the third degree can be found, which intersect the conicoid in the given curve and any given plane section.

14. Two curves of the fourth degree, in the same conicoid, intersect in eight points.

15. Two curves of the fourth degree, on the same surface of the third degree, intersect in not more than eight points.

16. Find the number of points in which two curves of the fifth degree on the same surface of the third degree intersect.

17. The eight points given by the equation

$$\frac{\alpha^2}{l}=\frac{\beta^2}{m}=\frac{\gamma^2}{n}=\frac{\delta^2}{r}$$

are so related that any conicoid passing through seven of them will pass through the eighth.

CHAPTER XIV.

TANGENT LINES, PLANES, AND CONES. POLES AND POLAR PLANES. NORMALS.

279. Def. If near any point P on a surface a second point Q be taken, and a straight line be drawn through P and Q; then, as Q moves up to P and ultimately coincides with it, the straight line PQ moves up to a limiting position PR, which is called a tangent line to the surface at P.

Since Q may generally move up to P in an infinite number of directions, there are generally an infinite number of tangent lines at any point.

280. *To find the equations of a tangent line to a surface at any point.*

Let the equation of the surface be $F(x, y, z) = 0$, and (x_0, y_0, z_0) be the point at which a tangent line is drawn, l, m, n its direction cosines.

The equations of the tangent line are

$$\frac{x - x_0}{l} = \frac{y - y_0}{m} = \frac{z - z_0}{n} = r.$$

At the points in which this line meets the surface the values of r are given by the equation

$$F(x_0 + lr,\ y_0 + mr,\ z_0 + nr) = 0,$$

or, $$F(x_0, y_0, z_0) + \{lF'(x_0) + mF'(y_0) + nF'(z_0)\}\, r + \&c. = 0,$$

since two points coincide with (x_0, y_0, z_0),

$$F(x_0, y_0, z_0) = 0,$$

and $$lF'(x_0) + mF'(y_0) + nF'(z_0) = 0.$$

The first of these equations represents the fact that (x_0, y_0, z_0) is a point on the surface, the second is a relation which must hold for every tangent line.

Subject to this condition the equations above will be the equations of a tangent line, *generally*.

Tangent lines at singular points of surfaces.

281. In particular cases, that is, for particular points in some surfaces, the equation derived from equating the coefficient of r to zero, may be satisfied identically for all values of $l:m:n$, which shews that *all* straight lines drawn through (x_0, y_0, z_0) meet the surface in two coincident points. In order to obtain a tangent line according to the definition in this case, we must take such a direction that the line may be a limiting position of a straight line joining this point with a third point which approaches and ultimately coincides with (x_0, y_0, z_0); this direction will, therefore, be given by equating the coefficient of r^2 to zero. The resulting equation is

$$\left(l\frac{d}{dx_0}+m\frac{d}{dy_0}+n\frac{d}{dz_0}\right)^2 F(x_0, y_0, z_0)=0,$$

and if this be not also satisfied identically for all values of $l:m:n$, it is the relation which must be satisfied in order that the equations given above may be those of the tangent line.

If the coefficient of r^n be the first, which is not identically zero, there are n points coincident with (x_0, y_0, z_0), and the relation between l, m, n is given by the equation

$$\left(l\frac{d}{dx_0}+m\frac{d}{dy_0}+n\frac{d}{dz_0}\right)^n F(x_0, y_0, z_0)=0.$$

Such points are called Singular or Multiple points of surfaces.

282. *To find the locus of all tangent lines to a surface through a given point on it.*

The equations of a tangent line to a surface at the point (x_0, y_0, z_0) are

$$\frac{x-x_0}{l}=\frac{y-y_0}{m}=\frac{z-z_0}{n},$$

subject to the condition

$$lF'(x_0)+mF'(y_0)+nF'(z_0)=0,$$

unless $\quad F'(x_0)=0,\ F'(y_0)=0, \text{ and } F'(z_0)=0;$

therefore, generally, the locus of the tangent lines is obtained by eliminating l, m, and n, and has the equation

$$(x-x_0)\,F'(x_0)+(y-y_0)\,F'(y_0)+(z-z_0)\,F'(z_0)=0.$$

The locus is, therefore, generally a plane, and this plane is called the *Tangent Plane* of the surface at the given point.

The direction-cosines of the normal to this plane being λ, μ, ν,

$$\frac{\lambda}{F'(x_0)}=\frac{\mu}{F'(y_0)}=\frac{\nu}{F'(z_0)}$$

$$=\pm\sqrt{\left\{\frac{1}{[F'(x_0)]^2+[F'(y_0)]^2+[F'(z_0)]^2}\right\}}.$$

283. If $F(x, y, z)$ be a rational algebraical function of the nth degree, the equation of the tangent plane admits of being written in a simplified form by means of a well-known property of homogeneous functions.

For, if we arrange the function in the form of the sum of a series of homogeneous functions, so that, separating the functional letter, $F=F_n+F_{n-1}+\ldots+F_1+c$, F_n denoting a function of the nth degree,

$$\left(x_0\frac{d}{dx_0}+y_0\frac{d}{dy_0}+\ldots\right)F_n\equiv nF_n;$$

hence the equation of the tangent plane appears in the form

$$xF'(x_0)+yF'(y_0)+zF'(z_0)$$

$$=\left(x_0\frac{d}{dx_0}+y_0\frac{d}{dy_0}+z_0\frac{d}{dz_0}\right)F(x_0, y_0, z_0)$$

$$\equiv\{nF_n+(n-1)\,F_{n-1}+\ldots+F_1\}\,(x_0, y_0, z_0)$$

$$=-\{F_{n-1}+2F_{n-2}+\ldots+(n-1)\,F_1\}\,(x_0, y_0, z_0)-nc,$$

since $$(F_n+F_{n-1}+\ldots+F_1)\,(x_0, y_0, z_0)+c=0.$$

Thus if $$F(x, y, z)=\frac{x^2}{2a}+\frac{y^2}{2b}-z,$$

the equation of the tangent plane at (x_0, y_0, z_0) is

$$\frac{x_0 x}{a} + \frac{y_0 y}{b} - z = z_0.$$

Tangent cone at a singular point.

284. At singular points the equation of the locus is

$$\left\{(x - x_0)\frac{d}{dx_0} + (y - y_0)\frac{d}{dy_0} + (z - z_0)\frac{d}{dz_0}\right\}^n F(x_0, y_0, z_0) = 0,$$

and it is called the *Tangent Cone* to the surface at the singular point, being generated by the motion of a straight line which constantly passes through the point (x_0, y_0, z_0).

285. *To find the equation of the tangent plane at any point of a surface given by the equation* $z = f(x, y)$.

Let a line be drawn through (x_0, y_0, z_0) whose equations are

$$x - x_0 = m(z - z_0), \quad y - y_0 = n(z - z_0),$$

the points in which this line meets the surface are those whose co-ordinates satisfy the equation

$$\begin{aligned} z - z_0 &= f\{x_0 + m(z - z_0), y_0 + n(z - z_0)\} - f(x_0, y_0), \\ &= (mp_0 + nq_0)(z - z_0) + \ldots\ldots\ldots \end{aligned}$$

and, if the line be a tangent line, two values of z are equal to z_0;

$$\therefore\ 1 = mp_0 + nq_0,$$

and, by eliminating m and n from this equation and the equations of the line, we obtain the locus of all the tangent lines through (x_0, y_0, z_0), whose equation will be

$$z - z_0 = p_0(x - x_0) + q_0(y - y_0),$$

which is therefore the equation of the tangent plane where such exists, i.e. unless p_0 and q_0 assume the indeterminate form $\frac{0}{0}$.

This equation is immediately deducible from the equation of Art. (282), by means of the equations

$$p_0 F'(z_0) + F'(x_0) = 0, \text{ and } q_0 F'(z_0) + F'(y_0) = 0.$$

286. *Geometrical explanation of the nature of the intersection of a surface with its tangent plane at any point.*

Every plane intersects a surface of the nth degree in a curve which is of the same degree; hence a tangent plane at any point intersects the surface in a curve of the nth degree, passing through the point of contact.

Now when a tangent plane exists, since it is the locus of the tangent lines at the point of contact, and each of these tangent lines contains two points, which coincide in the point of contact, it follows, that any line, drawn through the point of contact in the tangent plane, meets the curve of intersection in two points at the point of contact.

The point of contact is, therefore, a singular point in the curve of intersection.

This singular point may be either a conjugate point, as in the case of contact with an ellipsoid; or a multiple point, as in the case of a hyperboloid of one sheet; or a point through which two coincident lines pass, as in the case of a cylinder.

If the surface is of the second degree the curve of intersection is of the second degree, and, since it must contain a singular point, the only admissible lines of intersection are either an indefinitely small circle or ellipse, or else two straight lines which cross one another, or are coincident.

287. *If a plane intersect a surface in a curve which contains a singular point, the plane is generally a tangent plane to the surface at that singular point.*

For a straight line drawn in any direction in the plane, through a singular point, meets the surface in two points which ultimately coincide, and therefore generally satisfies the condition of being a tangent line to the surface.

If the point which is a singular point in the curve of intersection is also a singular point in the surface, the condition of passing through two coincident points is not sufficient to define a tangent line.

Thus, if at any point of a surface there be a conical tangent, there may be a multiple point in the curve of intersection of a plane intersecting the conical tangent, which will not make the cutting plane a tangent plane at the multiple point.

288. The form of the curve of intersection of a surface with the tangent plane at any point may be illustrated by taking the case of an anchor ring, supposed to be generated by the revolution of a circle about an axis in its plane not intersecting the circle.

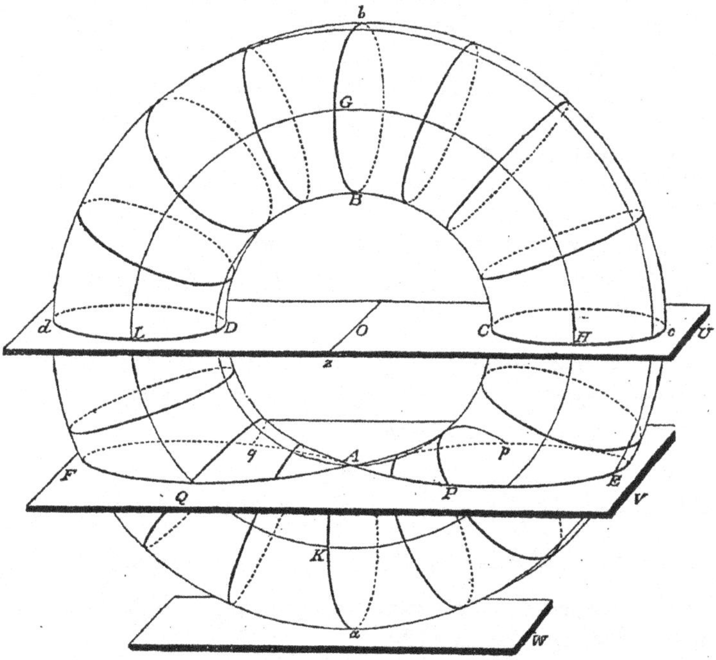

The figure represents the ring, with the generating circle in different positions as it revolves about the axis Oz.

The plane U is drawn through the axis Oz, intersecting the surface in the circles CHc, DLd.

Suppose this plane to move, parallel to itself, towards the position V, the closed curves in which it intersects the surface become elongated until they meet one another in the point A, forming for the position V of the plane a figure of eight, viz. $EPAqFQ$ which has a double point at A. Here we observe that the concavities of the circles AKa and $ACBD$, which are sections by planes perpendicular to V and to each other, lie in opposite directions with regard to the plane V, and that the tangent lines at A lie in that plane, which is therefore the tangent plane at A: and it is a tangent plane at no other point of the curve of intersection.

The sections by planes through A perpendicular to V change the directions of their concavities as they pass from the position AKa to $ACBD$, when they cross the tangents to the branches pAQ, PAq at the multiple point.

If the plane move past V to the position W the curve of intersection gradually assumes an oval form, which degenerates into a conjugate point at d.

It is clear also that a plane may meet the ring in the circle $GHKL$, in which case it is a tangent plane at every point of the curve in which it meets the surface, which is composed of two coincident circles, as may be seen by moving the plane inwards parallel to itself.

289. *To find the equations of the tangent line to the curve of intersection of a surface with its tangent plane at any point.*

Let the equation of the surface be

$$F(x, y, z) = 0,$$

and that of the tangent plane at (x_0, y_0, z_0)

$$(x - x_0) F'(x_0) + (y - y_0) F'(y_0) + (z - z_0) F'(z_0) = 0.$$

Let the equation of the tangent line to the curve of intersection at the point (x_1, y_1, z_1) be

$$\frac{x - x_1}{\lambda} = \frac{y - y_1}{\mu} = \frac{z - z_1}{\nu} = r.$$

The points in which it meets the curve are given by the values of r, which satisfy the equations

$$F(x_1 + \lambda r, \ y_1 + \mu r, \ z_1 + \nu r) = 0,$$

$$\text{or} \quad e^{r\left(\lambda \frac{d}{dx_1} + \mu \frac{d}{dy_1} + \nu \frac{d}{dz_1}\right)} F(x_1, y_1, z_1) = 0,$$

and $(x_1+\lambda r-x_0)\,F'(x_0)+(y_1+\mu r-y_0)\,F'(y_0)$
$$+(z_1+\nu r-z_0)\,F'(z_0)=0.$$

And since the tangent line meets the curve at two points generally, and in more than two at a singular point, we have for the general tangent line

$$\lambda F'(x_0)+\mu F'(y_0)+\nu F'(z_0)=0,$$
$$\lambda F'(x_1)+\mu F'(y_1)+\nu F'(z_1)=0;$$

which equations determine the ratios $\lambda:\mu:\nu$, except in the cases in which more than one system of values of $\lambda:\mu:\nu$ satisfy the equations; and this will happen,

I. when $F'(x_0)=0$, $F'(y_0)=0$, and $F'(z_0)=0$ simultaneously, which is the case when there is a tangent cone at the point (x_0, y_0, z_0);

II. when $F'(x_1)=0$, $F'(y_1)=0$, $F'(z_1)=0$, which is the case when (x_1, y_1, z_1) is a singular point in the surface;

III. when $\dfrac{F'(x_1)}{F'(x_0)}=\dfrac{F'(y_1)}{F'(y_0)}=\dfrac{F'(z_1)}{F'(z_0)}$, which will happen when (x_1, y_1, z_1) is the point of contact of the tangent plane.

In the last case, the tangent line meets the curve in more than two coincident points, and the condition for this is

$$\left(\lambda\frac{d}{dx_0}+\mu\frac{d}{dy_0}+\nu\frac{d}{dz_0}\right)^2 F(x_0, y_0, z_0)=0,$$

which, combined with the equation

$$\lambda F'(x_0)+\mu F'(y_0)+\nu F'(z_0)=0,$$

give two systems of values of the ratios $\lambda:\mu:\nu$, unless the differential coefficients of $F(x_0, y_0, z_0)$ of the second order vanish, in which case all lines drawn in the tangent plane through (x_0, y_0, z_0) meet the curve in three coincident points, and the multiple point is of a higher order of multiplicity, and if the sth differential coefficients are the first, which do not all vanish,

$$\left(\lambda\frac{d}{dx_0}+\mu\frac{d}{dy_0}+\nu\frac{d}{dz_0}\right)^s F(x_0, y_0, z_0)=0$$

is the equation which determines the s tangents to the multiple point, at the point of contact.

290. As an illustration of the nature of the intersection of a tangent plane with a surface which it touches, the case of triple tangent planes to surfaces of the third degree may be taken, discussed by Cayley*.

On every surface of the third degree there are 27 *straight lines, and* 45 *triple tangent planes, real or imaginary.*

A straight line meets a surface of the third degree generally in three points, which are given by an equation of the form

$$u + Du \,.\, r + \frac{1}{2} D^2u \,.\, r^2 + \frac{1}{6} D^3u \,.\, r^3 = 0.$$

Now the four constants in the equations of a straight line may be chosen so as to satisfy the equations $u = 0$, $Du = 0$, $D^2u = 0$, $D^3u = 0$, and all straight lines having such constants lie entirely in the surface, since the above equation is then satisfied for all values of r.

If a plane be drawn through such a straight line its line of intersection with the surface will be, in the general position, composed of that straight line and a conic forming a group of the third degree.

Now there are five positions of the plane for which the conic breaks up into two straight lines. For the equations of any surface of the third degree which contains the axis of x as one of its straight lines is of the form

$$F_3(y, z) + xF_2(y, z) + x^2F_1(y, z) = 0,$$

$F_3(y, z)$ denoting a function of y and z of the third degree, and $F_2(y, z)$, $F_1(y, z)$ of the second and first degrees.

If $z = \lambda y$ be the equation of a plane containing the axis of x the projection of the curve of intersection on the plane of xy has for its equation

$$y\{f_3(\lambda) + xf_2(\lambda) + x^2f_1(\lambda)\} = 0,$$

* *Cambridge and Dublin Mathematical Journal*, Vol. IV. p. 118.

where
$$f_3(\lambda) = a_3y^2 + 2a_2y + a_1,$$
$$f_2(\lambda) = 2b_2y + 2b_1,$$
$$f_1(\lambda) = c_1,$$

the subscripts of the letters a, b, c being the degree to which λ rises in the respective coefficients.

The curve of intersection consists therefore of the axis of x and a conic the equation of the projection of which is

$$a_3y^2 + 2b_2xy + c_1x^2 + 2a_2y + 2b_1x + a_1 = 0,$$

and the conic breaks up into two straight lines for values of λ, which satisfy the condition

$$a_3b_1^2 + c_1a_2^2 + a_1b_2^2 - a_3c_1a_1 - 2b_2a_2b_1 = 0;$$

which, being of the fifth degree in λ, shews that there are five positions of the plane, for which the conic becomes two straight lines.

In the general position of a plane through a straight line the plane is a tangent plane at the two points of intersection of the conic and straight line (Art. 279); in the five particular positions, the plane, intersecting the surface in three straight lines which form three double points, touches the surface at these three points, and it is therefore a triple tangent plane.

Through each of the three straight lines can be drawn, besides the plane in question, four other triple tangent planes, giving rise to 12 new triple tangent planes, and 24 new straight lines situated on the surface, making in all 27.

These are the only such straight lines which can be drawn on the surface, for any straight line on the surface must meet one of the three straight lines in any triple plane, since these three straight lines form the complete intersection of the plane with the surface; and the plane passing through such straight line and the line which it intersects must be one of the triple tangent planes containing that line, since it intersects the surface in two and therefore three straight lines.

Each triple tangent plane contains three lines, and five can be drawn through each of the 27 lines, therefore the whole number of triple tangent planes is $\dfrac{5 \cdot 27}{3} \equiv 45$.

Ruled Surfaces.

291. The student is already familiar with certain surfaces which are capable of being generated by straight lines, or through every point of which some straight line may be drawn which coincides, throughout its length, with the surface. For example, a plane, a cone, a cylinder, an hyperboloid of one sheet, an hyperbolic paraboloid.

Among these surfaces he is aware that any portion of a conical or cylindrical surface, if supposed perfectly flexible, might be developed into a plane without tearing or rumpling.

We shall now give some account of the general character of surfaces which have this property, distinguishing them from those which, although capable of being generated by the motion of a straight line, are incapable of development into a plane.

292. DEF. A *Ruled Surface* is a surface which is capable of generation by the motion of a straight line; or a surface through every point of which a straight line can be drawn, which lies entirely in the surface.

DEF. If a ruled surface be such that each generating line intersects that which is next consecutive, the surface is called a *Developable Surface.*

DEF. If a ruled surface be such that consecutive positions of the generating line do not intersect, the surface is called a *Skew Surface.*

Developable Surfaces.

293. *Explanation of the development of developable surfaces into a plane.*

Let Aa, Bb, Cc, ... be a series of straight lines taken in order, according to any proposed law, so as to satisfy the condition that each intersects the preceding, viz. in the points a, b, c, ...

Since Aa, Bb intersect in a, they lie in the same plane, similarly, the successive pairs of lines Bb and Cc, Cc and Dd, &c. lie in one plane; thus, a polygonal surface is formed by the successive plane elements AaB, BbC, &c.

This surface might be developed into one plane by turning the face AaB about Bb, until it formed a continuation of the plane BbC, and again turning the two so forming one face about

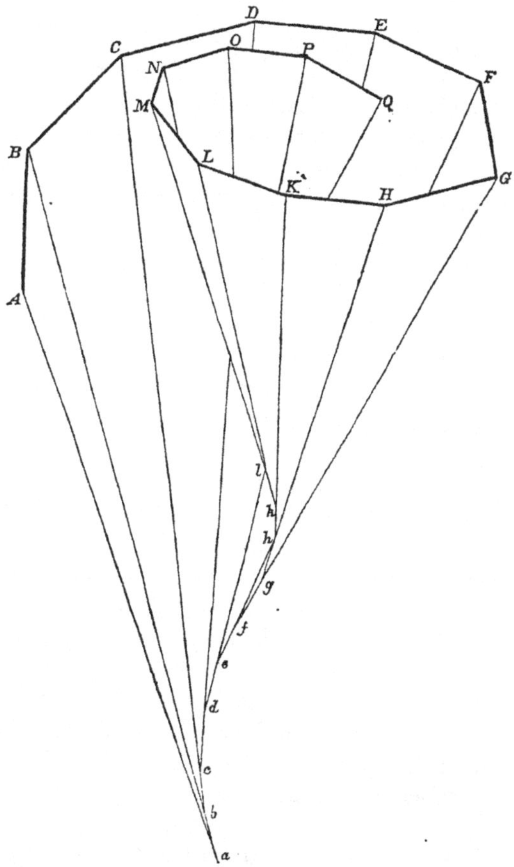

Cc until the three AaB, BbC, CcD were in one plane, and so on; the whole surface might, therefore, be developed into one plane without tearing or rumpling.

The same is true however near the lines Aa, Bb, ... are taken, hence, in the limit, we arrive at the property from which this class of surfaces derives its name, which as we have seen is derivable from the fact that two consecutive positions of the generating line always intersect.

Edge of Regression.

294. The polygon *abcd*, ... whose sides are in the direction of the lines *Bb*, *Cc*, ... becomes in the limit a curve which is generally of double curvature and is called the *Edge of Regression*, from the fact that the surface bends back at this curve so as to be of a cuspidal form, and every generating line of the system is a tangent to the edge of regression, which is therefore the envelope of all the generating lines.

In the case of a cylinder, the edge of regression is at an infinite distance.

295. *To find the general nature of the intersection of a tangent plane to a developable surface with the surface.*

The plane containing the element *DdE* of the surface represented by the figure evidently becomes in the limit a tangent plane to the developable surface at any point *D* in the generating line *Dd*, since it contains the two tangent lines *Dd*, and the limiting positions of lines joining such points as *D* and *E*, which ultimately coincide; and again, supposing *DdE* in the plane of the paper, *Ff* meets this plane in *e*, *Ggf* meets it in some point f', *Hhg* in g', &c., and similarly for *Cc*, *Bb*, ... on the other side.

The complete intersection of the surface and tangent plane is therefore the double line formed by the coincidence of *Dd*, *Ee*, and the limit of the polygon $a'b'c'def'g'$ ... which is a curve touching the double line *Dd* at the edge of regression.

Cor. *To find the nature of the contact of the edge of regression and the tangent plane.*

The plane containing the generating lines *Dd*, *Ee* contains the three angular points *c*, *d*, *e* of the polygon in the limit, therefore the tangent plane contains two consecutive elements of the curve edge of regression, and is what is called the osculating plane at that point.

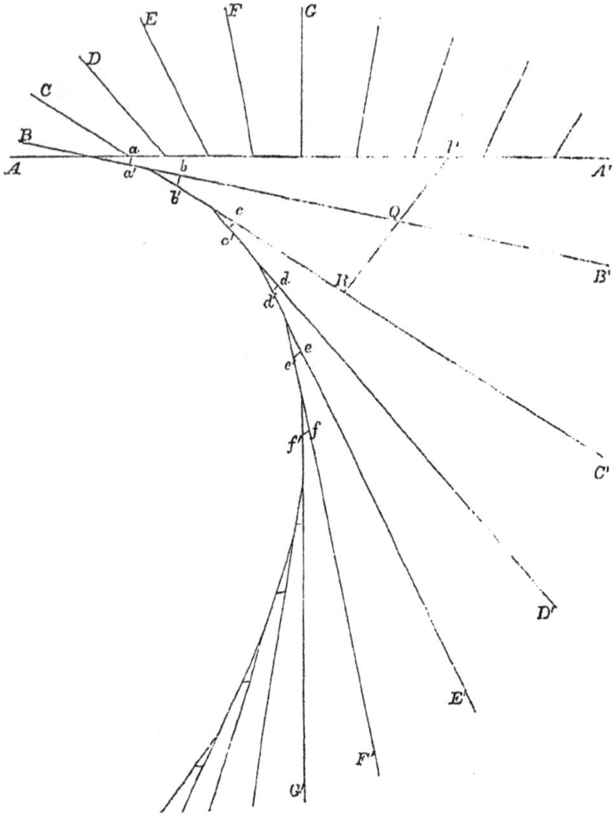

[To face page 221.]

296. *The shortest line which joins two points on a developable surface is that curve whose osculating plane contains the normal to the surface at every point.*

If the surface be developed into a plane, the shortest line must be developed into the straight line joining the two points. If on the polygonal surface in the figure of page 219, $ABCD...K$ be the polygon, which in the limit becomes the shortest line joining A and K, since on development this becomes a straight line, two consecutive sides EF, FG must be inclined at equal angles to the line Ff. Hence a straight line drawn through F, perpendicular to the line Ff in the plane bisecting the angle between the planes EFf, GFf will evidently lie in the plane EFG, and bisect the angle EFG. This line will be in the limit the normal to the surface, and the plane EFG will be the osculating plane.

Therefore the shortest line is the curve whose normal at every point lies in the osculating plane at that point.

Such a line is called a geodesic line of the surface, and it will be hereafter shewn that the property enunciated for developable surfaces is true for geodesic lines on all surfaces.

If the geodesic line, joining two given points, be drawn on a right circular cone, the equation of the projection upon the base can be shewn to be

$$\frac{1}{r}\sin(\gamma\sin\alpha) = \frac{1}{b}\sin(\theta\sin\alpha) + \frac{1}{a}\sin\{(\gamma-\theta).\sin\alpha\},$$

a, b being the distances of the given points from the axis, γ the angle between these distances, and α the semivertical angle of the cone.

Skew Surfaces and Curves of greatest density.

297. Let AA', BB', CC', DD', &c. be straight lines drawn according to some fixed law such that none intersects the next consecutive, let aa', bb', cc', dd', ... be the shortest distances. Suppose now that we take two of the generating lines as CC', DD', and imagine DD' twisted about c' so as to be parallel to CC', and united with it by means of a uniform elastic membrane: if now DD be returned to its original position, the portion of the membrane near dd', being unstretched,

will be denser than any other portion. If the same process be adopted for every line, the series of membranes will generate a surface which ultimately, as the lines approach nearer to one another, becomes a *skew* or twisted surface.

The curve which is the limit of the polygon formed by joining $a, b, c, d, \dots$ at which the membranes have the greatest density is called *the Curve of Greatest Density*.

Thus, in the figures at pages 121 and 124, the principal sections ABa and AOA' are curves of greatest density on the hyperboloid of one sheet and the hyperbolic paraboloid.

298. *To explain the nature of the contact of a tangent plane to a skew surface at any point.*

Let P be any point of a skew surface, AA' the generating line passing through P, suppose a plane to be drawn through P containing BB' the next consecutive position of the generating line, this plane will intersect the third line CC' in some point R, and if PR be joined it will meet BB' in Q, and PR will therefore be a tangent line at P having a contact of the second order at least, and, if the surface be of the second order, it must coincide entirely with the surface. The tangent plane at P is the plane containing AA' and PR'; for any change of position of P, R will change its position, and the tangent planes at points in AA' will always contain AA', but as the point of contact moves along AA', they will move about AA' through all positions.

The tangent plane therefore at any point contains the generating line and some other curve which may be a straight line, as in the case of a surface of the second degree, whereas there can be no developable surface of the second degree except a cylinder, since the curve of intersection of the tangent plane cannot exceed the double generating line.

299. *To shew that the equation of the tangent plane to a developable surface contains only one parameter.*

Since the general equations of a straight line involve four arbitrary constants, we must, in order to the generation of any ruled surface, have three relations connecting these constants, so that we may eliminate the four constants and obtain the equation of a surface the locus of all the straight lines. In developable surfaces, the generating straight lines are such that any two consecutive ones intersect, and the plane containing them is ultimately a tangent plane to the surface. The equation of this plane will then involve the four parameters, and by means of the three relations we may eliminate three, so that the general equation of the tangent plane to a developable surface will involve only one parameter, and we may write it in the form

$$z = \alpha x + \phi(\alpha) y + \psi(\alpha),$$

α being the parameter, and $\phi(\alpha)$, $\psi(\alpha)$ functions of that parameter, given in any particular case.

In skew surfaces, the equation of the tangent plane at any point will involve the parameter of the generating straight line passing through the point, but not containing the consecutive straight line, will involve some other parameter which fixes the tangent plane among all the planes containing that straight line.

We may also arrive at the conclusion that the equation of the tangent plane to a developable surface can only involve one parameter, from the consideration that if it involved two, we should by varying them infinitesimally, obtain the equations of three planes, which would ultimately intersect in a definite point, instead of in one straight line, so that the plane could in general have only one point of contact with the surface which it touched; and the surface would therefore not be developable.

300. *To find the form of the curve of intersection of a developable surface with a tangent plane.*

Let the equation of a plane be given in the form

$$z = \alpha x + \phi(\alpha) y + \psi(\alpha) \qquad (1)$$

containing one parameter α, by the variation of which the plane assumes different positions.

The equation of the plane in its next consecutive position is

$$z = (\alpha + d\alpha) x + \phi(\alpha + d\alpha) y + \psi(\alpha + d\alpha),$$

and the line of intersection has for its equations

$$0 = x + \phi'(\alpha)\, y + \psi'(\alpha) \text{ and } z = \alpha x + \phi(\alpha)\, y + \psi(\alpha).$$

If we eliminate α between these equations we obtain a surface which is the locus of all such lines.

If the equation of the plane involved two arbitrary parameters, the plane would not move in such a manner as to give with a consecutive position some definite line the locus of which would be a surface.

Assuming therefore (1) as the equation of a plane of the system, let the plane of xy correspond to the value $\alpha = 0$, let the axis of y be the line of intersection with the next consecutive, and the next to this pass through the origin;

$$\therefore\ \phi(0) = 0,\ \phi'(0) = 0,\ \psi(0) = 0,\ \psi'(0) = 0,\ \psi''(0) = 0.$$

Also in order that y may be determinate $\dfrac{\psi''(0)}{\phi''(0)}$ must vanish. Hence we may express $\phi(\alpha)$, $\psi(\alpha)$, as follows:

$$\phi(\alpha) = a\alpha^{m+1}(1 + \epsilon),$$
$$\psi(\alpha) = b\alpha^{n+2}(1 + \epsilon'),$$

ϵ, ϵ' vanishing simultaneously with α, and m being $< n + 1$.

The intersection of the surface with the plane of xy will be given by the elimination of α between

$$0 = \alpha\,\{x + a\alpha^m(1 + \epsilon)\, y + b\alpha^{n+1}(1 + \epsilon')\}$$

and $\quad 0 = x + (m + 1)\, a\alpha^m(1 + \eta)\, y + (n + 2)\, b\alpha^{n+1}(1 + \eta'),$

$\alpha = 0$ corresponds to Oy, and, for the curve of intersection, in the neighbourhood of O,

$$0 = ma\alpha^m y + (n + 1)\, b\alpha^{n+1},$$

and $$0 = mx - (n + 1 - m)\, b\alpha^{n+1};$$

$$\therefore\ \left\{\frac{mx}{(n+1-m)\, b}\right\}^{n+1-m} = \left\{-\frac{may}{(n+1)\, b}\right\}^{n+1};$$

$$\therefore\ y^{n+1} = Cx^{n+1-m},$$

or the curve is parabolic touching Oy at O.

301. *If a series of straight lines, generating a surface, be described according to a law such that the shortest distance between two consecutive lines is of an order superior to the first, it will be at least of the third.*

Since the four parameters, entering the equation of a line, must be capable of being eliminated, there must be three relations between those parameters besides the two equations of the line; hence if the equations be $x = mz + a$, $y = nz + b$, m, n, a, b must be functions of one parameter, which by its variation gives rise to different positions of the generating line.

The shortest distance between two consecutive lines of the system is

$$\frac{\Delta m\, \Delta b - \Delta n\, \Delta a}{\sqrt{\overline{\Delta m}|^2 + \overline{\Delta n}|^2 + (m\Delta n - n\Delta m)^2}},$$

and $$\Delta m = dm + \frac{1}{2} d^2m + \frac{1}{6} d^3m + \ldots\ldots$$

and similarly for Δn, Δa, and Δb;

$$\therefore\ \Delta m\, \Delta b - \Delta n\, \Delta a = dm\, db - dn\, da,$$

$$+ \frac{1}{2}\, (dm\, d^2b + db\, d^2m - dn\, d^2a - da\, d^2n),$$

+ terms of the following order higher than the third,

the denominator is of the first order, and if $dm\, db - dn\, da$ be not zero, the numerator is of the second order, but if

$$dm\, db - dn\, da = 0$$

always, we have also

$$dm\, d^2b + db\, d^2m - dn\, d^2a - da\, d^2n = 0\,;$$

or the numerator is of the fourth or higher order.

Hence the truth of the proposition which is due to M. Bouquet.

Tangent planes touching along a curve line.

302. We have seen (Art. 287) that, when a plane intersects a surface, at every point of the curve of intersection, through which

an arbitrary line drawn in the plane passes through two coincident points, the plane is a tangent plane to the surface, or such a point is a multiple point on the surface. If the curve of intersection consist of two or more coincident lines, this will occur at every point of such curves, hence, either the plane will be a tangent plane to the surface at every point of such multiple curve, or will contain a multiple line on the surface. Conversely, if a tangent plane touch along a curve line such a curve line will be a multiple line on the tangent plane. Thus in the case of the anchor ring, the plane which touches the anchor ring at every point has for its curve of intersection the two circles coincident in LKH; also the tangent plane to a cone contains two generating lines which ultimately coincide, and is therefore a tangent plane at every point of the generating lines which it contains.

Similarly, a surface of the fourth degree admits of the case of a double conic, or of a quadruple straight line, as in the case of two cones touching along a generating line.

A surface of the fifth degree might be composed of one of the third, and one of the second degree, in which case it is possible that a tangent plane might meet the former in a triple, and the latter in a double straight line.

303. *To find the conditions that a tangent plane may touch the surface at every point in which it meets it.*

Let the tangent plane at the point (x_0, y_0, z_0) in the surface whose equation is

$$z = f(x, y) \ldots\ldots\ldots\ldots\ldots\ldots\ldots\ldots (1)$$

have the property.

Sufficient conditions are that $p = p_0$ and $q = q_0$ throughout the curve common to the surface and tangent plane.

Since the curve of contact is a compound curve containing two coincident curves at least, therefore if (x, y, z) and $(x + dx, y + dy, z + dz)$ be two points common to the surface and tangent plane, two values of $dx : dy$ must be coincident at least.

The equation of the tangent plane is

$$z = p_0 x + q_0 y + c;$$

$$\therefore\ p_0\,dx + q_0\,dy = p\,dx + q\,dy + \frac{1}{2}\left(r\overline{dx}|^2 + 2s\,dx\,dy + t\overline{dy}|^2\right) + \&c.$$

Only one value of the ratio $dx : dy$ exists ultimately, unless

$$p = p_0 \text{ and } q = q_0,$$

and in this case the two values are given by

$$r\overline{dx}|^2 + 2s\,dx\,dy + t\overline{dy}|^2 = 0;$$

and since the roots are equal, because at every point two tangent lines coincide, we have the necessary condition, $rt - s^2 = 0$.

Or, since the tangent plane remains constant for all points common to it and the surface, p and q are constant when x and y receive small increments;

$$\therefore\ dp = 0 = r\,dx + s\,dy,$$

$$dq = 0 = s\,dx + t\,dy,$$

whence

$$rt - s^2 = 0 \ldots\ldots\ldots\ldots\ldots\ldots\ldots\ldots\ldots (2).$$

It is easily seen that the condition $rt - s^2 = 0$ is not *sufficient*, although *necessary*, since for the curve common to (1) and (2)

$$\overline{dp}|^2 + \overline{dq}|^2 = (r + t)\left(r\overline{dx}|^2 + 2s\,dx\,dy + t\overline{dy}|^2\right),$$

which is not necessarily $= 0$, for values of $dx : dy$ obtained from these equations.

It will be seen hereafter that $rt - s^2 = 0$ may be true for other points in the surface which are not in the curve of contact.

Thus, in the case of a developable surface always touched by a plane whose equation is $z = \alpha x + \phi(\alpha)\,y + \psi(\alpha)$,

$$p = \quad \alpha + \frac{d\alpha}{dx}\{x + \phi'(\alpha)\,y + \psi'(\alpha)\} = \alpha,$$

$$q = \phi(\alpha) + \frac{d\alpha}{dy}\{x + \phi'(\alpha)\,y + \psi'(\alpha)\} = \phi(\alpha);$$

$$\therefore\ q = \phi(p),$$

$$s = r\phi'(p), \quad \text{and } t = s\phi'(p).$$

Therefore $rt - s^2 = 0$ at every point of the surface, and we have shewn that the tangent plane does not necessarily touch at every point in which it meets the surface.

304. *To find the curve of greatest inclination to a given plane which can be traced on a surface.*

Let the equation of the surface be $F(x, y, z) = 0$, l, m, n the direction cosines of the given plane.

The tangent plane at any point of such a curve is

$$(x-x_0)F'(x_0) + (y-y_0)F'(y_0) + (z-z_0)F'(z_0) = 0,$$

and the direction cosine of the line of intersection with the given plane are proportional to

$$mF'(z_0) - nF'(y_0),\ nF'(x_0) - lF'(z_0),\ \text{and } lF'(y_0) - mF'(x_0).$$

The direction of the curve line having the property proposed, is perpendicular to this line; therefore the differential equations of the curve are

$$\{mF'(z) - nF'(y)\}\,dx + \{nF'(x) - lF'(z)\}\,dy + \{lF'(y) - mF'(x)\}\,dz = 0,$$

and $$F'(x)\,dx + F'(y)\,dy + F'(z)\,dz = 0.$$

The first of these equations with the equation of the surface and any point chosen on the surface through which the curve shall pass, are sufficient completely to determine the curve.

If the plane be that of xy, $l = m = 0$, and the equation becomes

$$F'(y)\,dx - F'(x)\,dy = 0,$$

which with $F(x, y, z) = 0$ is sufficient.

Cor. If the equations for obtaining $dx : dy : dz$ become identical, the direction of the line of greatest inclination will be indeterminate; in this case

$$\frac{mF'(z) - nF'(y)}{F'(x)} = \frac{nF'(x) - lF'(z)}{F'(y)} = \frac{lF'(y) - mF'(x)}{F'(z)}$$

$$= \frac{0}{lF'(x) + mF'(y) + nF'(z)}.$$

Therefore $F'(x) : F'(y) : F'(z) = l : m : n$ at every point of the surface, which can only happen when the surface is a plane parallel to the given plane.

Tetrahedral Co-ordinates.

305. *To find the equation of the tangent plane at a given point in a given surface referred to tetrahedral co-ordinates.*

Let $$\phi = F(\alpha, \beta, \gamma, \delta) = 0 \dots\dots\dots\dots (1)$$
be the equation of the given surface, $(\alpha_0, \beta_0, \gamma_0, \delta_0)$ the given point P, and
$$l\alpha + m\beta + n\gamma + r\delta = 0 \dots\dots\dots\dots (2)$$
the equation of the tangent plane required.

Since the tangent plane contains all the tangent lines to the surface, it must be satisfied by the co-ordinates of points taken in any direction on the surface, which ultimately coincide with the proposed point.

Hence (1) and (2) are satisfied by the co-ordinates of P and also by $\alpha_0 + d\alpha_0$, $\beta_0 + d\beta_0 \dots$ when $d\alpha_0$, $d\beta_0$, ... are indefinitely diminished;
$$\therefore\ l d\alpha_0 + m d\beta_0 + n d\gamma_0 + r d\delta_0 = 0,$$
$$F'(\alpha_0)\, d\alpha_0 + F'(\beta_0)\, d\beta_0 + F'(\gamma_0)\, d\gamma_0 + F'(\delta_0)\, d\delta_0 = 0,$$
and $$d\alpha_0 + d\beta_0 + d\gamma_0 + d\delta_0 = 0,$$
from the condition
$$\alpha_0 + \beta_0 + \gamma_0 + \delta_0 = 1, \text{ (Art. 8)};$$
and these equations are satisfied for an infinite number of values of the ratios
$$d\alpha_0 : d\beta_0 : d\gamma_0 : d\delta_0;$$
therefore, employing arbitrary multipliers A and B in order to eliminate $d\gamma_0$ and $d\delta_0$, we obtain the equations
$$l = AF'(\alpha_0) + B,\ m = AF'(\beta_0) + B,\ \&c. \dots\dots\dots (3),$$
and, multiplying these equations in order by α_0, β_0, γ_0, δ_0, observing that
$$\alpha_0 F'(\alpha_0) + \beta_0 F'(\beta_0) + \&c. \dots = 0,$$
and $$l\alpha_0 + m\beta_0 + \dots = 0,$$
we have $B = 0$.

Hence, multiplying equations (3) by α, β, γ, δ, (2) becomes identical with

$$\alpha F'(\alpha_0) + \beta F'(\beta_0) + \gamma F'(\gamma_0) + \delta F'(\delta_0) = 0,$$

which is the equation of the tangent plane.

306. *To find the relation between the cosines which a tangent line at any point of a surface, referred to tetrahedral co-ordinates, makes with the normals to the fundamental planes.*

Let λ, μ, ν, ρ be these cosines (Art. 62) for a tangent line at a point $(\alpha_0, \beta_0, \gamma_0, \delta_0)$; these are proportional to the values $d\alpha_0$, $d\beta_0$, $d\gamma_0$, $d\delta_0$ given by

$$F'(\alpha_0)\, d\alpha_0 + F'(\beta_0)\, d\beta_0 + F'(\gamma_0)\, d\gamma_0 + F'(\delta_0)\, d\delta_0 = 0,$$

and
$$d\alpha_0 + d\beta_0 + d\gamma_0 + d\delta_0 = 0,$$

the relations between λ, μ, ν, ρ are therefore

$$\lambda F'(\alpha_0) + \mu F'(\beta_0) + \nu F'(\gamma_0) + \rho F'(\delta_0) = 0,$$

and
$$\lambda + \mu + \nu + \rho = 0.$$

307. *To find the equation of the tangent cone at a singular point.*

If
$$F'(\alpha_0) = 0,\ F'(\beta_0) = 0, \ldots\ldots$$

the first of these equations is replaced by

$$\left(\lambda \frac{d}{d\alpha_0} + \mu \frac{d}{d\beta_0} + \nu \frac{d}{d\gamma_0} + \rho \frac{d}{d\delta_0}\right)^2 \phi_0 = 0,$$

and the equation of the locus of the tangent lines at the proposed point, which is a multiple point in this case, is

$$\left\{(\alpha - \alpha_0) \frac{d}{d\alpha_0} + (\beta - \beta_0) \frac{d}{d\beta_0} + \ldots\right\}^2 \phi_0 = 0,$$

which, since the coefficient of -2α is

$$\left(\alpha_0 \frac{d}{d\alpha_0} + \ldots\right) \frac{d\phi_0}{d\alpha_0} = (n-1) \frac{d\phi_0}{d\alpha_0} = (n-1) F'(\alpha_0) = 0,$$

becomes
$$\left(\alpha \frac{d}{d\alpha_0} + \beta \frac{d}{d\beta_0} + \ldots\right)^2 \phi_0 = 0,$$

the equation of the tangent cone in a homogeneous form.

And, similarly, if the differential coefficients of the $\overline{r-1}]^{\text{th}}$ and inferior orders all become zero at the proposed point, the equation of the tangent cone is

$$\left(\alpha \frac{d}{d\alpha_0} + \ldots\right)^r \phi_0 = 0.$$

308. *To find the class of a surface of the nth degree.*

This is to find the number of tangent planes which can be drawn to the surface so as to contain a given straight line.

Let the equation of the surface be $F(\alpha, \beta, \gamma, \delta) = 0$, and let $(\alpha', \beta', \gamma', \delta')$ $(\alpha'', \beta'', \gamma'', \delta'')$ be two points in the given straight line.

A tangent plane through the given straight line contains the given points, and if $(\alpha_0, \beta_0, \gamma_0, \delta_0)$ be the point of contact of any such tangent plane its equation will be

$$\alpha\ F'(\alpha_0) + \beta\ F'(\beta_0) + \gamma\ F'(\gamma_0) + \delta\ F'(\delta_0) = 0;$$

$$\therefore\ \alpha'\ F'(\alpha_0) + \beta'\ F'(\beta_0) + \gamma'\ F'(\gamma_0) + \delta'\ F'(\delta_0) = 0,$$

and $$\alpha''\ F'(\alpha_0) + \beta''\ F'(\beta_0) + \gamma''\ F'(\gamma_0) + \delta'\ F'(\delta_0) = 0.$$

These equations, which are of the $\overline{n-1}]^{\text{th}}$, and the equation

$$F(\alpha_0, \beta_0, \gamma_0, \delta_0) = 0,$$

which is of the nth degree, determine generally $n \,.\, (n-1)^2$ points, and the same number of tangent planes, hence the surface given by an equation of the nth degree is of the $n \,.\, \overline{n-1}]^2$ class.

It will be shewn hereafter how this number is diminished when there are multiple points and lines on the surface.

Four-Point Co-ordinates.

309. The proposition in the system of four-point co-ordinates, which corresponds to that in plane co-ordinate systems of finding the equation of the tangent plane at a given point of a surface, is to find the equation of the point of contact of any plane whose co-ordinates satisfy the tangential equations of the surface.

The property which we shall employ, is that the point of contact of a tangent plane is also a point in a contiguous tangent plane which moves up to and ultimately coincides with the former.

310. *To find the equation of the point of contact of a tangent plane to a given surface determined by an equation in four-point co-ordinates.*

Let $\phi \equiv F(\alpha, \beta, \gamma, \delta) = 0$........................(1)

be the equation of the surface, $\alpha_0, \beta_0, \gamma_0, \delta_0$ the co-ordinates of the given tangent plane.

Assume $l\alpha + m\beta + n\gamma + r\delta = 0$.....................(2)

to be the equation of the point of contact.

Then, $\alpha_0 + d\alpha_0, \beta_0 + d\beta_0 \ldots$ being the co-ordinates of a tangent plane to the surface which ultimately coincides with the former, these co-ordinates as well as the former satisfy the equations (1) and (2), and the relation between the point co-ordinates,

$$\frac{\alpha^2}{p_1^2} + \frac{\beta^2}{p_2^2} + \ldots - \frac{2\alpha\beta}{p_1 p_2} \cos CD - \ldots = 0, \quad \text{Art. (108)}.$$

Hence, for an infinite number of systems of values of

$$d\alpha_0 : d\beta_0 : d\gamma_0 : d\delta_0,$$

we have

$$l d\alpha_0 + \quad m d\beta_0 + \quad n d\gamma_0 + \quad r d\delta_0 = 0,$$

$$F'(\alpha_0)\, d\alpha_0 + F'(\beta_0)\, d\beta_0 + F'(\gamma_0)\, d\gamma_0 + F'(\delta_0)\, d\delta_0 = 0,$$

and
$$\lambda d\alpha_0 + \quad \mu d\beta_0 + \quad \nu d\gamma_0 + \quad \rho d\delta_0 = 0,$$

where

$$\lambda = \frac{\alpha_0}{p_1} - \frac{\beta_0}{p_2} \cos CD - \frac{\gamma_0}{p_3} \cos DB - \frac{\delta_0}{p_4} \cos BC,$$

and similarly for μ, ν, ρ. Art. (108).

If therefore we assume A and B so that

$$l = AF'(\alpha_0) + B\lambda, \text{ and } m = AF'(\beta_0) + B\mu,$$

we obtain in consequence of the indeterminateness of $d\gamma_0 : d\delta_0$

$$n = AF'(\gamma_0) + B\nu, \text{ and } r = AF'(\delta_0) + B\rho.$$

Multiplying these equations by α_0, β_0, γ_0, δ_0, and observing that

$$l\alpha_0 + m\beta_0 + \ldots = 0, \text{ and } \alpha_0 F'(\alpha_0) + \beta_0 F'(\beta_0) + \ldots = 0,$$

and also

$$\lambda\alpha_0 + \mu\beta_0 + \ldots = 1,$$

we obtain $B = 0$, and we derive

$$l\alpha + m\beta + \ldots = \lambda\{\alpha F'(\alpha_0) + \ldots\}.$$

Hence the equation of the point of contact is

$$\alpha F'(\alpha_0) + \beta F'(\beta_0) + \gamma F'(\gamma_0) + \delta F'(\delta_0) = 0.$$

311. *To find the tangential equation of the curve line in which the tangent plane to a surface given by four-point co-ordinates touches along a curve line.*

In the case of a singular tangent plane, for which there is not a single point but a curve of contact,

$$F'(\alpha_0) = 0, \; F'(\beta_0) = \ldots\ldots = 0 \quad \ldots\ldots\ldots\ldots\ldots (1).$$

If we take a plane $(\alpha_0 + d\alpha_0, \beta_0 + d\beta_0, \ldots)$ indefinitely near to $(\alpha_0, \beta_0, \gamma_0, \delta_0)$ the equation of the surface gives the relation

$$\left(d\alpha_0 \frac{d}{d\alpha_0} + d\beta_0 \frac{d}{d\beta_0} + \ldots\right)^2 \phi_0 = 0 \quad \ldots\ldots\ldots\ldots (2),$$

and if $(\alpha, \beta, \gamma, \delta)$ be any plane passing through the line of intersection of these planes, we may obtain as in Art. (275) or directly by geometrical considerations, if θ be the inclination of this plane to the first of the above planes and ϕ the angle between these two planes,

$$\alpha \sin\phi = (\alpha_0 + d\alpha_0)\sin(\theta + \phi) - \alpha_0 \sin\theta$$

$$= \alpha_0 \cos\theta \sin\phi + d\alpha_0 \sin\theta, \text{ ultimately;}$$

$$\therefore \frac{d\alpha_0}{\alpha - \alpha_0\cos\theta} = \frac{d\beta_0}{\beta - \beta_0\cos\theta} = \ldots$$

therefore, having regard to equations (1), we obtain from (2),

$$\left(\alpha \frac{d_0}{d\alpha_0} + \beta \frac{d}{d\beta_0} + \dots\right)^2 \phi_0 = 0, \text{ as in Art. 307,}$$

the relation which holds for all planes which touch the curve of contact.

This is therefore the tangential equation of the curve.

312. *To find the degree of a tangential surface of the nth class.*

Let the equation of the surface be $F(\alpha, \beta, \gamma, \delta) = 0$, and $(\alpha', \beta', \gamma', \delta')$ $(\alpha'', \beta'', \gamma'', \delta'')$ two tangent planes intersecting in a given line.

Any point in the given straight line and surface is in the given tangent planes, and if $(\alpha_0, \beta_0, \gamma_0, \delta_0)$ be the tangent planes of which this point is the point of contact,

$$\alpha F'(\alpha_0) + \beta F'(\beta_0) + \dots = 0,$$

$$\therefore \alpha' F'(\alpha_0) + \beta' F'(\beta_0) + \dots = 0,$$

$$\alpha'' F'(\alpha_0) + \beta'' F'(\beta_0) + \dots = 0.$$

These equations, which are of the $\overline{n-1}|^{\text{th}}$, and the equation of the surface, which is of the nth degree, determine generally $n \,.\, \overline{n-1}|^2$ tangent planes, and the same number of points of contact.

Hence, a surface given by a tangential equation of the nth degree is of the $n \,.\, \overline{n-1}|^2$ degree.

Polar Co-ordinates.

313. *To find the polar equation of the tangent plane to a surface at a given point.*

Let the equation of the surface be $\frac{1}{r} = u = f(\theta, \phi)$, and u_0, θ_0, ϕ_0 co-ordinates of the point of contact of the tangent plane.

The equation of the tangent plane is of the form

$$pu = \cos\alpha\cos\theta + \sin\alpha\sin\theta\cos(\phi - \beta), \quad \text{Art. (79)},$$

and the constants p, α, and β are to be determined from the consideration that the tangent plane contains not only the point of contact but adjacent points which have moved up to and ultimately coincided with that point.

Hence the values of $\dfrac{du}{d\theta}$ and $\dfrac{du}{d\phi}$ at the point of contact are the same for both tangent plane and surface, let v_0, w_0 be those values;

$$\begin{aligned}
\therefore\ pu_0 &= \cos\alpha\cos\theta_0 + \sin\alpha\sin\theta_0\cos(\phi_0 - \beta),\\
pv_0 &= -\cos\alpha\sin\theta_0 + \sin\alpha\cos\theta_0\cos(\phi_0 - \beta),\\
pw_0 &= - \sin\alpha\sin\theta_0\sin(\phi_0 - \beta);
\end{aligned}$$

$$\therefore\ p(u_0\sin\theta_0 + v_0\cos\theta_0) = \sin\alpha\cos(\phi_0 - \beta),$$

$$p(u_0\cos\theta_0 - v_0\sin\theta_0) = \cos\alpha;$$

the last three of these equations give readily the values of the constants: and the equation of the tangent plane is

$$\begin{aligned}
u = {} & (u_0\cos\theta_0 - v_0\sin\theta_0)\cos\theta\\
& + (u_0\sin\theta_0 + v_0\cos\theta_0)\cos(\phi - \phi_0)\cos\theta\\
& + w_0\operatorname{cosec}\theta_0\sin(\phi - \phi_0)\sin\theta.
\end{aligned}$$

This equation can also be written in the form

$$\begin{aligned}
\frac{r_0^2}{r} = {} & \frac{d}{d\theta_0}\left[r_0\{\sin\theta_0\cos\theta - \cos\theta_0\sin\theta\cos(\phi - \phi_0)\}\right]\\
& - \frac{dr_0}{d\phi_0}\operatorname{cosec}\theta_0\sin\theta\sin(\phi - \phi_0).
\end{aligned}$$

314. *To find the perpendicular distance from the pole upon the tangent plane.*

This may be obtained from the first three equations of the last article by squaring and adding, whence

$$p^2(u_0^2 + v_0^2 + w_0^2\operatorname{cosec}^2\theta) = 1,$$

$$\text{or,} \quad \frac{1}{p^2} = u_0^2 + \left(\frac{du_0}{d\theta_0}\right)^2 + \left(\frac{du_0}{d\phi_0}\right)^2 \operatorname{cosec}^2 \theta_0.$$

315. The following method serves to shew the geometrical signification of the partial differential coefficients, and may be useful as an exercise.

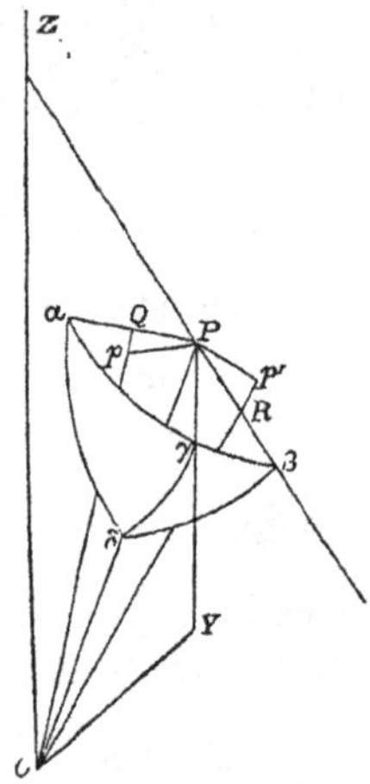

Let P be the point of contact, PR a tangent line passing through OZ, and PQ a tangent line in the plane through OP perpendicular to the plane POZ, take R and Q points very near to P; in OQ, OR take $Op = OP = Op'$, then $Pp = r \sin \theta d\phi$ and $Pp' = rd\theta$ ultimately, and $Qp, -Rp'$ are respectively the values of dr due to changes of θ and ϕ considering the other constant,

$$\therefore \frac{dr}{rd\theta} = -\frac{Rp'}{Pp'} = -\cot OPR,$$

$$\text{and} \quad \frac{dr}{r \sin \theta d\phi} = \frac{Qp}{Pp} = -\cot OPQ.$$

Draw OY perpendicular to the tangent plane QPR, and on a sphere, whose centre is P, let $\alpha\delta\beta$ be a spherical triangle, with its angular points in PQ, PO, PR, join $\delta\gamma$, γ being the intersection of PY and $\alpha\beta$, then $\delta\gamma$ is perpendicular to $\alpha\beta$, and $\alpha\delta\beta$ is a right angle. Hence $\cot \alpha\delta = \cot \delta\gamma \cos \alpha\delta\gamma$, and $\cot \beta\delta = \cot \delta\gamma \sin \alpha\delta\gamma$;

$$\therefore\ \cot^2 \alpha\delta + \cot^2 \beta\delta = \cot^2 \delta\gamma = \frac{r^2 - p^2}{p^2};$$

$$\therefore\ \frac{1}{p^2} = \frac{1}{r^2} + \frac{1}{r^4}\left(\frac{dr}{d\theta}\right)^2 + \frac{1}{r^4}\left(\frac{dr}{d\phi}\right)^2 \operatorname{cosec}^2 \theta.$$

Polar Planes and Poles.

316. *To find the locus of the points of contact of tangent lines, drawn to a given surface from a given point.*

Let $\phi \equiv F(x, y, z) = 0$ be the equation of the given surface, and α, β, γ the co-ordinates of the given point. Then, if (x, y, z) be one of the points of contact, the tangent plane to the surface at (x, y, z) must pass through (α, β, γ). This gives the condition

$$(\alpha - x)\frac{d\phi}{dx} + (\beta - y)\frac{d\phi}{dy} + (\gamma - z)\frac{d\phi}{dz} = 0,$$

which, combined with the equation of the surface, determines the required locus, or the *curve of contact.*

It has been shewn (Art. 283) that for points on the surface $x\frac{d\phi}{dx} + y\frac{d\phi}{dy} + z\frac{d\phi}{dz}$ is of the degree $n-1$, or lower by unity than the degree of the given surface; therefore by combining the above equation with that of the surface we obtain an equation of the $(n-1)^{\text{th}}$ degree.

The curve of contact for any conicoid is therefore a plane curve, and it is obvious that the equation of this plane is always real, whether the points of contact be real or imaginary.

317. Def. The polar plane of a given point with respect to a given conicoid is the plane on which lie the points of contact, real or imaginary, of the tangent lines drawn from the point to the conicoid; and the point from which the tangent lines are drawn is called the pole of the plane.

318. *To find the equation of the polar plane of a given point, with respect to a given conicoid, in three-plane co-ordinates.*

Let

$$ax^2 + by^2 + cz^2 + 2a'yz + 2b'zx + 2c'xy$$
$$+ 2a''x + 2b''y + 2c''z + d = 0$$

be the equation of the conicoid, (α, β, γ) the given point.

Then, by the last article, the equation of the polar plane is

$$\alpha\,(ax + c'y + b'z + a'') + \ldots + a''x + b''y + c''z + d = 0.$$

The polar plane of the origin is therefore

$$a''x + b''y + c''z + d = 0.$$

The equations determining the center may be deduced from this equation; for, since the polar line of the center in any plane section through it is at infinity, the polar plane will be at infinity, and the above equation must reduce to

$$\text{constant} = 0,$$

whence we have, for the center,

$$a\alpha + c'\beta + b'\gamma + a'' = 0, \;\; \&c.$$

as before determined, Art. (231).

319. *To find the equation of the polar plane of a given point, with respect to a given conicoid, in tetrahedral co-ordinates.*

Let $\phi \equiv F(\alpha, \beta, \gamma, \delta) = 0$ be the equation of the conicoid, $(\alpha_0, \beta_0, \gamma_0, \delta_0)$ the given point. Then, if $(\alpha, \beta, \gamma, \delta)$ be one of the points of contact, the tangent plane at $(\alpha, \beta, \gamma, \delta)$ will pass through $(\alpha_0, \beta_0, \gamma_0, \delta_0)$; or,

$$0 = \alpha_0 \frac{d\phi}{d\alpha} + \beta_0 \frac{d\phi}{d\beta} + \gamma_0 \frac{d\phi}{d\gamma} + \delta_0 \frac{d\phi}{d\delta} \equiv \alpha \frac{d\phi_0}{d\alpha_0} + \beta \frac{d\phi_0}{d\beta_0} + \gamma \frac{d\phi_0}{d\gamma_0} + \delta \frac{d\phi_0}{d\delta_0},$$

the equation of the polar plane.

320. *To find the equations of the center of a given conicoid in tetrahedral co-ordinates.*

The center being the pole of the plane at infinity, of which the equation is

$$\alpha + \beta + \gamma + \delta = 0, \quad (\text{Art. } 81),$$

this equation must coincide with

$$\alpha\frac{d\phi_0}{d\alpha_0}+\beta\frac{d\phi_0}{d\beta_0}+\gamma\frac{d\phi_0}{d\gamma_0}+\delta\frac{d\phi_0}{d\delta_0}=0.$$

The equations of the center are therefore, suppressing the suffix,

$$\frac{d\phi}{d\alpha}=\frac{d\phi}{d\beta}=\frac{d\phi}{d\gamma}=\frac{d\phi}{d\delta}.$$

321. *To find the co-ordinates of the polar plane of a given point, with respect to a given conicoid, in four-point co-ordinates.*

Let $\phi\equiv F(\alpha, \beta, \gamma, \delta)=0$ be the tangential equation of the conicoid, and $l\alpha+m\beta+n\gamma+r\delta=0$ of the given point.

Then if $(\alpha_0, \beta_0, \gamma_0, \delta_0)$ be a tangent plane to the conicoid, the equation of its point of contact is

$$\alpha\frac{d\phi_0}{d\alpha_0}+\beta\frac{d\phi_0}{d\beta_0}+\gamma\frac{d\phi_0}{d\gamma_0}+\delta\frac{d\phi_0}{d\delta_0}=0, \quad \text{(Art. 310)},$$

or $$\alpha_0\frac{d\phi}{d\alpha}+\beta_0\frac{d\phi}{d\beta}+\gamma_0\frac{d\phi}{d\gamma}+\delta_0\frac{d\phi}{d\delta}=0 \quad\ldots\ldots\ldots\ldots (1).$$

Also, since the tangent plane passes through the given point, we must have

$$l\alpha_0+m\beta_0+n\gamma_0+r\delta_0=0 \ldots\ldots\ldots\ldots\ldots\ldots (2).$$

Hence, if $(\alpha, \beta, \gamma, \delta)$ be the polar plane, $\alpha, \beta, \gamma, \delta$ must be determined so as to make (1) and (2) coincide, giving the equations

$$\frac{\frac{d\phi}{d\alpha}}{l}=\frac{\frac{d\phi}{d\beta}}{m}=\frac{\frac{d\phi}{d\gamma}}{n}=\frac{\frac{d\phi}{d\delta}}{r},$$

which determine the ratios of the co-ordinates.

The equation of the pole of a given plane $(\alpha_0, \beta_0, \gamma_0, \delta_0)$ will therefore be

$$\alpha\frac{d\phi_0}{d\alpha_0}+\beta\frac{d\phi_0}{d\beta_0}+\gamma\frac{d\phi_0}{d\gamma_0}+\delta\frac{d\phi_0}{d\delta_0}=0.$$

322. *To find the equation of the center of a given conicoid, in four-point co-ordinates.*

The center is the pole of the plane at infinity, of which the equations are

$$\alpha = \beta = \gamma = \delta.$$

The equation of the center will therefore be

$$\alpha \frac{d\phi_0}{d\alpha_0} + \beta \frac{d\phi_0}{d\beta_0} + \gamma \frac{d\phi_0}{d\gamma_0} + \delta \frac{d\phi_0}{d\delta_0} = 0,$$

when $$\alpha_0 = \beta_0 = \gamma_0 = \delta_0.$$

But $$\alpha \frac{d\phi_0}{d\alpha_0} + \beta \frac{d\phi_0}{d\beta_0} + \gamma \frac{d\phi_0}{d\gamma_0} + \delta \frac{d\phi_0}{d\delta_0}$$

$$\equiv \alpha_0 \frac{d\phi}{d\alpha} + \beta_0 \frac{d\phi}{d\beta} + \gamma_0 \frac{d\phi}{d\gamma} + \delta_0 \frac{d\phi}{d\delta}.$$

The equation of the center will therefore be

$$\frac{d\phi}{d\alpha} + \frac{d\phi}{d\beta} + \frac{d\phi}{d\gamma} + \frac{d\phi}{d\delta} = 0.$$

Enveloping Cones.

323. If the tangent planes drawn through a point to a conicoid be real, and therefore the intersection of its polar with the conicoid be also real, the cone whose vertex is the point, and base the curve of intersection, will be such that all its generating lines will touch the conicoid in real points. Such a cone is called an *Enveloping Cone.* The equation of this cone will in all cases be real, but if the curve of intersection be imaginary, the vertex will be the only real point lying on the cone.

The readiest method of finding the equation of such a cone requires a preliminary proof of the following proposition.

324. *Two conicoids which intersect in one plane curve will intersect also in another plane curve.*

The curve of intersection of two conicoids is met by an arbitrary plane in four, real or imaginary, points. Two of these will lie on the plane curve which is, by hypothesis, the partial intersection of the conicoids. Their remaining partial intersection will therefore be met by an arbitrary plane in two points, and will therefore be a plane curve.

It follows from this that if a proper conicoid can be described containing two plane curves of the second degree, an infinite number can be so described. For the two planes may be considered as one conicoid, and the two curves are therefore the intersection of two conicoids, whence, by Art. (258), an infinite number of conicoids can be drawn containing them.

Making the two move up to coincidence, we see that an infinite number of conicoids can be described touching another along a given plane curve. The equation of any of these conicoids, containing only one parameter, (Art. 258), will be determined if we make it pass through a fixed point. Hence if we take a point, and its polar plane with respect to a conicoid, the only conicoid which can be drawn, passing through the point and touching the conicoid along the curve of intersection with the polar, will be the corresponding enveloping cone.

Enveloping Cone.

325. *To find the equation of the enveloping cone of a given conicoid, whose vertex is at a given point.*

I. Three Plane Co-ordinates.

Let $F(x, y, z) \equiv F_2(x, y, z) + 2F_1(x, y, z) + F_0 = 0$ be the equation of the conicoid, (α, β, γ) the given point.

Then the equation of the polar of the given point will be

$$\alpha F'(x) + \beta F'(y) + \gamma F'(z) + 2\{F_1(x, y, z) + F_0\} = 0, \text{ Art. (318).}$$

The general equation of a conicoid touching the conicoid along the curve of intersection will be

$$F(x, y, z) = k\{\alpha F'(x) + \beta F'(y) + \gamma F'(z) + 2F_1(x, y, z) + 2F_0\}^2.$$

If this be the enveloping cone, the point (α, β, γ) must be a point on it, that is

$$F(\alpha, \beta, \gamma) = k\{2F_2(\alpha, \beta, \gamma) + 4F_1(\alpha, \beta, \gamma) + 2F_0\}^2$$
$$= 4k\{F(\alpha, \beta, \gamma)\}^2,$$

whence the equation of the cone is

$$4F(\alpha, \beta, \gamma)\, F(x, y, z)$$
$$= \{\alpha F'(x) + \beta F'(y) + \gamma F'(z) + 2F_1(x, y, z) + 2F_0\}^2.$$

II. Tetrahedral, or Quadriplanar, Co-ordinates.

Let $F(\alpha, \beta, \gamma, \delta) = 0$ be the equation of the conicoid, $(\alpha_0, \beta_0, \gamma_0, \delta_0)$ the given point. The equation of the cone will then be

$$F(\alpha, \beta, \gamma, \delta) = k\{\alpha F'(\alpha_0) + \beta F'(\beta_0) + \gamma F'(\gamma_0) + \delta F'(\delta_0)\}^2,$$

k being determined as before.

Hence

$$F(\alpha_0, \beta_0, \gamma_0, \delta_0) = k\{\alpha_0 F'(\alpha_0) + \ldots\ldots\}^2$$
$$\equiv k\{2F(\alpha_0, \beta_0, \gamma_0, \delta_0)\}^2,$$

and the equation of the cone is

$$4F(\alpha_0, \beta_0, \gamma_0, \delta_0)\, F(\alpha, \beta, \gamma, \delta)$$
$$= \{\alpha F'(\alpha_0) + \beta F'(\beta_0) + \gamma F'(\gamma_0) + \delta F'(\delta_0)\}^2.$$

III. Four-point Co-ordinates.

In this system an enveloping cone corresponds to a plane section in the system of tetrahedral co-ordinates. If $(\alpha_0, \beta_0, \gamma_0, \delta_0)$ be a given plane, the pole of this plane is, $F(\alpha, \beta, \gamma, \delta) = 0$ being the tangential equation of the conicoid,

$$\alpha F'(\alpha_0) + \beta F'(\beta_0) + \gamma F'(\gamma_0) + \delta F'(\delta_0) = 0,$$

and every tangent plane to the enveloping cone must therefore satisfy the equations

$$F(\alpha, \beta, \gamma, \delta) = 0,$$
$$\alpha F'(\alpha_0) + \beta F'(\beta_0) + \gamma F'(\gamma_0) + \delta F'(\delta_0) = 0.$$

The tangential equation of the curve of contact will be

$$4F(\alpha_0, \beta_0, \gamma_0, \delta_0)\, F(\alpha, \beta, \gamma, \delta)$$
$$= \{\alpha F'(\alpha_0) + \beta F'(\beta_0) + \gamma F'(\gamma_0) + \delta F'(\delta_0)\}^2.$$

Asymptotes.

326. Def. A straight line is an asymptote to a surface when it meets the surface in two points at least at an infinite distance, while the line itself remains at a finite distance.

Def. A plane is an asymptotic plane, when in it an infinite number of asymptotic lines can be drawn to the surface.

Def. A plane is a singular asymptotic plane when all straight lines drawn in it are asymptotic lines.

Def. An asymptotic surface is the surface which is enveloped by all asymptotic planes to the surface.

327. *To find the asymptotic lines, planes, and surface to a given surface.*

Let $\phi \equiv F(x, y, z) = 0$ be the equation of a surface,

(x_0, y_0, z_0) a point in an asymptotic line,

$$\frac{x-x_0}{\lambda} = \frac{y-y_0}{\mu} = \frac{z-z_0}{\nu} = r, \text{ its equations};$$

and let $F(\lambda, \mu, \nu)$ be arranged in a series of homogeneous functions of degrees $n, n-1, \ldots$ so that

$$F(\lambda, \mu, \nu) \equiv \phi_n + \phi_{n-1} + \ldots \phi_r.$$

The points in which the line meets the surface are given by the equation

$$F(x_0 + \lambda r, \; y_0 + \mu r, \; z_0 + \nu r) = 0,$$

or, if D denote the operation $x_0 \dfrac{d}{d\lambda} + y_0 \dfrac{d}{d\mu} + z_0 \dfrac{d}{d\nu}$,

$$r^n\phi_n + r^{n-1}(D\phi_n + \phi_{n-1}) + r^{n-2}\left(D^2\phi_n + D\phi_{n-1} + \frac{1}{2}\phi_{n-2}\right) + \ldots = 0,$$

and, since two of the roots are infinite,

$$\phi_n = 0, \quad \text{and} \quad \phi_{n-1} + D\phi_n = 0,$$

$\phi_n = 0$ denotes that the line is parallel to a generating line of the cone $F_n(x, y, z) = 0$, and

$$D\phi_n + \phi_{n-1} \equiv \left(x_0 \frac{d}{d\lambda} + y_0 \frac{d}{d\mu} + z_0 \frac{d}{d\nu}\right) \phi_n + \phi_{n-1} = 0 \ldots (1),$$

is the equation of a plane in which the asymptotic lines in any direction must be drawn, since the condition

$$\left(\lambda \frac{d}{d\lambda} + \mu \frac{d}{d\mu} + \nu \frac{d}{d\nu}\right) \phi_n = n\phi_n = 0$$

shews that the line lies entirely in the plane.

This is therefore the asymptotic plane containing the lines in the direction corresponding to any solution of $\phi_n = 0$.

Since the plane (1) is a function of λ, μ, ν, which are connected by the relations $\phi_n = 0$, and $\lambda^2 + \mu^2 + \nu^2 = 1$, the asymptotic surface may be found by eliminating λ, μ, ν, and $d\lambda$, $d\mu$, $d\nu$ between the equations (1), $\phi_n = 0$, and the differentials of the three equations.

If (x_0, y_0, z_0) be an arbitrary point, the equations $\phi_n = 0$, and (1), determine $n(n-1)$ directions in which asymptotic straight lines can be drawn through the point.

328. *To find the asymptotes of a central conicoid.*

Let the equation of the conicoid be $ax^2 + by^2 + cz^2 = 1$, the directions of the asymptotic lines are given by

$$a\lambda^2 + b\mu^2 + c\nu^2 = 0 \ldots\ldots\ldots\ldots\ldots\ldots\ldots (1).$$

The equation of the asymptotic plane containing lines whose directions are (λ, μ, ν) is

$$a\lambda x_0 + b\mu y_0 + c\nu z_0 = 0 \ldots\ldots\ldots\ldots\ldots\ldots\ldots (2).$$

To obtain the asymptotic surface, we have the above equations and the equations in the differentials,

$$a x_0\, d\lambda + b y_0\, d\mu + c z_0\, d\nu = 0,$$
$$a\lambda\, d\lambda + b\mu\, d\mu + c\nu\, d\nu = 0,$$
$$\lambda\, d\lambda + \mu\, d\mu + \nu\, d\nu = 0.$$

Eliminating by indeterminate multipliers

$$ax_0 + Aa\lambda + B\lambda = 0,$$

$$by_0 + Ab\mu + B\mu = 0,$$

$$cz_0 + Ac\nu + B\nu = 0.$$

$\therefore$ by (1) and (2), $B = 0$, and $\dfrac{x_0}{\lambda} = \dfrac{y_0}{\mu} = \dfrac{z_0}{\nu}$;

$\therefore$ by (1), $\qquad ax_0^2 + by_0^2 + cz_0^2 = 0,$

which is the asymptotic cone, every tangent plane to which is an asymptotic plane, lines in which parallel to the line of contact meet the surface in two points at infinity, viz. the points in which it meets the parallel generating lines, which are the lines of intersection of the surface and the plane (2).

329. *To find the asymptotic lines of closest contact in any asymptotic plane.*

If the surface be of a higher degree than the second, we can determine generally lines which meet the surface in three points at an infinite distance, the conditions that this should be the case are

$$\phi_n = 0 \quad\ldots\ldots\ldots\ldots\ldots\ldots \quad (1),$$

$$D\phi_n + \phi_{n-1} = 0 \quad\ldots\ldots\ldots\ldots\ldots\ldots \quad (2),$$

and $$D^2\phi_n + 2D\phi_{n-1} + 2\phi_{n-2} = 0 \quad\ldots\ldots\ldots\ldots\ldots\ldots \quad (3).$$

For any set of values of $\lambda : \mu : \nu$, which satisfy the equation (1), the intersection of the surface (3) with the plane (2) is the locus of the points through which the corresponding asymptotic lines have closer contact with the surface.

Since, if we write $x_0 + \lambda r$ for x_0, &c., and let

$$\Delta \equiv \lambda \frac{d}{d\lambda} + \mu \frac{d}{d\mu} + \nu \frac{d}{d\nu},$$

$D^2\phi_n$ becomes $(D + r\Delta)^2\phi_n \equiv D^2\phi_n + 2r\Delta\,(D\phi_n) + r^2\Delta^2\phi_n$

$$\equiv D^2\phi_n + 2r\,.\,(n-1)\,D\phi_n + r^2\,.\,n\,.\,(n-1)\,\phi_n,$$

and $D\phi_{n-1}$ becomes $(D + r\Delta)\,\phi_{n-1} \equiv D\phi_{n-1} + r\,.\,(n-1)\,\phi_{n-1},$

the equations (3) and (2) are simultaneously satisfied for all values of r, i. e. they intersect in two straight lines which are the asymptotic lines required.

330. If we applied this to the case of the conicoid the equation (3) would be identical with the equation of the surface, and the two particular asymptotic lines would be the generating lines in which the plane cuts the surface, since a line which meets a conicoid in these points lies entirely in the surface.

331. If we take as an example the surface

$$x^4 - y^2z^2 - 2a^2yz = 0,$$

the equations become

$$\lambda^4 - \mu^2\nu^2 = 0 \quad\ldots\ldots\ldots\ldots\ldots\ldots\ldots\ldots (1),$$

$$2x_0\lambda^3 - \mu\nu(\mu z_0 + \nu y_0) = 0 \quad\ldots\ldots\ldots\ldots\ldots\ldots (2),$$

$$6x_0^2\lambda^2 - 2y_0z_0\mu\nu - (\mu z_0 + \nu y_0)^2 = 2a^2\mu\nu \quad\ldots\ldots\ldots (3),$$

$$\text{or, } \lambda^2 = \pm\,\mu\nu, \text{ by (1);}$$

$$\therefore\ 2x_0\lambda \mp (\mu z_0 + \nu y_0) = 0.$$

Hence for the intersections of (2) and (3),

$$3(\mu z_0 + \nu y_0)^2 - 4y_0z_0\mu\nu - 2(\mu z_0 + \nu y_0)^2 = 4a^2\mu\nu;$$

$$\therefore\ \mu z_0 - \nu y_0 = \pm\, 2a\lambda, \text{ if } \lambda^2 = \mu\nu,$$

and (3) is evidently reducible to

$$12x_0^2\lambda^2 + (\mu z_0 - \nu y_0)^2 - 3(\mu z_0 + \nu y_0)^2 = \pm\, 4a^2\lambda^2,$$

which represents a hyperboloid of one or two sheets.

If we take $\lambda^2 = -\mu\nu$, we obtain imaginary asymptotic lines.

332. Mr Walton* has defined an asymptotic plane as a plane touching a surface at an infinite distance and passing within a finite distance from the origin of co-ordinates.

If the asymptotic plane correspond to an ordinary tangent plane, the tangent lines drawn from different points in the plane to the point of contact become parallel when this point moves to an infinite distance. Hence the ordinary asymptotic plane is

* *Cambridge and Dublin Mathematical Journal*, Vol. III. p. 28.

one in which all lines drawn in a certain direction meet the surface in two points at an infinite distance.

As in the tangent plane there are generally two directions, real or imaginary, in which the tangent lines pass through three consecutive points, so among the parallel asymptotic lines there are generally two lines which possess the property of Art. (329).

A singular asymptotic plane is one which touches along a line at infinity, if considered as a limit of a tangent plane; and, if considered as the locus of asymptotic lines, it is a plane such that lines drawn in *any* direction in it meet the surface in two points at an infinite distance.

The analytical condition is obtained by considering that for some solutions of $\phi_n = 0$ the equation $D\phi_n + \phi_{n-1} = 0$ must give an equation which is independent of the values λ, μ, ν.

Thus for the surface $xyz = a^3$,

$$\lambda\mu\nu = 0, \text{ and } x_0\mu\nu + y_0\nu\lambda + z_0\lambda\mu = 0,$$

we have three singular asymptotic planes $x_0 = 0$, $y_0 = 0$, and $z_0 = 0$, which are the only asymptotic planes.

333. *Asymptotic surfaces which have a contact of a higher degree than the second.*

If any relation which satisfies $\phi_n = 0$, makes $D\phi_n + \phi_{n-1} = 0$ independently of any relation between x_0, y_0, z_0, we may find a locus of straight lines drawn in the direction corresponding to that relation which shall pass through three points at an infinite distance: the equation of the locus is

$$D^2\phi_n + 2D\phi_{n-1} + 2\phi_{n-2} = 0,$$

which, the reasoning of Art. (329) shews, is the equation of a cylindrical surface or of two planes.

The existence of such asymptotic surfaces shews that there is a singular point at an infinite distance.

Thus in the surface whose equation is

$$z(x+y)^2 - a(x^2 - y^2) + b^2 z = 0,$$

we have the equations

$$\nu(\lambda+\mu)^2=0 \ldots\ldots\ldots\ldots\ldots\ldots\ldots\ldots\ldots(1),$$

$$\{2\nu(x_0+y_0)+z_0(\lambda+\mu)-a(\lambda-\mu)\}(\lambda+\mu)=0 \ldots\ldots\ldots(2),$$

$$2z_0(x_0+y_0)(\lambda+\mu)+\nu(x_0+y_0)^2-2a(x_0\lambda-y_0\mu)+b^2\nu=0 \ldots(3),$$

$\nu=0$ gives the asymptotic plane

$$z_0(\lambda+\mu)-a(\lambda-\mu)=0,$$

and the particular lines

$$0\,.\,(x_0+y_0)^2+\lambda y_0-\mu x_0=0,$$

one of which is at an infinite distance.

$\lambda+\mu=0$ satisfies (2) identically, and the asymptotic surface is

$$\nu(x_0+y_0)^2-2a\lambda(x_0+y_0)+b^2\nu=0,$$

which gives two planes if $a^2\lambda=$ or $>b^2$, and it is easily seen, that the straight lines in which these planes intersect the plane

$$2\lambda z-\nu(x-y)=0,$$

lie entirely in the surface.

334. If the equation $D\phi_n+\phi_{n-1}=0$ be of such a form that the terms involving x_0, y_0, z_0 disappear, this indicates that the tangent lines for those particular directions are at an infinite distance; in this case we can find an asymptotic surface.

For let λ', μ', ν' be values of λ, μ, ν which are near those which make ϕ_n and $\phi_{n-1}=0$, and reduce $D\phi_n$ to zero identically, and suppose ϕ_s^2 to be the factor of ϕ_n which gives rise to a factor ϕ_s in $D\phi_n$; then at the points in which

$$\frac{x}{\lambda'}=\frac{y}{\mu'}=\frac{z}{\nu'}=r$$

meets the surface we have, ϕ' being the corresponding value of ϕ,

$$\phi_n' r^n+\phi'_{n-1}r^{n-1}+\ldots=0,$$

or

$$\phi'_{n-2s}\phi_s'^2 r^n+\phi'_{n-1}r^{n-1}+\ldots=0;$$

therefore at an infinite distance,

$$\phi_{n-2s}\phi_s^2 r^{2s}+\phi_{n-1}r^{2s-1}=0,$$

which reduces to form

$$u_{2s} + u_{2s-1} = 0,$$

which is the equation of the asymptotic surface.

Thus $$z(x+y)^2 - az^2 + bx^2 = 0,$$
being the surface,

$$\nu'(\lambda' + \mu')^2 r^3 + a\nu'^2 r^2 + b\lambda'^2 r^2 = 0,$$

when λ', μ' make $\lambda' + \mu'$ nearly $= 0$;

$$\therefore \nu'(\lambda' + \mu')^2 r^2 - a\nu'^2 r = 0,$$

and the equation of the asymptotic surface is

$$\nu'(x+y)^2 - a\nu' z = 0,$$

or $$(x+y)^2 - az = 0, \text{ a parabolic cylinder.}$$

Normals.

335. DEF. The normal to a surface at an ordinary point of a surface is the straight line drawn through the point perpendicular to the tangent plane at that point.

DEF. The normal cone at a singular point of a surface is the locus of the normals to the tangent planes to the conical tangent at that point.

336. *To find the equations of the normal at any point of a surface.*

Let the equation of the surface be $F(x, y, z) = 0$.

The equation of the tangent plane at any point, which is not a singular point, is

$$(x - x_0) F'(x_0) + (y - y_0) F'(y_0) + (z - z_0) F'(z_0) = 0.$$

The direction cosines of the normal are proportional to

$$F'(x_0),\ F'(y_0),\ F'(z_0).$$

Therefore its equations are

$$\frac{x - x_0}{F'(x_0)} = \frac{y - y_0}{F'(y_0)} = \frac{z - z_0}{F'(z_0)},$$

which is a determinate line, except in the cases in which

$$F'(x_0)=0,\ F'(y_0)=0,\ \text{and } F'(z_0)=0.$$

337. If the surface be given by the equation $z=f(x, y)$, the equation of the tangent plane is

$$z-z_0=p_0(x-x_0)+q_0(y-y_0),$$

and the equations of the normal are

$$x-x_0+p_0(z-z_0)=0,\ \text{and } y-y_0+q_0(z-z_0)=0.$$

338. *To find the equation of the normal cone at a singular point of a surface.*

Let $\phi\equiv F(x, y, z)=0$ be the equation of the surface, and let u, v, w, u', v', and w' be the values of $\frac{d^2\phi}{dx^2}, \frac{d^2\phi}{dy^2}, \frac{d^2\phi}{dz^2}, \frac{d^2\phi}{dy\,dz}, \frac{d^2\phi}{dz\,dx}$ and $\frac{d^2\phi}{dx\,dy}$ at the singular point (x_0, y_0, z_0).

Any of the sides of the normal cone is perpendicular to each of two consecutive tangent lines to the surface at the singular point, or the normal cone is reciprocal to the tangent cone.

Let λ, μ, ν be the direction cosines of a side of the normal cone, and l, m, n those of a tangent line at (x_0, y_0, z_0);

$$\therefore\ l^2u+m^2v+n^2w+2u'mn+2v'nl+2w'lm=0,$$

$$l^2+m^2+n^2=1,\ \text{and } l\lambda+m\mu+n\nu=0\text{.........(1)},$$

and similar equations are true for the next consecutive tangent line;

$$\therefore\ (lu+mw'+nv')\,dl+\text{......}=0,$$
$$l\,dl+\text{......}=0,$$
$$\lambda\,dl+\text{......}=0,$$

hence, employing the arbitrary multipliers A and B, we obtain

$$lu+mw'+nv'+Al+B\lambda=0,$$
$$lw'+mv+nu'+Am+B\mu=0,$$

and $$lv'+mu'+nw+An+B\nu=0;$$

therefore multiplying by l, m, n and adding, the equations (1) give $A = 0$,

hence, writing $vw - u'^2 = p$, $v'w' - uu' = p'$, &c.,

$$l\,(uvw + 2u'v'w' - uu'^2 - vv'^2 - ww'^2) + A\,(\lambda p + \mu r' + \nu q') = 0,$$

....

....

also $$l\lambda + m\mu + n\nu = 0\,;$$

$$\therefore\ \lambda^2 p + \mu^2 q + \nu^2 r + 2\mu\nu p' + 2\nu\lambda q' + 2\lambda\mu r' = 0,$$

and the equation of the normal cone is therefore

$$p\,(x - x_0)^2 + \ldots\ldots + 2p'\,(y - y_0)\,(z - z_0) + \ldots = 0.$$

339. Cor. The condition that the normal cone should reduce to two planes is

$$N \equiv pqr + 2p'q'r' - pp'^2 - qq'^2 - rr'^2,$$

$$\equiv p\,(qr - p'^2) + r'\,(p'q' - rr') + q'\,(p'r' - qq') = 0,$$

but
$$qr - p'^2 = (uw - v'^2)\,(uv - w'^2) - (v'w' - uu')^2$$
$$= u\,(uvw + 2u'v'w' - uu'^2 - v'w' - w'w'^2)$$
$$\equiv uT \text{ suppose.}$$

Similarly, $p'q' - rr' = w'T$, and $p'r' - qq' = v'T$;

$$\therefore\ N = (pu + r'w' + q'v')\,T = T^2\,;$$

therefore when $N = 0$, $T = 0$, or the tangent cone degenerates into two planes, as it ought to be, from the nature of the normal cone.

340. *To find the normal to a surface given by the tangential equation.*

Let the equation of the surface be $F(\alpha, \beta, \gamma, \delta) = 0$, and let $(\alpha_0, \beta_0, \gamma_0, \delta_0)$ be a tangent plane, also $(\alpha, \beta, \gamma, \delta)$ a plane containing the normal; this is perpendicular to the tangent plane, and by Art. (276) the condition of perpendicularity is

$$\lambda_0 \frac{\alpha}{p_1} + \mu_0 \frac{\beta}{p_2} + \nu_0 \frac{\gamma}{p_3} + \rho_0 \frac{\delta}{p_4} = 0 \ldots\ldots\ldots\ldots\ldots (1),$$

where

$$\lambda_0 = \frac{\alpha_0}{p_1} - \frac{\beta_0}{p_2} \cos CD + \frac{\gamma_0}{p_3} \cos DB - \frac{\delta_0}{p_4} \cos BC,$$

and similarly for μ_0, ν_0, ρ_0.

Since this plane also contains the point of contact

$$\alpha F'(\alpha_0) + \beta F'(\beta_0) + \gamma F'(\gamma_0) + \delta F'(\delta_0) = 0 \text{ .... (2).}$$

The equations (1) and (2) determine the normal.

The equation (1) gives the direction of the normal since it represents a point at infinity in the direction perpendicular to the tangent plane, see Arts. (107), (110).

341. *To find the number of normals which can be drawn from a given point to a surface of the nth degree.*

Let $F(x, y, z) = 0$ be the surface. The number of normals will be the same from whatever point they be drawn, the number will therefore be found by investigating the number of normals which can be drawn from a point at an infinite distance, which we may assume in Ox produced.

The number will therefore be equal to the number of normals parallel to Ox, together with the number of normals to a plane section at an infinite distance.

If (x_0, y_0, z_0) be the foot of a normal parallel to Ox, $F'(y_0) = 0$, $F'(z_0) = 0$, which combined with the equation $F(x_0, y_0, z_0) = 0$ gives $n.(n-1)^2$ solutions.

Again, any plane section of the surface will be of the nth degree, and the number of normals drawn to any curve $f(x, y) = 0$ of the nth degree is, in like manner, the number of normals parallel to Ox, together with the normals which can be drawn at points at an infinite distance, whose number is n; now, the number of normals parallel to Ox are given by the number of solutions of $f'(y_0) = 0$, and $f(x_0, y_0) = 0$, which are $n.(n-1)$, hence, the number of normals to the plane section at an infinite distance is n^2.

Therefore, the number of normals which can be drawn to the surface from any point

$$= n.\overline{n-1}|^2 + n^2 = n^3 - n^2 + n.$$

Applications.

342. *Tangent plane of a central conicoid.*

The equation being $ax^2 + by^2 + cz^2 = 1$, that of the tangent plane at (x_0, y_0, z_0) is $ax_0x + by_0y + cz_0z = 1$, and if l, m, n be direction cosines of the normal to this plane

$$\frac{ax_0}{l} = \frac{by_0}{m} = \frac{cz_0}{n} = \sqrt{\left(\frac{ax_0^2 + by_0^2 + cz_0^2}{\frac{l^2}{a} + \frac{m^2}{b} + \frac{n^2}{c}}\right)};$$

therefore the equation of the tangent plane may be written in the form

$$lx + my + nz = \sqrt{\left(\frac{l^2}{a} + \frac{m^2}{b} + \frac{n^2}{c}\right)}.$$

1. For the ellipsoid

$$\frac{x^2}{a^2} + \frac{y^2}{b^2} + \frac{z^2}{c^2} = 1,$$

$$\frac{x_0^2}{a^2} + \frac{y_0^2}{b^2} + \frac{z_0^2}{c^2} = 1.$$

The tangent plane

$$\frac{xx_0}{a^2} + \frac{yy_0}{b^2} + \frac{zz_0}{c^2} = 1$$

meets the ellipsoid when

$$\frac{(x - x_0)^2}{a^2} + \frac{(y - y_0)^2}{b^2} + \frac{(z - z_0)^2}{c^2} = 0,$$

or in the single point (x_0, y_0, z_0).

2. For the hyperboloid of one sheet

$$\frac{x^2}{a^2} + \frac{y^2}{b^2} - \frac{z^2}{c^2} = 1,$$

$$\frac{x_0^2}{a^2} + \frac{y_0^2}{b^2} - \frac{z_0^2}{c^2} = 1.$$

The tangent plane

$$\frac{xx_0}{a^2} + \frac{yy_0}{b^2} - \frac{zz_0}{c^2} = 1$$

meets the hyperboloid in points which satisfy the equation

$$\left(\frac{x^2}{a^2}+\frac{y^2}{b^2}\right)\left(\frac{x_0^2}{a^2}+\frac{y_0^2}{b^2}\right)-\left(\frac{xx_0}{a^2}+\frac{yy_0}{b^2}\right)^2$$

$$=\left(1+\frac{z^2}{c^2}\right)\left(1+\frac{z_0^2}{c^2}\right)-\left(1-\frac{zz_0}{c^2}\right)^2,$$

$$\text{or}\quad \left(\frac{xy_0-yx_0}{ab}\right)^2=\left(\frac{z-z_0}{c}\right)^2;$$

or in two straight lines determined by the equations

$$\frac{xy_0-yx_0}{ab}=\pm\frac{z-z_0}{c},$$

$$\text{and}\qquad \frac{xx_0}{a^2}+\frac{yy_0}{b^2}-\frac{zz_0}{c^2}=1.$$

343. *Tangent plane and normal to the paraboloid.* The equation being

$$\frac{x^2}{a^2}\pm\frac{y^2}{b^2}=\frac{2z}{c},$$

that of the tangent plane is

$$\frac{(x-x_0)\,x_0}{a^2}\pm\frac{(y-y_0)\,y_0}{b^2}=\frac{z-z_0}{c},$$

$$\text{or}\quad \frac{xx_0}{a^2}\pm\frac{yy_0}{b^2}=\frac{z+z_0}{c},$$

those of the normal are

$$\frac{a^2(x-x_0)}{x_0}=\frac{b^2(y-y_0)}{\pm y_0}=\frac{c(z-z_0)}{-1}$$

$$=\frac{(x-x_0)\,x_0\pm(y-y_0)\,y_0+(z-z_0)\,2z_0}{\dfrac{x_0^2}{a^2}\pm\dfrac{y_0^2}{b^2}-\dfrac{2z_0}{c}};$$

therefore, the equations of the normal may be written

$$\frac{a^2(x-x_0)}{x_0}\mp\frac{b^2(y-y_0)}{y_0}=0,$$

$$\text{and}\qquad x_0(x-x_0)\pm y_0(y-y_0)+2z_0(z-z_0)=0.$$

The last equation shews that the normal to a paraboloid at any point lies in the tangent plane to a surface whose equation is

$$x^2 \pm y^2 + 2z^2 = x_0^2 \pm y_0^2 + 2z_0^2$$

drawn through the same point.

344. *Tangent plane and normal to the helicoid.*

The helicoid is generated by the motion of a straight line, subjected to pass through a given axis to which it is perpendicular, and about which it twists with an angular velocity proportional to the velocity of the point of intersection.

The equation of the helicoid is

$$z = c \tan^{-1} \frac{y}{x},$$

if the axis of z be the line to which the generating line is perpendicular,

$$p = -\frac{cy}{x^2 + y^2}, \quad q = \frac{cx}{x^2 + y^2}.$$

The equation of the tangent plane at (x_0, y_0, z_0) is

$$(x_0^2 + y_0^2)(z - z_0) = c(-xy_0 + yx_0),$$

and the equations of the normal are

$$\frac{x - x_0}{-y_0} = \frac{y - y_0}{x_0} = \frac{c(z - z_0)}{x_0^2 + y_0^2}.$$

The tangent plane cuts the surface in the points which satisfy the equation

$$(x_0^2 + y_0^2)(z - z_0) = cx_0 x\left(\tan\frac{z}{c} - \tan\frac{z_0}{c}\right),$$

or in an infinite number of straight lines, one of which coincides with the generating line.

The normal is a tangent line to a circular cylinder whose axis is that of the surface and which passes through the point at which the normal is drawn.

345. *Tangent plane to the anchor-ring.*

Let the plane containing the centers of the generating circles be taken for the plane of xy and the axis of rotation for the axis of z, and ρ be the distance of any point (x, y, z) from the axis, c that of the center of the generating circle, a its radius,

$$\rho^2 = x^2 + y^2, \quad \text{and } z^2 + (\rho - c)^2 = a^2.$$

The equation of the anchor-ring is

$$(x^2 + y^2 + z^2 + c^2 - a)^2 - 4c^2(x^2 + y^2) = 0.$$

The equation of the tangent plane at (x_0, y_0, z_0) is

$$\{x_0(x - x_0) + y_0(y - y_0)\}(\rho_0 - c) + z_0(z - z_0)\rho_0 = 0;$$

$$\text{or, } (\rho_0 - c)(x_0 x + y_0 y) + \rho_0 z_0 z = \rho_0^2(\rho_0 - c) + \rho_0 z_0^2$$

$$= \{\rho_0 a^2 + c(\rho_0 - c)\}.$$

To find the form of the curve EAF in figure, page 213,

let $x_0 = c - a$, $y_0 = 0$, $z_0 = 0$, $\rho_0 = c - a$, then the equation of the tangent plane is $x = c - a$, and the form of the curve of intersection is given by the equation

$$\{y^2 + z^2 + 2c(c - a)\}^2 = 4c^2\{y^2 + (c - a)^2\};$$

$$\text{or, } (y^2 + z^2)^2 = 4ac\, y^2 - 4c(c - a)z^2.$$

This curve will be the lemniscate of Bernoulli if $c = 2a$.

To find a tangent plane which touches the surface along a curve-line, the ratios of the coefficients in the equation of the tangent plane must be constant;

$$\text{or, } \frac{x_0(\rho_0 - c)}{z_0\rho_0}, \ \frac{y_0(\rho_0 - c)}{z_0\rho_0} \text{ and } \frac{a^2 + c(\rho_0 - c)}{z_0} \text{ are constant,}$$

therefore either $\rho_0 = c$, or $y_0 = mx_0$, m constant.

If $\rho_0 = c$, $z_0 = \pm a$ satisfying the condition.

If $y_0 = mx_0$, x_0, y_0, and z_0 must all be constant.

Hence, the only tangent planes are those whose equations are $z = \pm a$.

346. As an illustration of singular points of surfaces we may take the case of a conjugate curve line corresponding to a conjugate point in two dimensions. If $F(x, y, z) = 0$ be the

equation of a surface containing such an isolated curve (x_0, y_0, z_0) being any point in the curve, any line through this point meets the surface in two coincident points for all directions, if therefore (λ, μ, ν) be the direction of the line

$$\lambda F'(x_0) + \mu F'(y_0) + \nu F'(z_0) = 0,$$

for all values of λ, μ, ν;

$$\therefore\ F'(x_0) = 0, \quad F'(y_0) = 0, \text{ and } F'(z_0) = 0.$$

These equations and $F(x_0, y_0, z_0) = 0$ are true for all points in the line.

The equations are therefore true for $x_0 + dx_0$, &c., if u, v, w, u', v', w' be the values of $\frac{d^2F_0}{dx_0^2}\ldots$

$$u\, dx_0 + w'dy_0 + v'dz_0 = 0,$$

$$w'dx_0 + v\, dy_0 + u'dz_0 = 0,$$

$$v'dx_0 + u'dy_0 + wdz_0 = 0\,;$$

and therefore $$\begin{vmatrix} u, & w', & v' \\ w', & v, & u' \\ v', & u', & w \end{vmatrix} = 0.$$

This is the condition that the locus of the tangent lines at (x_0, y_0, z_0), whose equation is

$$u\,(x - x_0)^2 + \ldots\ldots + 2u'(y - y_0),\ (z - z_0) + \ldots\ldots = 0,$$

may separate into two real or impossible planes. But for a conjugate line it is manifest that these must be impossible, and we shall have the further conditions,

$$vw > u'^2, \quad wu > v'^2, \quad uv > w'^2, \text{ (Art. 225)}.$$

EXAMPLE.

The surface whose equation is

$$a^2\{2(y^2+z^2) - x^2\}^2 = (y^2+z^2)(y^2+z^2-a^2)^2$$

will be found to have a conjugate line in the plane of yz.

S

342. *If a cone be described with any point of a conicoid as its vertex, and a focal as its base the normal to the conicoid at the vertex is an axis of the cone.*

Let the equation of the conicoid be

$$\frac{x^2}{a^2}+\frac{y^2}{b^2}+\frac{z^2}{c^2}=1,$$

and let $x'y'z'$ be the co-ordinates of P the vertex; let a tangent line through P meet the dirigent cylinder in E, E', and let F, F' be the foci corresponding to the directrices through E, E'; and Q, Q' the feet of these directrices.

Then, since EE' passes through a fixed point P, QQ' will pass through a fixed point $(x'y')$ the projection of P, the tangents at Q, Q' to the dirigent conic will intersect on the polar of $x'y'$ with respect to the dirigent conic, that is, on the straight line whose equation is

$$\frac{a^2-c^2}{a^4}xx'+\frac{b^2-c^2}{b^4}yy'=1 \text{ .... (1)}.$$

Now, since F, F' are the poles, with respect to the trace of the surface on xy, of the tangents at Q, Q', FF' will pass through the pole of (1) with respect to that trace, or through a point whose co-ordinates are

$$\frac{a^2-c^2}{a^2}x', \quad \frac{b^2-c^2}{b^2}y'.$$

But, the equations of the normal at P being

$$\frac{x-x'}{\frac{x'}{a^2}}=\frac{y-y'}{\frac{y'}{b^2}}=\frac{z-z'}{\frac{z'}{c^2}},$$

it appears that FF' always passes through the point where the normal meets xy. Also, by Art. (207), EPE' makes equal angles with FP, $F'P$, hence the normal at P, being perpendicular to EPE', makes equal angles with FP, $F'P$.

That is, any plane through the normal will cut the cone in two straight lines making equal angles with the normal, or the normal is an axis of the cone.

The two other axes of the cone will be normals to two other conicoids through the point confocal with the given one, and the axes of the cone being mutually at right angles, the confocal surface will also cut each other at right angles.

Hence, through any point may be drawn three surfaces having a given focal curve, and these surfaces will cut each other at right angles at all points of intersection. These properties may readily be proved otherwise.

Changing the signs of a^2, b^2, c^2 we obtain the theorem for all central conicoids, and the non-central surfaces being the limits of these, the proposition will be equally true for them.

XV.

(1) The tangent planes to an ellipsoid at points lying on a plane section will intersect any fixed plane in straight lines which touch a conic section.

(2) The locus of the intersection of two tangent planes to the cone

$$\frac{x^2}{a} + \frac{y^2}{b} + \frac{z^2}{c} = 0,$$

which are at right angles, is the cone

$$(b + c)\,x^2 + (c + a)\,y^2 + (a + b)\,z^2.$$

(3) Find the equation of the tangent plane upon the principle that no other plane can pass between it and the surface in the neighbourhood of the point through which it is drawn.

(4) If two planes be drawn at right angles to each other touching the central conicoid

$$ax^2 + by^2 + cz^2 = 1,$$

and having their line of intersection in a given direction (l, m, n); shew that the locus of their line of intersection is the right circular cylinder

$$x^2 + y^2 + z^2 = (lx + my + nz)^2 + \frac{m^2 + n^2}{a} + \frac{n^2 + l^2}{b} + \frac{l^2 + m^2}{c}.$$

(5) If the non-central conicoid

$$\frac{y^2}{a} + \frac{z^2}{b} = 2x,$$

be taken, the locus is

$$2l\,(lx + my + nz) - 2x = a\,(n^2 + l^2) + b\,(l^2 + m^2).$$

(6) The locus of the intersection of three tangent planes to the conicoid

$$ax^2 + by^2 + cz^2 = 1,$$

which are mutually at right angles, is

$$x^2 + y^2 + z^2 = \frac{1}{a} + \frac{1}{b} + \frac{1}{c},$$

and to the conicoid

$$\frac{y^2}{a} + \frac{z^2}{b} = 2x, \text{ is } x = -\frac{a+b}{2}.$$

(7) The locus of the intersection of three tangent lines to the ellipsoid

$$\frac{x^2}{a^2} + \frac{y^2}{b^2} + \frac{z^2}{c^2} = 1,$$

mutually at right angles, is

$$(b^2 + c^2)\, x^2 + (c^2 + a^2)\, y^2 + (a^2 + b^2)\, z^2 = b^2c^2 + c^2a^2 + a^2b^2.$$

(8) If p, p' be the perpendiculars from the center on parallel tangent planes to two confocal conicoids,

$$p^2 - p'^2$$

is a constant quantity.

(9) If three conicoids be drawn, through a given point $(x'y'z')$, confocal with the ellipsoid

$$\frac{x^2}{a^2} + \frac{y^2}{b^2} + \frac{z^2}{c^2} = 1,$$

the locus of the intersection of three tangent planes to them, mutually at right angles, is the sphere

$$x^2 + y^2 + z^2 = x'^2 + y'^2 + z'^2.$$

(10) If three planes be drawn, mutually at right angles, and each passing through a tangent line of a plane curve of the second degree, the locus of their intersection is a sphere.

(11) The tangent plane to the surface $xyz = a^3$ cuts off a tetrahedron of constant volume from the co-ordinate planes.

(12) If two surfaces of the second degree have two common generating lines of the same system, they will have two other common generating lines, and touch each other in four points.

(13) The tangent plane to the surface whose equation referred to tetrahedral co-ordinates is

$$l\beta\gamma + m\gamma\alpha + n\alpha\beta + l'\alpha\delta + m'\beta\delta + n'\gamma\delta = 0,$$

at the point A is $l'\delta + n\beta + m\gamma = 0$.

(14) If the tangent planes at A, B, C, D form a tetrahedron $abcd$, find the equations of the lines Aa, &c., and shew that they will meet in a point if $ll' = mm' = nn'$.

(15) If Aa, Bb intersect, then also Cc, Dd will intersect.

(16) Shew that the surface, whose equation is

$$mn\beta\gamma + nl\gamma\alpha + lm\alpha\beta + lr\alpha\delta + mr\beta\delta + nr\gamma\delta = 0,$$

satisfying the conditions of (14), can never be a ruled surface; and that it will be an elliptic paraboloid, if

$$\frac{1}{l^2} + \frac{1}{m^2} + \frac{1}{n^2} + \frac{1}{r^2} = \frac{1}{mn} + \frac{1}{nl} + \frac{1}{lm} + \frac{1}{lr} + \frac{1}{mr} + \frac{1}{nr}.$$

(17) The straight lines in which the tangent planes at A, B, C, D to the surface in (16) meet the opposite faces of the fundamental tetrahedron will lie on the plane, $l\alpha + m\beta + n\gamma + r\delta = 0$.

(18) If Aa, Bb, Cc, Dd meet the above surface again in the points a', b', c', d', the tangent planes at A, a' and the plane BCD intersect in the same straight line, and the four straight lines so determined lie on the plane $l\alpha + m\beta + n\gamma + r\delta = 0$.

(19) In the surface $l\alpha\beta = m\gamma\delta$ which passes through the edges BC, CA, AB, DB, find the points in AC, BD, and in BC, AD, at which the tangent planes are parallel, and thence shew that the center of the surface lies on the line joining the middle points of AB, CD.

(20) This surface will be a paraboloid if $l = m$.

(21) If the straight line joining the middle points of AB, CD meet this surface in P, Q, the tangent planes at P, Q are parallel to AB and CD.

(22) The surface $l\alpha^2 + m\beta^2 + n\gamma^2 + r\delta^2 = 0$, will be a paraboloid, if $\frac{1}{l} + \frac{1}{m} + \frac{1}{n} + \frac{1}{r} = 0$, and will be elliptic, or hyperbolic, according as $lmnr$ is negative or positive.

(23) If this condition be satisfied, and if a, b, c be the middle points of DA, DB, DC, a', b', c' of BC, CA, AB, this surface will touch the planes $b'c'a$, $c'a'b$, $a'b'c$, abc; and also the three planes $bcb'c'$, $cac'a'$, $aba'b'$; and the points of contact of the former are the angular points of a tetrahedron whose faces intersect the corresponding faces of $ABCD$ in four straight lines lying in one plane; and this plane passes through the points of contact of the latter, and is parallel to the axis of the paraboloid.

(24) The surface

$$l\alpha\gamma + m\alpha\delta + n\beta\gamma + r\beta\delta = 0,$$

of which the edges AB, CD are generating lines, will be a paraboloid if $l+r=m+n$; and if $l+m+n+r=0$, the straight line joining the middle points of AB, CD will lie on the surface.

(25) If two arbitrary points be taken on each of four straight lines meeting in a point, the only conicoids which can be described through the eight points are cones, or combinations of planes.

(26) Investigate the condition that the general equation of the second degree may represent a cone from the consideration that every plane will have the same pole with respect to it.

(27) O is a fixed point, P a point such that the polar planes of O, P with respect to a given conicoid are at right angles, shew that the locus of P is the plane diametral to all chords of the conicoid perpendicular to the polar plane of O.

(28) In any conicoid passing through the sides of a quadrilateral $ABCD$ in space, the polar plane of the center of gravity of the tetrahedron $ABCD$ will be parallel to AC and BD.

(29) The polar plane of any point on a directrix will pass through the corresponding focus, and the line joining the point to the focus will be at right angles to the polar plane.

(30) If O be a fixed point on a conicoid, OP, OQ, OR any three chords mutually at right angles, the pole of the plane PQR will lie on a fixed plane.

(31) The surface, whose equation is $\frac{l}{\alpha}+\frac{m}{\beta}+\frac{n}{\gamma}+\frac{r}{\delta}=0$, has a tangent cone at each of the angular points of the fundamental tetrahedron. Any two of these cones have a common tangent plane, and a common plane section containing the edge of the tetrahedron opposite to the common generating line: also the six planes of these common sections meet in a point.

(32) The points on a conicoid, the normals at which intersect the normal at a fixed point, lie on a cone of the second degree, having its vertex at the fixed point.

(33) From different points of the straight line $\frac{x}{a}=\frac{y}{b}$, $z=0$, asymptotic straight lines are drawn to the hyperboloid $\frac{x^2}{a^2}+\frac{y^2}{b^2}+\frac{z^2}{c^2}=1$; shew that they will all lie in the planes $\frac{x}{a}-\frac{y}{b}=\pm\frac{z}{c}\sqrt{2}$.

(34) Find the tangent planes to the two surfaces

$$(1)\quad (x^2+y^2+z^2)^2=4a^2(x^2+y^2);$$

$$(2)\quad b^3z=(x^2+y^2)^2-a^2(x^2+y^2);$$

which touch them along a curved line.

(35) Find the tangent cone at the origin to the surface

$$(x^2+y^2+ax)^2-(c^2-a^2)(x^2+z^2)=0;$$

and shew that as a diminishes and ultimately vanishes, the tangent cone contracts, and ultimately becomes a straight line, and as a increases up to c, it expands and finally becomes a plane.

(36) Shew that the asymptotic planes to the surface

$$z(x^2+y^2)-ax^2-by^2=0,$$

are parallel to the plane of xy, and that the locus of straight lines in these planes having contact of the second order at infinity is

$$z=a; \text{ or } z=b;$$

and that the axis of z is an evanescent asymptotic cylinder.

(37) If a globe be placed upon a table, the breadth of the elliptic shadow cast by a fixed luminous point is independent of the position of the globe.

(38) If an ellipsoid having its least axis vertical, be substituted for the globe, determine the condition of the shadow of the globe being circular. It may be shewn that the locus of the luminous point must be an hyperbola, and that the radius of the circular shadow is independent of the mean axis of the ellipsoid.

CHAPTER XV.

METHOD OF RECIPROCAL POLARS.

343. If we take any plane passing through a given point, the polar line of the point with respect to the section of a conicoid by this plane will be the intersection of the polar plane and the plane of section; for the points of contact of all tangent lines, real or impossible, from the point to the conicoid, lie on the polar plane, and therefore the points of contact of the two tangent lines drawn from the point of the section lie on the intersection of the polar plane and the plane of section, which is therefore the polar line.

Hence, any straight line through a point will be harmonically divided by the conicoid and the polar plane of the point.

If P be any point, and Q any point on the polar plane of P, then, taking any plane section through PQ, since Q is on the polar line of P, P is on the polar line of Q, and therefore on the polar plane of Q.

Conversely, any plane passing through P will have its pole on the polar plane of P.

Hence, any plane passing through PQ will have its pole on the line of intersection of the polar planes of P and Q, and the polar plane of any point lying on PQ will pass through the same line of intersection. The relation between PQ and this line of intersection being thus reciprocal, the straight lines are said to be reciprocal to each other with respect to the conicoid.

It is manifest that reciprocal straight lines cannot intersect unless both are tangent lines to the conicoid.

If a cone be described, with vertex P, meeting the conicoid in one and therefore (Art. 324) in two plane sections, these planes will intersect in a straight line lying in the polar plane of P; for any plane through P intersects the two planes in two straight lines which meet in a point on the polar line of P, and therefore on the polar plane of P.

Let Q be the pole of the plane through P and the line of intersection of the two planes, then by taking plane sections through PQ, it is obvious that a cone can be described with vertex Q meeting the conicoid in the same plane sections.

If therefore we take any two plane sections of a conicoid, there will generally be two points lying on the straight line reciprocal to the intersection of the planes, with which, as vertices, two cones may be described containing the plane sections.

If the two planes be parallel, their line of intersection being at infinity, the reciprocal straight line will pass through the center.

Of two reciprocal straight lines, one will always meet the conicoid in real, and the other in impossible points, the tangent planes at the two real points passing through the reciprocal straight line.

These reciprocal properties are particular cases of the general method of reciprocal polars, to which we now proceed.

344. *Reciprocal points and planes.*

Suppose a system A of planes such that, corresponding to every plane P of the system, is determined one, and only one, point p' of a system A'; then, if $(\alpha', \beta', \gamma', \delta')$ be the co-ordinates of p', the equation of P must be of the form

$$(a_1\alpha' + b_1\beta' + c_1\gamma' + d_1\delta')\,\alpha + (a_2\alpha' + \ldots)\,\beta + (a_3\alpha' + \ldots)\,\gamma + (a_4\alpha' + \ldots)\,\delta = 0;$$

$a_1, b_1, c_1, d_1, a_2 \ldots$ being constants for all the planes.

Since this equation may be written in the form

$$(a_1\alpha + a_2\beta + a_3\gamma + a_4\delta)\,\alpha' + (b_1\alpha + \ldots)\,\beta' + \ldots = 0,$$

we see that to every point $(\alpha, \beta, \gamma, \delta)$, or p, of the system A corresponds a determinate plane P' of the system A'.

Also, if a point p lie on the plane P, the plane P' passes through the point p'. If we take two planes P, Q, we determine two points p', q'; if r be any point on (P, Q), R' will pass through p', q'; if R pass through (P, Q), r' will lie on $p'q'$. The straight lines (P, Q), $p'q'$ may thus be considered corresponding straight lines, and the relations between the planes, straight lines, and points of the system A, and the points, straight lines, and planes, corresponding respectively to them in the system A', are reciprocal.

If we take a series of planes in the first system touching a surface S, we shall get a corresponding series of points in the second system lying on some surface S'. Now suppose three of the planes PQR to move up to coincidence at a point p on S, the points p', q', r' will move up to coincidence on S', and the plane passing through them will be a tangent plane at the point corresponding to the ultimate position of P. Hence S and S' are reciprocal surfaces in the sense that each may be generated from the other, either as the locus of points corresponding to the tangent plane of the other, or as enveloped by planes corresponding to points on the other.

If S be a cone with vertex p, then since all the tangent planes to S pass through p, all the points of S' will lie on the plane P', and since any tangent to S has an infinite number of points of contact, there will be at each point of S' an infinite number of tangent planes. S' will therefore in this case be a plane curve, every point of S' corresponding to a tangent plane to S, and every tangent line to S' corresponding to a generating line of S.

If S be any developable surface, every tangent plane to S will have an infinite number of points of contact lying in a straight line, and accordingly at every point of S' will be an infinite number of tangent planes intersecting in one straight line. S' will therefore in this case be a curve of double curvature, a tangent line to S' corresponding to a generating line of S. Also, it is immediately seen that the developable generated by the tangent lines to S' corresponds to the edge of regression of S.

345. The equation found in Art. (319) for the polar plane of a given point with respect to a given conicoid shews that the relation between a point and its polar plane is a particular case of the general relation between a plane and its corresponding point described in the preceding article. If the point be limited by lying on a surface of the nth degree, which is met by an arbitrary straight line in n points, the reciprocal surface

will be of the nth class, since it will have n tangent planes, passing through the straight line which is the reciprocal of any arbitrary straight line, and which is therefore itself arbitrary. The relation between these two surfaces is reciprocal by the properties already discussed.

Such surfaces are said to be *polar reciprocals*, the one of the other, with respect to the conicoid, which is called the *auxiliary conicoid.*

If the auxiliary conicoid is not specified it is always supposed to be a sphere, and any change in the radius of the sphere not altering the species of the reciprocal surface, but only its dimensions, one surface is in this case said to be the polar reciprocal of the other with respect to the point which is the center of the sphere.

The polar reciprocal of a conicoid is always a conicoid, since a conicoid is both of the second degree, and the second class. Also, since the equation of the auxiliary conicoid involves nine constants, which will enter into the equation of the polar reciprocal, we may, by properly choosing the auxiliary conicoid, make the polar reciprocal coincide with any assigned conicoid, so that from any proposition which is generally true for a conicoid, we may form one true for all conicoids which are the reciprocal polars of the former, that is, for all conicoids whatever.

If we restrict the auxiliary conicoid to real surfaces, this will not be strictly true, for, in that case, the polar reciprocal of every ruled surface must be a ruled surface, since to a straight line every point of which lies in the one surface, corresponds a straight line, any plane passing through which will be a tangent plane to the reciprocal surface, and which must therefore be wholly on the surface. Thus the polar reciprocal of an ellipsoid may be any ellipsoid, elliptic paraboloid, or hyperboloid of two sheets; and of a hyperboloid of one sheet, any hyperboloid of one sheet, or hyperboloic paraboloid. The polar reciprocals of cones and cylinders will of course be plane curves, and in the latter case the plane will pass through the center of the auxiliary conicoid.

346. *On the species of polar reciprocals of conicoids.*

The polar reciprocal of an umbilical surface, namely, an ellipsoid, hyperboloid of two sheets, or elliptic paraboloid, will be an ellipsoid, an elliptic paraboloid, or a hyperboloid of two sheets, according as the center of the auxiliary conic lies within, upon, or without the surface. For to any tangent plane to the surface (A) passing through the center of the auxiliary conicoid (B_1), corresponds a point on the reciprocal surface (A'), its pole with respect to (B), at an infinite distance. Hence if the center of (B) lie without (A), there will be a plane section of (A') at infinity, corresponding to the enveloping cone from the center of (B). If the center of B lie upon A, the plane at infinity, which is the polar plane of the center of B, will touch A'. If the center of B be within A, the enveloping one becomes impossible, and the plane at infinity will not meet A' in real points. Hence, since the polar reciprocal cannot be a ruled surface, the results will be as stated.

The polar reciprocal of either of the skew surfaces will be an hyperboloid of one sheet, or an hyperbolic paraboloid, according as the center of the auxiliary conic does not, or does, lie upon the surface, the reasoning being precisely similar to that for the umbilical surfaces.

347. *On the reciprocal polar of a conicoid with respect to a given point.*

If O be the given point, and a sphere be described with center O, and radius k, then if OP be drawn perpendicular to any plane, the pole of the plane with respect to the sphere will be a point Q on OP, such that $OQ \cdot OP = k^2$; since we take any plane through OP, we shall get a circle and straight line, and Q will be the pole of the straight line with respect to the circle. Hence the construction for the reciprocal polar of a conicoid with respect to a given point is as follows. Through the given point let fall a perpendicular on any tangent plane to the conicoid, and on this perpendicular take a point such that the rectangle under the whole perpendicular and the part of it intercepted between this point and the given point is constant; the locus of the points so determined is the reciprocal polar with respect to the point. We might of course equally take the reciprocal construction, namely, draw a plane perpendicular to any radius vector of the conicoid from the given point, and at a distance from the given point such that the rectangle under this distance and the radius vector is constant; the planes so determined will touch the reciprocal pole.

The reciprocal polar of a sphere with respect to a point will be a prolate surface of revolution of which the point is the focus, and the line joining the point and the center of the sphere the axis; for, taking any plane through this line, the section of the reciprocal polar by this plane will be a conic of which the point is the focus and the line before mentioned the major axis. The reciprocal of the polar of the point, with respect to the sphere, will be the center, and the reciprocal of the center, the directrix plane of the surface of revolution, exactly as in two dimensions. Properties of conicoids of revolution having a common focus may be immediately obtained in this manner. These are, however, generally at once deducible from the corresponding properties of plane curves. It is shewn (Art. 210), that the enveloping cone from a point on a focal curve of a conicoid is a right cone. If we take then the polar reciprocal of the conicoid with respect to this point, the tangent planes to the asymptotic cone of the reciprocal surface will be perpendicular to generating lines of the enveloping cone of the conicoid, and the asymptotic cone will therefore also be a right cone, or the surface will be one of revolution. This result is of course true whether the asymptotic cone employed in the proof be real or impossible. Conversely, the reciprocal polar of a surface of revolution with respect to a point will be a conicoid of which the point is a focus.

Hence, from a sphere may be obtained, by successive reciprocations, any of the umbilical conicoids, but, as before shewn, the ruled surfaces cannot be obtained in this manner.

348. *To find the equation of the reciprocal polar of a conicoid with respect to any point on a focal curve.*

Let the equation of the conicoid be

$$\frac{x^2}{a^2}+\frac{y^2}{b^2}+\frac{z^2}{c^2}=1,$$

let the given point be (α, β) lying on the focal curve,

$$\frac{x^2}{a^2-c^2}+\frac{y^2}{b^2-c^2}=1, \quad z=0;$$

and let the radius of the auxiliary sphere be k.

Then if we take a tangent plane to the conicoid at the point (x', y', z') the corresponding point in the reciprocal polar will be given by the equations

$$\frac{x-\alpha}{\frac{px'}{a^2}}=\frac{y-\beta}{\frac{py'}{b^2}}=\frac{z}{\frac{pz'}{c^2}}=\frac{k^2}{\varpi},$$

p being the perpendicular from the center, and ϖ from the point (α, β) on the tangent plane.

Hence, since $\frac{\varpi}{p}=1-\frac{\alpha x'}{a^2}-\frac{\beta y'}{b^2}$, the equations for the point on the reciprocal become

$$\frac{x-\alpha}{\frac{x'}{a^2}}=\frac{y-\beta}{\frac{y'}{b^2}}=\frac{z}{\frac{z'}{c^2}}=\frac{-k^2}{\frac{\alpha x'}{a^2}+\frac{\beta y'}{b^2}-1}$$

$$=\alpha(x-\alpha)+\beta(y-\beta)+k^2=\sqrt{\frac{a^2(x-\alpha)^2+b^2(y-\beta)^2+c^2z^2}{\frac{x'^2}{a^2}+\frac{y'^2}{b^2}+\frac{z'^2}{c^2}}}.$$

The equation of the reciprocal polar is therefore

$$a^2(x-\alpha)^2+b^2(y-\beta)^2+c^2z^2-\{\alpha(x-\alpha)+\beta(y-\beta)+k^2\}^2=0.$$

349. *The reciprocal of a conicoid, with respect to a point on a modular focal curve, is a conicoid of revolution capable of generation by the modular method; and, with respect to a point on an umbilical focal curve, is one capable of generation by the umbilical method.*

Removing the origin to its center, the equation will become

$$(a^2-\alpha^2)x^2-2\alpha\beta xy+(b^2-\beta^2)y^2+c^2z^2=\frac{k^4}{1-\frac{\alpha^2}{a^2}-\frac{\beta^2}{b^2}}.$$

If we again transform to the principal axes, the equation will become

$$h^2x^2+l^2y^2+c^2z^2=\frac{k^4}{1-\frac{\alpha^2}{a^2}-\frac{\beta^2}{b^2}},$$

h^2, l^2 being the roots of the equation

$$(s - a^2 + \alpha^2)(s - b^2 + \beta^2) = \alpha^2\beta^2,$$

and since $\dfrac{\alpha^2}{a^2 - c^2} + \dfrac{\beta^2}{b^2 - c^2} = 1$, we see that c^2 is one root, and the other will therefore be $a^2 + b^2 - c^2 - (\alpha^2 + \beta^2)$.

Now $\alpha^2 + \beta^2$ is less than $a^2 - c^2$, and greater than $b^2 - c^2$; the second root is therefore $> b^2 < a^2$, and the reciprocal surface is an oblate spheroid for points lying on the modular conic.

Similarly, for points on the umbilical conic, the reciprocal surface is a prolate conicoid of revolution, and will be either a spheroid, paraboloid, or hyperboloid of two sheets according as the point lies within, upon, or without the ellipsoid.

Corresponding results may be deduced for the other umbilical conicoids. Hence, the reciprocal polar of an umbilical conicoid is, an oblate spheroid for a point on the modular focal, and a prolate conicoid of revolution for a point on the umbilical focal.

For the hyperboloid of one sheet, we shall have, changing the sign of c^2, as the equation of the reciprocal surface in its simplest form,

$$(a^2+b^2+c^2-\alpha^2-\beta^2)\,x^2-c^2y^2-c^2z^2=\frac{k^4}{1-\dfrac{\alpha^2}{a^2}-\dfrac{\beta^2}{b^2}}=\frac{-k^4a^2b^2}{c^2(a^2+b^2+c^2-\alpha^2-\beta^2)},$$

since α, β are connected by the equations

$$\frac{\alpha^2}{a^2+c^2}+\frac{\beta^2}{b^2+c^2}=1.$$

Now $\alpha^2 + \beta^2 > b^2 + c^2 < a^2 + c^2$, and the surface is therefore a hyperboloid of revolution of one sheet. The same will be the case for the other focal, which is also modular in this surface.

The hyperbolic paraboloid, being the limit of the hyperboloid of one sheet, the reciprocal surface, for any point on either focal, will also be a hyperboloid of one sheet.

These results may be generally summed up as follows: The reciprocal of a conicoid with respect to a point on a focal curve is a surface of revolution, of the species capable of generation by the method corresponding to the focal curve on which the point is placed. See Arts. 198, 204.

350. *On cones and cylinders which are the reciprocal polars of circles with respect to a point.*

If O be the center of the auxiliary sphere, V the point corresponding to the plane of the circle, the tangent planes to the reciprocal cone are perpendicular to the lines from O to points on the circle; hence the cone which is the reciprocal of the circle, and the cone whose vertex is O and base the circle will be reciprocal cones, Art. (215), and since OV is perpendicular to the plane of the circle, that is to one of the dirigent planes of the cone vertex O and base the circle, O will be a point on a focal line of the reciprocal cone, Art. (216).

Hence from properties of circles in one plane we may obtain reciprocal properties of cones having a common vertex and focal line.

If O be in the plane of the circle, the vertex of the reciprocal cone goes off to infinity, and we obtain a cylinder, one of the focal lines of which passes through O. Conversely, the reciprocal of a conic section with respect to a focus is a right circular cylinder.

Hence if we take the reciprocal with respect to O of the section of the reciprocal cone made by a plane through O perpendicular to V, we obtain a cylinder passing through the given circle, and having its generating lines parallel to OV. This is therefore a right circular cylinder, and O must be a focus of the section of which it is the reciprocal.

That is, the focal lines of a cone pass through the foci of the sections of the cone made by planes perpendicular to them.

If O be the vertex of a right circular cone containing the circle, the tangent planes to the reciprocal cone will all be equally inclined to the axis of the former cone, and the reciprocal cone will therefore also be a right circular cone, whose axis will pass through O.

The same reasoning applies to the case of any conic section, and hence when a conicoid is reciprocated with regard to a point on a modular focal curve, the reciprocal of the umbilical focal is a right cone, and *vice versâ*.

351. *Examples of the use of the method of Reciprocal Polars.*

In the following examples of the use of this method, the original proposition, and the one deduced from it, are placed side by side, and the student should endeavour in each case to obtain for himself the reciprocal proposition, and satisfy himself of its being equally general with the one from which it is deduced.

i. Any three conicoids intersect in eight real or impossible points.	Any three conicoids have eight real or impossible common tangent planes.
ii. An infinite number of conicoids can be described passing through the curve of intersection of any two.	An infinite number of conicoids can be described touching along a curve the developable surface generated by common tangent planes to any two.
iii. Two cones of the second degree can be described, each containing two given plane sections of a conicoid.	Two enveloping cones of the same conicoid will intersect each other in two plane curves.
iv. If two conicoids have one common plane section, they will also have another.	If two conicoids have one common enveloping cone, they will also have another.
v. If three conicoids have one common plane section, the second planes of intersection of the surfaces, taken two and two, will intersect in the same straight line.	If three conicoids have one common enveloping cone, the second enveloping cones of these surfaces, taken two and two, will have their vertices in the same straight line.
vi. Any line through a point is divided into two equal parts by two parallel tangent planes to a conicoid, and the parallel plane through the center.	Any straight line through a point is divided harmonically by a conicoid, and the polar plane of the point with respect to the conicoid.
vii. Two conicoids, each touching another along a plane curve, will themselves intersect in two plane curves, whose planes pass through the line of intersection of the planes of contact.	Two conicoids, each touching another conicoid along a plane curve will have two common enveloping cones, whose vertices lie on the same straight line with the poles of the planes of contact.
viii. Two conicoids, touching each other at two points, will have two plane curves of intersection passing through those points.	Two conicoids, touching each other at two points, will have two common enveloping cones, whose vertices lie on the line of intersection of the tangent planes at those points.

ix. If a conicoid pass through seven fixed points, the polar plane of any other fixed point will pass through a fixed point.	If a conicoid touch seven given planes, the locus of its center will be a plane.
x. Two circles, lying in the same plane, cannot have more than two real common points.	Two cones, having a common vertex, and a common focal line, cannot have more than two real common tangent planes.
xi. Two spheres have two common enveloping cones.	If two prolate surfaces of revolution have a common focus, their points of intersection will lie on plane sections.
xii. Any enveloping cone of a sphere is a cone of revolution.	If, in a prolate surface of revolution, a cone be described with a focus of the surface as vertex, and a plane section as base, it will be a cone of revolution.
xiii. The vertex of a common enveloping cone of two spheres lies on the line joining their centers.	If two prolate surfaces of revolution have a common focus, their planes of intersection will pass through the line of intersection of their directrix planes.
xiv. If a sphere be inscribed in a right circular cone, the section of the cone made by any plane touching it will have the point of contact as a focus, and its directrix will lie in the plane of contact of the sphere and cone.	If a paraboloid of revolution be described passing through a given ellipse, and a right circular cylinder be described also passing through the ellipse, the axis of the cylinder will be parallel to that of the paraboloid, and will pass through the pole of the plane of the ellipse.
xv. If a sphere be inscribed in a conicoid of revolution, the section of the conicoid by any plane touching the sphere will have the point of contact for a focus, and the corresponding directrix will lie on the plane of contact of the sphere and conicoid.	If with any point on the focal hyperbola of an ellipsoid as focus be described a paraboloid of revolution enveloping the ellipsoid, the axis of this paraboloid will be parallel to the generating lines of one of the right circular cylinders which envelope the ellipsoid, and the axis of this cylinder will pass through the pole of the plane of contact of the ellipsoid and paraboloid.

Reciprocating this last with respect to an arbitrary point, we obtain the following proposition, which is due, we believe, to M. Chasles.

If two conicoids touch each other along a plane curve, and a tangent plane be drawn to one of them at an umbilicus, the section of the other made by this plane will have the umbilicus for a focus, and its corresponding directrix in the place of contact of the two conicoids.

For, reciprocating with respect to a point O, the surface corresponding to the paraboloid will pass through O (Art. 346); also O will be a point on its focal curve (Art. 347); O will then be an umbilicus (Art. 192). Also the tangent plane at O will correspond to the point at infinity on the paraboloid, and therefore to the vertex of the right circular cylinder; hence to the right circular cylinder will correspond a plane section of the reciprocal of the ellipsoid by the tangent plane at O. Since the cylinder is a right circular cylinder, the corresponding curve will be a conic whose focus is O, and since the axis of the cylinder passes through the pole of the plane of contact of the ellipsoid and paraboloid, the directrix of the corresponding conic will lie in the plane of contact of the two corresponding surfaces.

We have not been able to discover M. Chasles' own proof of this proposition. An algebraical proof by Mr E. W. Walker will be found in the *Quarterly Journal of Mathematics*, 1861, Vol. IV.

352. *To find the degree of the polar reciprocal of a surface of the nth degree.*

To every point on a surface corresponds a tangent plane to the reciprocal surface, and conversely; hence the number of points in which the reciprocal surface is met by a straight line will be equal to the number of tangent planes which can be drawn to the given surface containing a given straight line. This number, by Art. (308), is $n(n-1)^2$, which is therefore the degree of the reciprocal surface. The class of the reciprocal surface will be the degree of the original surface, or will be n. The class of a surface of the degree $n(n-1)^2$ $(\equiv m)$, will in general be $m(m-1)^2$, which can only be equal to n, when $n=1$, or 2. It will hereafter be explained how the existence of singular points and lines causes the reduction of the class of a surface.

353. *To shew that if two tetrahedrons be so related that each angular point of one is the pole of a face of the other, with respect to any conicoid, the lines of intersection of corresponding faces will*

lie on one conicoid, as also will the lines joining corresponding angular points.

Let $ABCD$, $abcd$, be the two tetrahedrons, and let the equation of the conicoid, referred to $ABCD$, be

$$p\alpha^2 + q\beta^2 + r\gamma^2 + s\delta^2 + l\beta\gamma + m\gamma\alpha + n\alpha\beta + l'\alpha\delta + m'\beta\delta + n'\gamma\delta = 0.$$

Then the equation of bcd is

$$2p\alpha + n\beta + m\gamma + l'\delta = 0;$$

and the equations of the line of intersection of BCD, bcd, are

$$n\beta + m\gamma + l'\delta = 0, \quad \alpha = 0 \quad \text{.................. (1)},$$

$$\text{of } CDA,\ cda, \quad n\alpha + l\gamma + m'\delta = 0, \quad \beta = 0 \quad \text{.......... (2)},$$

$$\text{of } DAB,\ dab, \quad m\alpha + l\beta + n'\delta = 0, \quad \gamma = 0 \quad \text{......... (3)},$$

$$\text{of } ABC,\ abc, \quad l'\alpha + m'\beta + n'\gamma = 0, \quad \delta = 0 \quad \text{......... (4)}.$$

The equations of any line intersecting (1) and (2) may then be written

$$k\alpha + n\beta + m\gamma + l'\delta = 0, \quad n\alpha + k'\beta + l\gamma + m'\delta = 0.$$

The condition that this line may intersect (3) is, therefore,

$$\begin{vmatrix} k & n & l' \\ n & k' & m' \\ m & l & n' \end{vmatrix};$$

and that it may intersect (4), is

$$\begin{vmatrix} k & n & m \\ n & k' & l \\ l' & m' & n' \end{vmatrix};$$

and these conditions are obviously the same.

Eliminating k and k' from the equations of the line, and either of these determinants, we find the equation of the surface to be

$$mnl'\alpha^2 + nlm'\beta^2 + lmn'\gamma^2 + l'm'n'\delta^2 + (mm' + nn')(l\beta\gamma + l'\alpha\delta) + \ldots = 0.$$

Since the faces of the fundamental tetrahedron meet this surface in straight lines, they will be tangent planes to it; since the equation involves five constants, it may be considered to be the general equation of a conicoid touching the faces of the fundamental tetrahedron.

The second part of the proposition may be proved algebraically, but, reciprocating the whole system with respect to any conicoid, we obtain two new tetrahedrons and a new conicoid forming a system similarly related, and the reciprocals of the straight lines above considered will be the straight lines joining the vertices of the new tetrahedrons. These straight lines will therefore also lie on one conicoid, the reciprocal of the former one.

If $ll' = mm' = nn'$, the equation obtained for the conicoid reduces in that case to two coincident planes.

In this case the lines of intersection of corresponding faces lie in one plane, and the lines joining corresponding angular points will pass through one point.

It follows from this, that if any two tetrahedrons be so related that the lines joining corresponding vertices meet in a point, the lines of intersection of corresponding faces will lie in one plane, and conversely.

XVI.

(1) The eccentricity of any section of a prolate conicoid is $e \cos \alpha$, e being the eccentricity of any principal section of the conicoid, and α the inclination of the cutting plane to the axis.

(2) If a conicoid touch the faces of the fundamental tetrahedron $ABCD$ in a, b, c, d; shew that if Aa, Bb intersect each other, Cc and Dd will also intersect each other.

(3) If a series of straight lines be drawn through a point O, such that the straight lines, reciprocal to them with respect to a given conicoid, are respectively perpendicular to them, these straight lines will lie on a cone of the second degree, and the reciprocal straight lines will be tangent lines to a parabola.

(4) If two plane sections A, B of a conicoid be taken, and if O, O' be the vertices of the two cones which can be described passing through A, B (Art. 343), then if two conicoids be described touching the former conicoid along A, B respectively, and passing through O, the tangent plane to each of these at O will be the polar plane of O', and they will also have a common plane section in the polar plane of O. Also state the reciprocal of this theorem.

(5) If three cones of the second degree have a common vertex, and a common focal line, the lines of intersection of the common tangent planes to them, taken two and two, will lie in a plane.

(6) The reciprocal polar of the surface $ax^2+by^2+cz^2=1$ with respect to the surface $a'yz+b'zx+c'xy=1$, is

$$\frac{(b'z+c'y)^2}{a}+\frac{(c'x+a'z)^2}{b}+\frac{(a'y+b'x)^2}{c}=1.$$

(7) If two hyperboloids of revolution of two sheets, or two prolate spheroids, have a common focus, and equal minor axes, they will have a common enveloping cylinder, one of whose focal lines will pass through the common focus.

(8) The reciprocal polar of the surface

$$ax^2+by^2+cz^2=1$$

with respect to a point (α, β, γ) is

$$abc\,\{\alpha(x-\alpha)+\beta(y-\beta)+\gamma(z-\gamma)+k^2\}^2$$
$$=bc\,(x-\alpha)^2+ca\,(y-\beta)^2+ab\,(z-\gamma)^2.$$

(9) Any two surfaces of the family represented by the equation

$$a^2x^2+b^2y^2-(\alpha x+\beta y+\gamma z+k^2)^2+c^2(x^2+y^2+z^2)=0,$$

for different values of c, possess the property that the line joining the points of contact of any common tangent plane subtends a right angle at the origin.

(10) The reciprocal polar of the surface

$$ax^2+by^2+cz^2+2a'yz+2b'zx+2c'xy=1,$$

with respect to a sphere of radius k, and center (α, β, γ) is

$$(abc+2a'b'c'-aa'^2-bb'^2-cc'^2)\,\{\alpha(x-\alpha)+\beta(y-\beta)+\gamma(z-\gamma)+k^2\}^2$$
$$=\lambda(x-\alpha)^2+\mu(y-\beta)^2+\nu(z-\gamma)^2+2\lambda'(y-\beta)(z-\gamma)+\ldots$$

where λ, μ, ν denote $bc-a'^2$, &c., and λ', μ', ν' denote $b'c'-aa'$, &c.

(11) From the equation of the reciprocal surface in (10), obtain the equations of the focal curves of the given conicoid in the form

$$\frac{a'+\lambda yz-x(\mu'y+\nu'z-\lambda'x)}{\rho yz-\lambda'}=\ldots\ldots$$

ρ denoting $abc+2a'b'c'-aa'^2-bb'^2-bc'^2$.

(12) Also, if the given conicoid be one of revolution, determine the foci, from the consideration that the reciprocal polar, with respect to a focus, will be a sphere.

(13) The reciprocal polar of the hyperboloid $\alpha\beta = k\gamma\delta$ with respect to the conicoid $\alpha^2 + \beta^2 = \gamma^2 + \delta^2$ is $k\alpha\beta = \gamma\delta$.

(14) The reciprocal polar of the conicoid

$$mn\beta\gamma + nl\gamma\alpha + lm\alpha\beta + lr\alpha\delta + mr\beta\delta + nr\gamma\delta = 0,$$

with respect to the conicoid

$$a\alpha^2 + b\beta^2 + c\gamma^2 + d\delta^2 = 0,$$

is $$\left(\frac{a\alpha}{l}\right)^2 + \ldots - \frac{ab\alpha\beta}{lm} - \ldots = 0.$$

(15) If BC, CA, AB be three chords of a conicoid, each of which subtends a right angle at a fixed point S, the plane ABC will touch a prolate conicoid of revolution, of which S is a focus, and the polar plane of S the corresponding directrix plane.

(16) If three tangent planes to a cone of the second degree intersect in three straight lines VP, VQ, VR, and if P, Q, R be points such that QR, RP, PQ each subtends a right angle at a fixed point O, the plane PQR will envelope a conicoid.

CHAPTER XVI.

GENERAL THEORY OF POLARS AND TANGENT LINES.

354. THE methods which we shall adopt in the discussion of Polars of Surfaces represented by rational algebraical equations, are those employed by Joachimstal and Cayley in a discussion concerning tangents in Crelle's *Journal*, Vol. XXXIV., and by Salmon, in the *Quarterly Journal of Mathematics*, on tangent lines to surfaces.

355. *Method of determining the position of any point in a straight line, with reference to two other known points in the line.*

Let P, P' be any two points $(\alpha, \beta, \gamma, \delta)$ and $(\alpha', \beta', \gamma', \delta')$, and R any other point in the line joining them, whose algebraical distances from P and P', estimated in the directions which correspond to a position of R between P and P', are in the ratio $\mu : \lambda$.

The co-ordinates of R will be $\dfrac{\lambda\alpha + \mu\alpha'}{\lambda + \mu}$, ... so that, when the ratio $\mu : \lambda$ is given, the position of R is completely determined, observing that when the ratio is negative, R will be in PP' produced if the ratio be numerically greater than unity, and in $P'P$ produced, if less.

356. *When a straight line is drawn through any two points, to find the positions of the points of intersection with a given surface.*

Suppose $\phi \equiv F(\alpha, \beta, \gamma, \delta) = 0$ to be the equation, in a rational homogeneous form, of a surface of the nth degree; P, P' two points whose co-ordinates are $\alpha, \beta, \gamma, \delta$, and $\alpha', \beta', \gamma', \delta'$; R any point in the straight line drawn through P, P' determined by the ratio $\mu : \lambda$, as in the preceding article.

The points of intersection with the surface are given by the equation

$$F(\lambda\alpha+\mu\alpha',\quad \lambda\beta+\mu\beta',\quad \lambda\gamma+\mu\gamma',\quad \lambda\delta+\mu\delta')=0.$$

If we expand the function by the ordinary methods, writing for operations,

$$D\equiv\alpha'\frac{d}{d\alpha}+\beta'\frac{d}{d\beta}+\gamma'\frac{d}{d\gamma}+\delta'\frac{d}{d\delta},$$

$$\text{and } D'\equiv\alpha\frac{d}{d\alpha'}+\beta\frac{d}{d\beta'}+\gamma\frac{d}{d\gamma'}+\delta\frac{d}{d\delta'},$$

the values of the ratio $\mu:\lambda$ are given by either of the equivalent equations $\lambda^n e^{\frac{\mu D}{\lambda}}\phi=0$, or $\mu^n e^{\frac{\lambda D'}{\mu}}\phi'=0$, each expansion terminating at the $n+1^{\text{th}}$ term, and the corresponding positions of the n points of intersection of the straight line and surface are thus determined.

We may observe, from the identity of the two equations, that we obtain the following identical operations,

$$\lfloor r\, D'^{n-r}=\lfloor n-r\, D^r.$$

Poles and Polars.

357. *Description of the Polars of a surface with respect to a point.*

The surfaces, which are represented by the equations

$$D\phi=0,\quad D^2\phi=0,\ \ldots\ldots\ D^{n-1}\phi=0,$$

or by the equivalent equations

$$D'^{n-1}\phi'=0,\quad D'^{n-2}\phi'=0,\ \ldots\ldots\ D'\phi'=0,$$

are called the 1st, 2nd, ... $\overline{n-1}|^{\text{th}}$ Polars of the surface $\phi=0$ with respect to the point $(\alpha',\beta',\gamma',\delta')$, which is called the Pole.

The particular Polar $D^{n-1}\phi=0$, or $D'\phi'=0$ is also called the Polar Plane, and $D^{n-2}\phi=0$, or $D'^2\phi'=0$, the Polar Conicoid of the surface.

358. When the equation of the surface is given in the common co-ordinates, as $f(x,y,z)=0$, it may be reduced to the homogeneous form by writing $\frac{x}{t}$, $\frac{y}{t}$, $\frac{z}{t}$ for x, y, z, and the

equation of the polars will be obtained by means of the operation

$$D \equiv x'\frac{d}{dx} + y'\frac{d}{dy} + z'\frac{d}{dz} + t'\frac{d}{dt},$$

and in the equations $D\phi = 0$, ... so obtained making $t = 1 = t'$. The equation $t = 0$ will then represent the plane at infinity.

If the function $f(x, y, z)$ be arranged in homogeneous functions of ascending degrees, as $u_0 + u_1 + u_2 + \ldots\ldots = 0$, this equation reduces to the homogeneous form $u_0 t^n + u_1 t^{n-1} + u_2 t^{n-2} + \ldots\ldots = 0$.

If $x' = 0$, $y' = 0$, $z' = 0$,

$D^r\phi \equiv n \cdot (n-1) \ldots (n-r+1)\, u_0 t'^r t^{n-r} + (n-1) \ldots (n-r)\, u_1 t'^r t^{n-r-1} + \ldots$

Hence, the equation of the rth polar of the origin is

$$u_0 + \frac{n-r}{n} u_1 + \frac{(n-r)(n-r-1)}{n \cdot (n-1)} u_2 + \ldots\ldots = 0,$$

that of the polar plane of the origin is

$$nu_0 + u_1 = 0,$$

and that of the polar conicoid,

$$\frac{1}{2} n \cdot (n-1)\, u_0 + (n-1)\, u_1 + u_2 = 0.$$

If $n = 2$, the result of Art. (318) for the polar plane of the origin is obtained with respect to a conicoid, to which the polar conicoid reduces.

359. *Geometrical properties of Polars.*

If ρ, ρ_m be the distances of P and R_m from P',

$$\mu_m : \lambda_m :: \rho - \rho_m : \rho_m,$$

$D^r\phi' = 0$ is the locus of a point P, such that, if $\frac{\mu_1}{\lambda_1}, \frac{\mu_2}{\lambda_2}, \ldots\ldots$ be the n values of $\frac{\mu}{\lambda}$ corresponding to the intersections $R_1, R_2 \ldots$ of a line $P'P$ with the surface $\phi = 0$,

$$\Sigma\left(\frac{\mu_1 \mu_2 \ldots\ldots \mu_r}{\lambda_1 \lambda_2 \ldots\ldots \lambda_r}\right) = 0.$$

Hence, the $(n-r)^{\text{th}}$ polar is the locus of P, such that

$$\Sigma\left\{\left(\frac{1}{\rho} - \frac{1}{\rho_1}\right)\left(\frac{1}{\rho} - \frac{1}{\rho_2}\right) \ldots\ldots \left(\frac{1}{\rho} - \frac{1}{\rho_r}\right)\right\} = 0.$$

Thus, the polar plane with respect to P' is the locus of R, such that $\Sigma\left(\frac{1}{\rho}-\frac{1}{\rho_r}\right)=0$,

$$\text{or } \frac{n}{\rho}=\frac{1}{\rho_1}+\frac{1}{\rho_2}+\ldots\ldots+\frac{1}{\rho_n},$$

which gives the well-known harmonic property of the polar of a conicoid with respect to a point.

360. *Connexion between Diameters and Polar Surfaces.*

When the point, with respect to which the polars are taken, is at an infinite distance, the condition for a polar plane becomes $\Sigma(\rho-\rho_1)=0$, or $\Sigma(PR)=0$; therefore, the polar plane is a polar diametral plane corresponding to a system of parallel chords drawn in the direction of the infinitely distant point.

The condition for the polar conicoid becomes

$$\Sigma(\rho-\rho_1)(\rho-\rho_2)=0, \quad \text{or } \Sigma(PR_1 . PR_2)=0,$$

and the conicoid is a polar diametral conicoid for a system of parallel chords.

And generally, the polar surface of any degree with respect to a point at an infinite distance in a given direction is the polar diametral surface of the same degree corresponding to a system of chords in that direction.

361. *If tangent planes be drawn to a surface at the points in which a straight line meets it, and from any point Q in this line any other straight line be drawn meeting the surface in P_1, P_2... and the tangent planes in p_1, p_2......*

$$\Sigma\left(\frac{1}{QP}\right)=\Sigma\left(\frac{1}{Qp}\right).$$

If three straight lines be drawn through a point Q, meeting two surfaces in the same points R_1, R_2, ... S_1, S_2, ... and T_1, T_2, ... R, S, T the corresponding points in the polar planes with respect to Q are the same for the two surfaces; hence, the polar planes for these surfaces are the same. This is true when the three straight lines become ultimately coincident, in which case the surfaces touch one another at R_1, R_2, ...

Suppose now a straight line QR to meet a surface in R_1, R_2, ... and at these points tangent planes to be drawn, the surface and the system of tangent planes form two surfaces which have the same polar plane.

Hence, if any other line through Q meet the surface in P_1, P_2 ...... and the tangent planes in p_1, p_2 ......

$$\Sigma\left(\frac{1}{QP}\right) = \Sigma\left(\frac{1}{Qp}\right).$$

Properties of Polars.

362. *Every polar of a surface, with respect to a given pole, is a polar, with respect to the same pole, of every polar of a higher degree than its own.*

For $$D^{k+l}\phi = D^k(D^l\phi).$$

363. *Every line, drawn through a pole to a point in the curve of intersection of the first polar with the surface, meets the surface in two coincident points.*

For the equation $\lambda^n e^{\frac{\mu D}{\lambda}}\phi = 0$ has two values of μ equal to zero if $\phi = 0$, and $D\phi = 0$ simultaneously.

364. *If a surface have a multiple point of the mth degree, that point will be a multiple point of the $\overline{m-1}|^{\text{th}}$ degree on the first polar, with respect to any point not on the surface.*

Let P' be the pole, R the multiple point, P any point in $P'R$; m values of $\mu : \lambda$, in the equation $\lambda^n e^{\frac{\mu D}{\lambda}}\phi = 0$, corresponding to the multiple point R, are equal; hence, the equation

$$\frac{d}{d\mu}\lambda^n e^{\frac{\mu D}{\lambda}}\phi = 0, \text{ or } \lambda^{n-1} e^{\frac{\mu D}{\lambda}} D\phi = 0,$$

has $m-1$ equal values of $\mu : \lambda$; i. e. the first polar of $\phi = 0$ has a multiple point of the $\overline{m-1}|^{\text{th}}$ degree at R.

This is also obvious from the property of the polar given in Art. (359), for, if a straight line through the pole meet the surface in a multiple point of the mth degree, whose distance is ρ_1, the points in which it meets the first polar are given by the equation

$$\left(\frac{1}{\rho} - \frac{1}{\rho_1}\right)^m \left(\frac{1}{\rho} - \frac{1}{\rho_{m+1}}\right) \ldots\ldots \left(\frac{m}{\frac{1}{\rho} - \frac{1}{\rho_1}} + \ldots\ldots\right) = 0,$$

which has $m-1$ roots equal to ρ_1, hence the multiple point in the first polar is of the $\overline{m-1}|^{\text{th}}$ degree.

The rth polar has a multiple point at R of the $\overline{m-r}|^{\text{th}}$ degree.

365. *If a tangent cone on a double point of a surface becomes two non-coincident tangent planes, the first polar touches the line of intersection at the double point.*

In this case, if P' be the pole, P any point in the plane through P' and the line of intersection, there are two coincident positions of $P'P$, such that for *each* position $\lambda^n e^{\frac{\mu D}{\lambda}} \phi = 0$ gives two equal values of the ratio $\mu : \lambda$; and therefore one of these values satisfies the equation $\lambda^{n-1} e^{\frac{\mu D}{\lambda}} D\phi = 0$ for each of two coincident positions of $P'P$; that is, two coincident points in the line of intersection on the double point lie in the first polar, or the line of intersection is a tangent to the first polar.

Or, if ρ_1 be the distance of the double point from the pole, since there are two equal values of ρ_1, one value of ρ in the first polar is ρ_1, and, if the double point has two non-coincident tangent planes, the line of intersection is a tangent line; therefore, for two coincident directions the same is true, and $\rho = \rho_1$ for two points, coincident with the double point, in the line of intersection, which therefore touches the first polar.

366. *If the two tangent planes at a double point are coincident, the first polar has a tangent plane at that point coincident with them.*

For $P'P$ intersects the surface in two coincident points for any direction indefinitely near the multiple point; hence, the first polar has a point in the plane coincident with them, not only at the multiple point, but at the adjacent points; the plane is therefore a tangent plane to the first polar.

367. *If r generating lines of a conical tangent coincide, $r-1$ of the conical tangent of the first polar will also coincide.*

For, $P'P$, passing near the multiple generating line through r ultimately coincident points, will pass through $r-1$ ultimately coincident points of the first polar.

368. *If a surface have a multiple line of the mth degree, the first polar contains the same line as a multiple line of the $\overline{m-1}|^{\text{th}}$ degree.*

For if P be any point on the multiple line, $P'P$ has m equal values for the surface, and therefore $m-1$ for the first polar.

369. The propositions of Articles 364, 365, and 366, can be shewn directly as follows.

Let the angle D of the tetrahedron be taken for the multiple point and the angle A for the point Q with respect to which the polar is taken.

The equation of the surface will be

$$\phi \equiv \phi_m \delta^{n-m} + \phi_{m+1} \delta^{n-m-1} \ldots\ldots + \phi_n.$$

I. $\phi_m = 0$ is the conical tangent, the first polar corresponding to $(p_1, 0, 0, 0)$ is

$$\Delta\phi \equiv p_1 \frac{d\phi}{d\alpha} \equiv p_1 \frac{d\phi_m}{d\alpha} \delta^{n-m} + \ldots\ldots = 0,$$

and the conical tangent is $p_1 \dfrac{d\phi_m}{d\alpha} = 0$,

which shews that D is a multiple point of the $\overline{m-1}|^{\text{th}}$ degree.

II. If the multiple point have two non-coincident tangent planes, $m = 2$, and $\phi_2 = \psi_1 \chi_1$,

$$\frac{d\phi_2}{dx} = \psi_1 \frac{d\chi_1}{d\alpha} + \chi_1 \frac{d\psi_1}{d\alpha};$$

$\therefore\ \psi_1 = 0,\ \chi_1 = 0$ is a tangent line to the first polar.

III. If the planes be coincident $\phi_2 = \psi_1^2$,

$$\frac{d\phi_2}{d\alpha} = 2\psi_1 \frac{d\psi_1}{dx};$$

$\therefore\ \psi_1 = 0$ is a tangent plane to the first polar.

370. *If the pole be on the surface, the polar plane will be a tangent plane, at the pole, to the surface, and also to all the corresponding polars.*

For $D'\phi' = 0$ is the equation of the tangent plane at

$$(\alpha', \beta', \gamma', \delta').$$

Also, since the polar plane is the polar plane for the rth polar, in which the pole evidently lies, the polar plane is also a tangent plane at the pole, to the rth polar as well as to the surface.

This is easily seen also from the equation of the surface, $u_1 + u_2 + \ldots\ldots = 0$, in which the origin, a point on the surface, is taken for the pole, since the equation of the rth polar is

$$\left(1 - \frac{r}{n}\right) u_1 + \left(1 - \frac{r}{n}\right)\left(1 - \frac{r}{n-1}\right) u_2 + \ldots\ldots = 0.$$

$u_1 = 0$, the equation of the polar plane, is also the equation of the tangent plane to the surface and r^{th} polar.

371. *The locus of poles, whose polar planes pass through a given point, is the first polar with respect to that point.*

The polar plane of P' is $D'\phi' = 0$. If this plane pass through P_1, we have the equation $\left(\alpha_1 \frac{d}{d\alpha'} + \ldots\right)\phi' = 0$, therefore P' must lie in the surface, whose equation is $\left(\alpha_1 \frac{d}{d\alpha} + \ldots\right)\phi = 0$, which is the first polar with respect to P_1.

372. *Every plane is a polar plane corresponding to $(n-1)^3$ poles.*

Take three arbitrary points P_1, P_2, P_3, in the plane, the first polars of these points are of the $\overline{n-1}|^{\text{th}}$ degree. The first polar of P_1 is the locus of all points which are poles of planes through P_1, and therefore contains all poles of the given plane; the three surfaces which are first polars of P_1, P_2, P_3, each contain the poles of the given plane, and, therefore, since every common point is the pole of the plane containing P_1, P_2, P_3, there are $\overline{n-1}|^3$ such poles.

373. *The first polars of all points in a straight line have a common curve of intersection.*

The $\overline{n-1}|^3$ poles of any plane through two of the points lie on the curve of intersection of the polars of the two points, and this curve must therefore be the locus of the poles of all such planes; any point in the line of intersection of the planes must therefore have its first polar passing through the curve of intersection of the first polars of the two points taken.

Such a curve is a *Polar Curve* corresponding to the line.

COR. 1. If two lines intersect, their polar curves lie on the first polar of the point of intersection.

COR. 2. If any number of planes pass through a point, their poles lie on the first polar of the point.

COR. 3. A tangent line to the surface touches its polar curve on the point of contact with the surface.

On the Degree of a Reciprocal Surface.

374. The properties of the polars of a point with respect to a surface have been employed by Salmon, in the *Cambridge and Dublin Journal*, Vol. II., to explain the reduction of the class of a surface or the degree of its reciprocal, in the case of certain singularities in the surface; and we give some of the theorems relating to this reduction in order to introduce the student to some method of dealing with the subject.

375. *To estimate the effect of an ordinary double point of a surface upon the class of the surface, or the degree of its reciprocals.*

The number of tangent planes which can be drawn through a given line may be found by constructing the polar curve of the line, which is the intersection of the first polars corresponding to any two points in the line; the intersections of this curve with the surface gives $n . \overline{n-1}|^2$ points, and a plane drawn through any one of these points and the given line will generally be a tangent plane to the surface, since all the lines drawn from the point in that plane will be generally tangent lines.

But (Art. 364), it is seen that, if there be an ordinary double point on the surface, the first polars of any kind pass through the double point, and therefore the polar curve of the line passes through the point; hence, the lines drawn from the double point in the plane containing it and the given line, although they meet the surface in two coincident points, are not generally tangent lines.

Two of the planes, therefore, corresponding to the two points in which the polar curve meets the surface at the double point, are not tangent planes to the surface.

The number of tangent planes is therefore diminished by two, for each ordinary double point of the surface.

376. *To estimate the effect on the class when the conical tangent reduces to two planes.*

If the tangent planes at a double point be not coincident, the first polars touch their line of intersection; hence, to the number of coincident points of intersection of the three surfaces in the ordinary case is added one, since the intersection of each polar surface with the given surface touches the line of intersection.

If the tangent planes at a double point coincide, each tangent plane contains three coincident points in the three surfaces, and the whole number, by which the class is reduced, is therefore six.

377. The surface of the third degree, whose equation is

$$\frac{a}{\alpha}+\frac{b}{\beta}+\frac{c}{\gamma}+\frac{d}{\delta}=0,$$

has four double points, one at each angle of the fundamental tetrahedron. Hence the class of surfaces of the third degree, which is in general $3\,.\,2^2 \equiv 12$, is reduced for this surface by two for each double point: the surface is therefore of the fourth class.

If we reciprocate the surface with reference to

$$\alpha^2+\beta^2+\gamma^2+\delta^2=0,$$

let $(\alpha', \beta', \gamma', \delta')$ be a point in the reciprocal surface;

$$\therefore\ \alpha'\alpha+\beta'\beta+\gamma'\gamma+\delta'\delta=0$$

is the equation of a tangent plane to the given surface, and is therefore identical with

$$\frac{a\alpha}{\alpha''^2}+\frac{b\beta}{\beta''^2}+\frac{c\gamma}{\gamma''^2}+\frac{d\delta}{\delta''^2}=0,$$

$$\text{where}\ \frac{a}{\alpha''}+\frac{b}{\beta''}+\frac{c}{\gamma''}+\frac{d}{\delta''}=0;$$

$$\therefore \frac{\alpha'\alpha''^2}{a} = \frac{\beta'\beta''^2}{b} = \frac{\gamma'\gamma''^2}{c} = \frac{\delta'\delta''^2}{d};$$

$$\therefore (a\alpha')^{\frac{1}{3}} + (b\beta')^{\frac{1}{3}} + (c\gamma')^{\frac{1}{3}} + (d\delta')^{\frac{1}{3}} = 0$$

is the equation of the reciprocal surface, which is of the fourth degree, which is therefore the class of the surface as reduced above.

378. The surface $l\alpha\beta\gamma = \delta^3$ has double points at A, B, C, and the tangent surface at A reduces to two tangent planes, $\beta = 0$ and $\gamma = 0$, the class is therefore reduced by these for each double point, and the degree of the reciprocal will be

$$3 \,.\, 2^2 - 3 \,.\, 3 \equiv 3.$$

379. The wave surface, whose equation is

$$(a^2x^2 + b^2y + \delta z^2)(x^2 + y^2 + z^2) - a^2(b^2 + c^2)x^2 - b^2(c^2 + a^2)y^2$$
$$- c^2(a^2 + b^2)z^2 + a^2b^2c^2 = 0,$$

has four double points, real or imaginary, in each principal plane, and if we write $\frac{x'}{u}$ for x_1, $\frac{y'}{u}$ for y, &c. the symmetrical form shews that there are also four in the plane at infinity; hence there are sixteen double points, and the degree of the reciprocal surface will be $4 \,.\, 3^2 - 16 \,.\, 2 \equiv 4$.

380. *To estimate the effect of a double straight line in a surface.*

The polars contain the line singly, and the number of points which correspond to the multiple line which is common to the three surfaces is $5n - 8$ (Art. 272).

Now if in tetrahedral co-ordinates CD be taken for the multiple line, the equation of the surface may be written in the form $P\alpha^2 + 2Q\alpha\beta + R\beta^2 = 0$, in which P, Q, R are of the $\overline{n-2}|^{\text{th}}$ degree. There will therefore be certain points at which the tangent planes will be coincident, which will be determined by the intersection of the surface $PR = Q^2$ with the straight line, the number of points being $2(n-2)$.

Now at each of these points there will be an additional point common to the surfaces, and the whole number by which the class of the surface will be diminished will be $7n - 12$.

381. *To estimate the effect of a multiple line of the rth degree of multiplicity in the surface.*

The polars contain the line, each in the $\overline{r-1}|^{th}$ degree of multiplicity: and the number of points which correspond to the common multiple lines is (Art. 272)

$$(r-1)^2 n + 2r . (r-1)(n-1) - 2r . \overline{r-1}|^2$$

$$\equiv (r-1)\{(3r-1)n - 2r^2\}.$$

Now, if the line be taken for one of the edges, as CD, of the fundamental tetrahedron, the equation of the surface may be written in the form $F_r(\alpha, \beta) \equiv P\alpha^r + Q\alpha^{r-1}\beta + ... = 0$, in which the coefficients are of the $(n-r)^{th}$ degree.

The equation of the tangent planes to the surface at any point $(0, 0, \gamma', \delta')$ will be $P'\alpha^r + Q'\alpha^{r-1}\beta + ... = 0$, where P', Q'... are the values of P, Q... when $0, 0, \gamma', \delta$ have been substituted for $\alpha, \beta, \gamma, \delta$.

Now the points in CD at which there are two coincident tangent planes will be obtained by eliminating α, β between the equations

$$\frac{d}{d\alpha}F_r(\alpha, \beta) = 0, \quad \text{and } \frac{d}{d\beta}F_r(\alpha, \beta) = 0,$$

and the eliminant will be of the degree $r-1$ in each of the coefficients.

The degree of the resulting equation in γ', δ', will therefore be $2(r-1)(n-r)$.

And since the polars touch the line at each of these points, $2(r-1)(n-r)$ additional points will lie in the multiple line.

Hence the total number of points corresponding to the line, each of which is a point which gives an improper tangent plane, is

$$(r-1)\{3(r-1)n - 2r^2 + 2(n-r)\}\ .$$

$$\equiv (r-1)\{(3r+1)n - 2r(r+1)\}.$$

This is therefore the number by which the degree $n . \overline{n-1}|^2$ is reduced in the reciprocal surface.

Thus, if a surface contain a multiple line of the $\overline{n-1}|^{th}$ degree, which must be a straight line, the degree of the reciprocal surface will be

$$n \,.\, \overline{n-1}|^2 - (n-2)\{(3n-2)\,n - 2\,(n-1)\,n\}$$
$$\equiv n\,\{\overline{n-1}|^2 - (n-2)\,n\} \equiv n,$$

or the reciprocal surface is of the same degree as the original surface.

382. As an example of such a reduction, we will take the surface whose equation is $\alpha\beta^{n-1} - \gamma\delta^{n-1} = 0$, in which a straight line is of the $\overline{n-1}|^{\text{th}}$ degree of multiplicity, and the reciprocal surface will be found to be of the nth degree.

383. *To estimate the effect of a line of the rth degree of multiplicity, the line being the intersection of two surfaces of the kth and l^{th} degrees, $K=0$, and $L=0$.*

The number of points which correspond to the multiple line is (Art. 273)

$$lk\,\{\overline{nr-1}|^2 + 2\,(n-1)\,r\,.\,(r-1) - (k+l)\,r\,.\,\overline{r-1}|^2\}$$

The number of coincident tangent planes is obtained from the equation $F_r(K, L) = 0$, whose coefficients are of the degree necessary to make each term of the nth degree.

The eliminant of

$$\frac{d}{dK} F_r(K, L) = 0, \text{ and } \frac{d}{dL} F_r(K, L) = 0,$$

is of the degree $(n-kr)\,(r-1) + (n-lr)\,(r-1)$

$$\equiv (r-1)\,\{2n - (k+l)\,r\},$$

which gives a surface meeting the line in points whose number is

$$(r-1)\,\{2n - (k+l)\,r\}\,kl.$$

The degree of the reciprocal is therefore reduced by

$$kl\,(r-1)\,[n\,.\,(r-1) + 2\,(n-1)\,r + 2n - (k+l)\,\{r\,(r-1)+r\}]$$
$$\equiv kl\,(r-1)\,\{(3r+1)\,n - 2r + (k+l)\,r^2\}.$$

On the Relation of Straight Lines to Surfaces.

384. Since the methods employed in this chapter afford peculiar facilities in the examination of the positions of straight lines satisfying certain conditions relative to a given surface and

its singular points, we shall here follow Salmon in his application of them to the contact of straight lines with surfaces, in the *Quarterly Journal*, Vol. I. page 329, repeating some propositions which have been already discussed.

385. *To find the condition that a straight line may touch a surface at a given point.*

Let P', $(\alpha', \beta', \gamma', \delta')$ be the given point, P, $(\alpha, \beta, \gamma, \delta)$ any point in space, and R, $\left(\frac{\lambda\alpha+\mu\alpha'}{\lambda+\mu}, \ldots\right)$ any point in PP', and $\phi=0$ the equation of the surface.

The values of $\lambda : \mu$ for the positions of R in which PP' meets the surface are given by the equation $\mu^n e^{\frac{\lambda D'}{\mu}}=0$, and if PP' meet the surface in two points which coincide with P', two values of λ are zero, or $\phi'=0$, and $D'\phi'=0$.

These are *necessary* conditions that PP' should touch at P', and unless $D'\phi'=0$ is satisfied for all values of $\alpha, \beta, \gamma, \delta$, i.e. unless $\frac{d\phi'}{d\alpha'}, \frac{d\phi'}{d\beta'}, \ldots$ are all zero, the locus of P such that PP' is a tangent line is $\left(\alpha\frac{d}{d\alpha'}+\ldots\right)\phi'=0$, which is the equation of the tangent plane at P'.

$D'\phi'=0$ is not a *sufficient* condition for tangency if the differential coefficients are zero, for in that case PP' meets the surface in two coincident points for all positions of P, or P' is a multiple point, in this case P may be determined so that PP' meets the surface in three coincident points if its co-ordinates satisfy the equation $D'^2\phi'=0$, and unless this equation be satisfied for all values of $\alpha, \beta, \gamma, \delta$ or $\frac{d^2\phi'}{d\alpha'^2}, \ldots \frac{d^2\phi'}{d\alpha' d\beta'}, \ldots$ are all zero,

$$\left(\alpha\frac{d}{d\alpha'}+\ldots\right)^2\phi'=0$$

is the locus of such positions of P', and is the equation of the tangent cone at the double point.

The argument is easily continued in the case of triple ... r^{ple} singular points.

386. *To find the locus of the points of contact of all tangent lines which can be drawn from a given point to a surface.*

Let P' be the given point, P the point of contact in a tangent line to the surface, drawn through P'; in this case two positions of R at least coincide in P, and the equation $\lambda^n e^{\frac{\mu D}{\lambda}} \phi = 0$ has two values of $\mu : \lambda = 0$, i.e. $\phi = 0$ and $D\phi = 0$, the intersection of the two surfaces represented by these equations is the locus of the points of contact including the singular points of the surface which may lie in the curve of intersection for which $P'P$ is not a proper tangent. Hence, the locus is the intersection of the surface and its first polar with respect to the given point.

387. *To find the tangent lines at an ordinary point of a surface, which meet the surface in three coincident points on the point of contact.*

Let P' be the point of contact, P any point in such a tangent line, $\mu^n e^{\frac{\lambda D'}{\mu}} \phi' = 0$ must have three values of $\lambda = 0$, hence $\phi' = 0$, $D'\phi' = 0$, and $D'^2\phi' = 0$. The intersections of the surface $D'^2\phi' = 0$, with the tangent plane $D'\phi' = 0$, give the positions of the two tangent lines required, which are obviously the tangents to the section made by the tangent plane.

388. If the surface be of a higher degree than the second and if $D'^2\phi' = 0$ is identically satisfied at P' for all values of $\alpha, \beta, \gamma, \delta$, we can obtain three straight lines which meet the surface in four coincident points, viz. the intersection of the cone $D'^3\phi' = 0$ with the tangent plane $D'\phi' = 0$. And so on to the general case.

389. If the equation $D'\phi' = 0$ be satisfied identically, or the point P' be a singular point, the tangent lines which meet the surface in four coincident points are given by equations $D'^2\phi' = 0$, and $D'^3\phi' = 0$, and these are two conical surfaces, if they be not also identically satisfied, the first being the conical tangent to the double point and the second determining the six particular generating lines of the tangent cone which satisfy the required condition.

If the singular point have a conical tangent of the rth degree, the number of tangent lines meeting the curve in $r+2$ coincident points is $r(r+1)$.

390. *To find the equation of the conical surface, whose vertex is a given point, which envelopes a given surface.*

Let P' be the given point, P any point in any tangent line, the equation $\lambda^n e^{\frac{\mu D}{\lambda}} \phi = 0$ must contain equal values of $\lambda : \mu$, hence the equations

$$\lambda^n e^{\frac{\mu D}{\lambda}} \phi = 0 \text{ and } \lambda^{n-1} e^{\frac{\mu D}{\lambda}} D\phi = 0$$

must have a common root.

The eliminant of these equations is the equation of the locus of P and is the equation of the envelope.

If the surface be of the second order,

$$\lambda^2\phi + \lambda\mu D\phi + \tfrac{1}{2}\mu^2 D^2\phi = 0 \text{ and } (\lambda + \mu D)\, D\phi = 0$$

have their roots equal, or the equation of the envelope is

$$2\phi D^2\phi = (D\phi)^2, \text{ or } 4\phi\phi' = (D\phi)^2.$$

391. *To find the number of tangents which can be drawn from a given point, to meet a surface in three consecutive points.*

Let P' be the given point, then, if three positions of R coincide in P, $\lambda^n e^{\frac{\mu D}{\lambda}} \phi = 0$ must have three values of the ratio $\mu : \lambda = 0$; the conditions for this are $\phi = 0$, $D\phi = 0$, $D^2\phi = 0$, and the surfaces represented by these equations, viz. the given surface and its first and second polar with respect to the given point, intersect in $n(n-1)(n-2)$ points, or $n(n-1)(n-2)$ tangents can be drawn to touch in three consecutive points.

A point of contact of this kind corresponds to a cuspidal edge of the conical envelope.

392. *To find the number of tangent lines which can be drawn to a surface at a given point, so as to touch at another point as well.*

Let P' be the given point, P the other point of contact.

In order that the line may touch at P', two values of the ratio $\lambda : \mu$ in $\mu^n e^{\frac{\lambda D'}{\mu}} \phi' = 0$ are zero, hence $\phi' = 0$, and $D'\phi' = 0$; the equation which contains the remaining values of $\lambda : \mu$ is

$$u = \tfrac{1}{2}\mu^{n-2} D'^2\phi' + \ldots + \lambda^{n-2}\phi = 0,$$

and since two of these values are equal, $\frac{du}{d\lambda}=0$ and $\frac{du}{d\mu}=0$ have a common root, the eliminant V of these equations will be of the same order as $\phi^{n-3}.(D'^2\phi')^{n-3}$, or $(n+2)(n-3)$, and this is therefore the number of double tangents which can be drawn touching at a given point. The surface represented by this eliminant, $V=0$, intersects the tangent plane in the straight lines, which are the tangent lines having the double contact.

393. If the point P' be a multiple point of the rth degree, in order that the line $P'P$ shall touch the surface at P', $r+1$ values of $\lambda:\mu$ must be zero, and the equation which gives the remaining values is

$$u \equiv \frac{1}{\lfloor r+1}\mu^{n-r-1}D'^{r+1}\phi' + \ldots + \lambda^{n-r-1}\phi = 0,$$

the eliminant V of $\frac{du}{d\lambda}=0$ and $\frac{du}{d\mu}=0$ is of the same degree as

$(D'^{r+1}\phi')^{n-r-2}\phi^{n-r-2}$ or of the $(n+r+1)(n-r-2)^{th}$ degree, or of the

$$\{n(n-1)-(r+1)(r+2)\}^{th} \text{ degree.}$$

The intersection of $V=0$ with $D'^r\phi=0$ gives all such tangent lines.

394. *To find the locus of tangents which can be drawn from a singular point to a surface.*

Let P' be the singular point, suppose of the rth degree, r of the values of $\lambda:\mu$ in $\mu^n e^{\frac{\lambda D'}{\mu}}\phi'=0$ are equal to zero,

$$\therefore\ \phi'=0,\ D'\phi'=0,\ \ldots\ D'^{r-1}=0,$$

and the equation which gives the remaining values is

$$u \equiv \frac{1}{\lfloor r}\mu^{n-r}D'^r\phi' + \ldots + \lambda^{n-r}\phi = 0,$$

and, if the line $P'P$ is a tangent, two of the roots are equal, and $\frac{du}{d\lambda}=0$, $\frac{du}{d\mu}=0$, have a common root, and the eliminant V is of the degree of $(D'^r\phi')^{n-r-1}\phi^{n-r-1}$, which is $(n+r)(n-r-1)$.

$V=0$ is the equation of a surface containing all tangent lines through P' to the surface.

395. *To find the number of double tangent lines which can be drawn from a fixed point not in the surface.*

Take P' the fixed point, P one of the points of contact; then two positions of R coincide in P, and two other positions coincide, therefore from the equation $\lambda^n e^{\frac{\mu D}{\lambda}} \phi = 0$ we obtain $\phi = 0$ and $D\phi = 0$, also $u \equiv \frac{1}{2}\lambda^{n-2}D^2\phi + \ldots\ldots + \mu^{n-2}\phi' = 0$ contains equal roots; hence V the eliminant of $\frac{du}{d\lambda} = 0$ and $\frac{du}{d\mu} = 0$ is of the degree of $(D^2\phi)^{n-3}$, or of the $(n-2)(n-3)$ degree, and the number of solutions of $V=0$, $\phi = 0$, $D\phi = 0$, is double the number required, since each double tangent corresponds to two, hence

$$\frac{1}{2}n(n-1)(n-2)(n-3)$$

is the number of double tangents.

This is the number of double sides of the conical envelope whose vertex is P'.

Since our object is only to introduce the student to the methods which have been found to offer the greatest facilities in dealing with tangent lines to surfaces subject to given conditions, we forbear to follow out the whole of the problems relating to this subject so ably given by Salmon.

We shall therefore confine ourselves to a reference to the article by Salmon in which the subject is very nearly exhausted.

XVII.

1. The polars of any order, of all surfaces of the nth degree passing through $\phi(n) - 1$ given arbitrary points, have a common curve of intersection.

2. The polars of any order r, of all surfaces of the nth degree passing through $\phi(n) - 2$ points, have $(n-r)^3$ common points.

3. Prove that the surface reciprocal to the surface whose equation is $(x^2+y^2+z^2)^2 = a^2(x^2+y^2)$ is of the fourth degree, and explain the reduction of class.

4. Prove that the reciprocal surface of the surface whose equation is

$$\frac{l}{\alpha^n}+\frac{m}{\beta^n}+\frac{n}{\gamma^n}+\frac{r}{\delta^n}=0$$

is of the degree $n\,\overline{n+1}|^2$.

5. If P, Q be two points, P_{n-1}, Q_{n-1} their first polars with respect to S, prove that the first polars of P with respect to Q_{n-1} and of Q with respect to P_{n-1} are the same surface.

6. Prove also, that the xth polars of P with respect to Q_y and the yth of Q with respect to P_x are the same surface.

7. If P_x has a double point Q, Q_{n-x+1} has a double point P.

8. If the polar conicoid of

$$\frac{l}{\alpha}+\frac{m}{\beta}+\frac{n}{\gamma}+\frac{r}{\delta}=0,$$

with respect to P, resolve into two planes, there are four positions of P given by equations similar to

$$\frac{\alpha}{-l}=\frac{\beta}{m}=\frac{\gamma}{n}=\frac{\delta}{r}.$$

The corresponding conicoid is the planes

$$\alpha=0, \text{ and } \frac{\beta}{m}+\frac{\gamma}{n}+\frac{\delta}{r}=0,$$

and the plane polar is

$$\frac{3\alpha}{l}-\frac{\beta}{m}-\frac{\gamma}{n}-\frac{\delta}{r}=0.$$

9. The conditions that the first polar of

$$\frac{l}{\alpha}+\frac{m}{\beta}+\frac{n}{\gamma}+\frac{r}{\delta}=0,$$

with respect to $(\alpha', \beta', \gamma', \delta')$, may be a sphere, are

$$a^2\,mn+a'^2\,lr=b^2\,ln+b'^2\,mr=c^2\,lm+c'^2\,nr\equiv\rho;$$

and the locus of all such poles, corresponding to all surfaces of this form, is the curve

$$a^2\beta\gamma+a'^2\alpha\delta=b^2\gamma\alpha+b'^2\beta\delta=c^2\alpha\beta+c'^2\gamma\delta.$$

For a particular surface

$$\frac{\alpha}{\dfrac{c'^2}{m}+\dfrac{b'^2}{n}+\dfrac{a^2}{r}-\dfrac{\rho}{mnr}}=\ldots\ldots\ldots\ldots\ldots\ldots$$

CHAPTER XVII.

FUNCTIONAL AND DIFFERENTIAL EQUATIONS OF FAMILIES OF SURFACES.

396. In finding the equation of a surface generated by the motion of a curve of given species, it is obvious that the curve must satisfy such a number of conditions as will enable us, on expressing them analytically, to eliminate from the equations all the constants which distinguish the curve in any particular position, and to obtain a final equation involving only the current co-ordinates, and constants which are the same for all positions of the curve. Thus, if the equations of the curve involve n parameters, which vary with the position of the curve, we shall require $(n-1)$ fixed and independent conditions, to be satisfied by each curve, in order that the locus of the curve may be a surface. Then $(n-1)$ equations, expressing these conditions, together with the equations of the curve, will give $(n+1)$ independent equations satisfied by the co-ordinates of any point of the curve, from which the n parameters may be eliminated, and a final equation obtained, which is the equation of the locus of the curve.

397. If, however, a less number of conditions be given, although there is in this case no determinate locus, we may find a functional, or a partial differential equation, which must be satisfied by all the different surfaces generated by curves satisfying these conditions. Thus, if $(n-2)$ conditions be given, we may eliminate any $(n-1)$ of the n parameters, obtaining an equation of the form $u=\alpha$, where u is a determinate function of x, y, z, and such constants as do not depend on the position of the curve, and α is one of the parameters. Similarly eliminating other $(n-1)$, we may obtain an equation $v=\beta$. The equations

$$u=\alpha, \quad v=\beta,$$

may then be considered the equations of the generating curve, and, in order that this may generate a determinate surface, another condition will be necessary, which will give a relation between the parameters of the form $\beta = f(\alpha)$.

Hence, the equation of any surface so generated must be of the form

$$v = f(u),$$

where u, v are determinate functions of x, y, z, and constants, and f denotes an arbitrary function.

Eliminating the arbitrary function, we obtain a linear partial differential equation of the first order; and either of these equations may be considered as the general equation of a family of surfaces, generated by a curve of given species, fulfilling a number of conditions insufficient to determine a locus, but such that if any new condition be imposed, a particular surface belonging to this family may be found.

Again, if only $(n-3)$ be given, we may obtain two final equations of the form

$$\beta = f_1(x, y, z, \alpha); \quad \gamma = f_2(x, y, z, \alpha);$$

f_1, f_2 denoting known functions, and α, β, γ being parameters. For a determinate surface, two new independent conditions will be necessary, which will supply two equations of the form

$$\beta = \phi(\alpha), \quad \gamma = \psi(\alpha).$$

The general equation of this family of surfaces will then be found, if we eliminate α from the equations

$$\phi(\alpha) = f_1(x, y, z, \alpha); \quad \psi(\alpha) = f_2(x, y, z, \alpha).$$

This is not generally possible, ϕ, ψ being, for the family, arbitrary, and determinate only for a particular surface. In certain cases, however, it happens, that on eliminating $(n-2)$ of the constants, one of the remaining two also disappears, leaving an equation of the form $u = \alpha$, as in the former case. When this occurs, we may also obtain an equation of the form

$$f(x, y, z, \beta, \gamma) = 0,$$

where f is determinate. The general equation of the family of surfaces will then be

$$f\{x, y, z, \phi(u), \psi(u)\} = 0,$$

involving two arbitrary functions, from which may be obtained a partial differential equation not involving the functions. This differential equation will generally be of the third order, but occasionally of the second.

Similarly, if $(n-4)$ conditions be given, and if the equation be such that it is possible to deduce one equation of the form $u=\alpha$, we may obtain a general equation for the family of surfaces so generated which will involve three arbitrary functions, and so on for fewer given conditions.

398. *On the functional equations of families of Ruled Surfaces.*

The equations of a straight line involve four parameters; hence, if a straight line moves so as to satisfy three conditions, each condition being such as to give rise to one independent relation among the parameters, a determinate locus will be generated. Intersection with a given curve, or tangency of a given surface, are examples of such conditions.

If a straight line move so as to satisfy only two such conditions, we shall be able to obtain a functional or differential equation, which will include the whole family of surfaces so generated. Of this kind are conical surfaces with a given vertex, or cylindrical surfaces with a given direction of generating line; the condition that a straight line may pass through a given point, at a finite or infinite distance, giving rise to two relations among the parameters.

If a straight line move so as to satisfy only one such condition, the general equation of the family of surfaces generated cannot usually be obtained. If however the one condition be that it move parallel to a fixed plane, the exceptional case mentioned in the last Article arises, since this condition gives rise at once to an equation of the form $u=\alpha$, $u=0$ being the equation of the given plane.

399. *To find the general functional equation of cylindrical surfaces having their generating lines parallel to a given straight line.*

Let l, m, n be proportional to the direction-cosines of the given straight line.

Then the equations of the generating line may be written

$$\frac{x}{l} = \frac{y - \beta}{m} = \frac{z - \gamma}{n},$$

in which β, γ are the two parameters.

Hence,

$$l\beta = ly - mx, \quad l\gamma = lz - nx,$$

and the general functional equation will be

$$ly - mx = f(lz - nx),$$

or

$$F(ly - mx, \quad lz - nx) = 0.$$

400. *To find the general functional equation of conical surfaces having a given vertex.*

Let (a, b, c) be the given vertex, then the equations of the generating line may be written

$$\frac{x - a}{l} = \frac{y - b}{m} = \frac{z - c}{n},$$

in which the ratios $l : m : n$ are the parameters.

The functional equation will then be

$$\frac{x - a}{y - b} = f\left(\frac{x - a}{z - c}\right),$$

or

$$F\left(\frac{x - a}{y - b}, \frac{x - a}{z - c}\right) = 0.$$

This shews that if an equation $u = 0$ represent a cone, the first member of the equation may, by a change of origin, be reduced to a homogeneous function of x, y, z.

401. *To find the general functional equation of conoidal surfaces, having a given axis and a given directing plane.*

DEF. A conoidal surface, or conoid, is any surface generated by a straight line moving so as to intersect a given straight line, the *axis*, and to remain parallel to a given plane, the *directing plane*. If the axis be perpendicular to the directing plane, the surface is called a *right conoid*.

Let the equations of the axis be

$$\frac{x - a}{l} = \frac{y - b}{m} = \frac{z - c}{n} = r,$$

and (l', m', n') the direction-cosines of the directing plane: also, let the equations of the generating line be

$$\frac{x-a-lr}{\lambda}=\frac{y-b-mr}{\mu}=\frac{z-c-nr}{\nu}.$$

The equations will then involve the three parameters r, and the ratios $\lambda:\mu:\nu$, connected by the equation

$$l'\lambda+m'\mu+n'\nu=0.$$

We shall then have the equations

$$l'(x-a)+m'(y-b)+n'(z-c)=(ll'+mm'+nn')\,r,$$

and
$$\frac{m(x-a)-l(y-b)}{n(x-a)-l(z-c)}=\frac{m\lambda-l\mu}{n\lambda-l\nu}.$$

The general functional equation will accordingly be

$$l'(x-a)+m'(y-b)+n'(z-c)=F\left\{\frac{m(x-a)-l(y-b)}{n(x-a)-l(z-c)}\right\}.$$

If the axis be taken as the axis of x, and the directing plane as that of yz, this equation reduces to

$$x=F\left(\frac{y}{z}\right).$$

The general equation of a right conoid is of the form

$$l(x-a)+m(y-b)+n(z-c)=F\left\{\frac{m(x-a)-l(y-b)}{n(x-a)-l(z-c)}\right\}.$$

402. *To find the general functional equation of surfaces generated by a straight line moving parallel to a fixed plane.*

Let the equations of the generating line in any position be

$$\frac{x-\alpha}{\lambda}=\frac{y-\beta}{\mu}=\frac{z-\gamma}{\nu},$$

and let l, m, n be the direction-cosines of the fixed plane.

We shall then have

$$l\lambda+m\mu+n\nu=0,$$

and therefore

$$lx+my+nz=l\alpha+m\beta+n\gamma,$$

for any point on the line.

Also we may write one of the equations of the line in the form

$$z = \frac{\nu}{\lambda} x + \gamma - \frac{\nu}{\lambda} \alpha,$$

which shews that the general functional equation is

$$z = x\phi(lx + my + nz) + \psi(lx + my + nz).$$

It may also be taken as

$$z = y\phi_1(lx + my + nz) + \psi_1(lx + my + nz).$$

These equations will be found to lead to the same differential equation, which we shall hereafter find in a different manner. They may also be seen to coincide as follows.

Taking the former

$$z = x\phi(u) + \psi(u),$$

$$lz = (u - my - nz)\phi(u) + l\psi(u),$$

$$z\{l + n\phi(u)\} = u\phi(u) + l\psi(u) - my\phi(u);$$

$$z = -\frac{my\phi(u)}{l + n\phi(u)} + \frac{u\phi(u) + l\psi(u)}{l + n\phi(u)},$$

which is of the second form.

If the given plane be that of zx, the equation reduces to

$$z = x\phi(y) + \psi(y).$$

This is the exceptional case before mentioned (Art. 397). This family of surfaces includes conoids as a particular case.

403. *To find the general functional equation of surfaces of revolution.*

These may be considered as generated by the motion of a circle whose center lies on a fixed straight line, to which its plane is perpendicular, and whose circumference meets a fixed curve in the same plane as the axis.

Let (a, b, c) be any fixed point on the axis, (l, m, n) the direction-cosines of the axis. Then with center (a, b, c) we can describe a sphere passing through the generating circle in any position, and the equations of the circle may be taken to be

$$(x - a)^2 + (y - b)^2 + (z - c)^2 = r^2,$$

$$lx + my + nz = p.$$

Hence the required functional equation will be

$$(x-a)^2+(y-b)^2+(z-c)^2=f(lx+my+nz).$$

If the axis of z be taken for the axis, the general equation becomes

$$x^2+y^2+z^2=f(z),$$

which may be written

$$x^2+y^2=\phi(z) \text{ or } z=\psi(x^2+y^2).$$

Differential Equations of Families of Surfaces.

404. From the functional equation of a family of surfaces, the arbitrary functions can be eliminated by differentiation, and a partial differential equation obtained which must be satisfied at every point of any one of the family of surfaces. In the case however of a family of ruled surfaces, satisfying given conditions, the differential equation may be obtained directly as follows.

Let $F(x, y, z)=0$ be the equation of any ruled surface, and let

$$\frac{x'-x}{\lambda}=\frac{y'-y}{\mu}=\frac{z'-z}{\nu}=r$$

be the equation of any generating line. The equation

$$F(x+\lambda r,\ y+\mu r,\ z+\nu r)=0$$

must then be identically true, and the following equations must hold at every point of the surface,

$$\left.\begin{aligned} &F(x, y, z)=0,\\ &\lambda\frac{dF}{dx}+\mu\frac{dF}{dy}+\nu\frac{dF}{dz}=0,\\ &\dots\dots\dots\dots\dots\dots\\ &\left(\lambda\frac{d}{dx}+\mu\frac{d}{dy}+\nu\frac{d}{dz}\right)^n F=0, \end{aligned}\right\}\dots\dots\dots\dots (A),$$

if the equation be of the nth degree. From these equations, if $\lambda:\mu:\nu$ be given as functions of x, y, z, we may obtain a series of differential equations, any one of which must be true at any point of the surface, and may be considered a differential equation of the surface.

405. *To find the differential equation of cylindrical surfaces, having a given direction of the generating lines.*

If l, m, n be the direction-cosines of any generating line, we must have

$$l\frac{dF}{dx}+m\frac{dF}{dy}+n\frac{dF}{dz}=0, \quad\text{..............} \quad (1),$$

and if this condition be satisfied at every point, it may readily be shewn that the other equations of the system typified by

$$\left(l\frac{d}{dx}+\dots\right)^r F=0,$$

will also be satisfied. The equation (1) is therefore the differential equation required. We may also write it in the form

$$n=l\frac{dz}{dx}+m\frac{dz}{dy}.$$

406. *To find the differential equation of conical surfaces, having a given vertex.*

If (α, β, γ) be the co-ordinates of the given vertex, we shall have

$$\frac{\alpha-x}{\lambda}=\frac{\beta-y}{\mu}=\frac{\gamma-z}{\nu},$$

and we must have, at every point of the surface,

$$(\alpha-x)\frac{dF}{dx}+(\beta-y)\frac{dF}{dy}+(\gamma-z)\frac{dF}{dz}=0,$$

the differential equation required, whence it follows, that

$$\left\{(\alpha-x)\frac{d}{dx}+\dots\right\}^r F=0;$$

the operations being performed on F alone.

The differential equation may be written

$$z-\gamma=(x-\alpha)\frac{dz}{dx}+(y-\beta)\frac{dz}{dy}.$$

407. *To find the general differential equation of conoidal surfaces, having a given axis, and a given directing plane.*

Let the equations of the axis be $\frac{x-a}{l}=\frac{y-b}{m}=\frac{z-c}{n}$, and let l', m', n' be the direction-cosines of the directing-plane, λ, μ, ν those of any generating line. We shall then have, at any point of this generating line,

$$(x-a)(m\nu-n\mu)+(y-b)(n\lambda-l\nu)+(z-c)(l\mu-m\lambda)=0;$$

and since $l'\lambda+m'\mu+n'\nu=0$, we obtain, putting

$$\rho\equiv ll'+mm'+nn'.\ \rho'\equiv l'(x-a)+m'(y-b)+n'(z-c),$$

$$\frac{\lambda}{\rho(x-a)-l\rho'}=\frac{\mu}{\rho(y-b)-m\rho'}=\frac{\nu}{\rho(z-c)-n\rho'};$$

therefore the differential equation is

$$\{\rho(x-a)-l\rho'\}\frac{dF}{dx}+\{\rho(y-b)-m\rho'\}\frac{dF}{dy}+\{\rho(z-c)-n\rho'\}\frac{dF}{dz}=0.$$

The coefficients being linear functions of x, y, z, it follows, from this equation, that the other equations of the system (A) will hold.

If the axis be the axis of z, and the directing plane the plane of xy, this equation reduces to

$$x\frac{dF}{dx}+y\frac{dF}{dy}=0,\ \text{or}\ x\frac{dz}{dx}+y\frac{dz}{dy}=0.$$

408. *To find the general differential equation of surfaces generated by a straight line moving parallel to a fixed plane.*

Let l, m, n be the direction-cosines of the fixed plane, λ, μ, ν those of any generating line; we shall then have

$$l\lambda+m\mu+n\nu=0,$$

and the differential equation satisfied by such surfaces will be found by eliminating λ, μ, ν, from this equation, and

$$\lambda\frac{dF}{dx}+\mu\frac{dF}{dy}+\nu\frac{dF}{dz}=0,$$

$$\lambda^2\frac{d^2F}{dx^2}+\ldots+2\mu\nu\frac{d^2F}{dy\,dz}+\ldots=0.$$

The equation will accordingly be

$$\left(n\frac{dF}{dy}-m\frac{dF}{dz}\right)^2\frac{d^2F}{dx^2}+\ldots$$

$$+2\left(l\frac{dF}{dy}-m\frac{dF}{dx}\right)\left(m\frac{dF}{dz}-n\frac{dF}{dy}\right)^2\frac{d^2F}{dy\,dz}+\ldots=0,$$

which may be written

$$\left(m+n\frac{dz}{dy}\right)^2\frac{d^2z}{dx^2}-2\left(m+n\frac{dz}{dy}\right)\left(l+n\frac{dz}{dx}\right)\frac{d^2z}{dx\,dy}$$

$$+\left(l+n\frac{dz}{dx}\right)^2\frac{d^2z}{dy^2}=0.$$

If the fixed plane be taken as that of xy, these equations reduce to

$$\left(\frac{dF}{dy}\right)^2\frac{d^2F}{dx^2}-2\frac{dF}{dx}\frac{dF}{dy}\frac{d^2F}{dx\,dy}+\left(\frac{dF}{dx}\right)^2\frac{d^2F}{dy^2}=0,$$

and $$\left(\frac{dz}{dy}\right)^2\frac{d^2z}{dx^2}-2\frac{dz}{dx}\frac{dz}{dy}\frac{d^2z}{dx\,dy}+\left(\frac{dz}{dx}\right)^2\frac{d^2z}{dy^2}=0.$$

The remaining equations of the system (A) may be shewn to be satisfied in this case, but with the general form of the equation the work is tedious. In the case when the equation of the surface is $z=f(x, y)$, and the fixed plane becomes that of xy, we shall have

$$\nu=0,\ \lambda\frac{dz}{dx}+\mu\frac{dz}{dy}=0,\ \lambda^2\frac{d^2z}{dx^2}+2\lambda\mu\frac{d^2z}{dx\,dy}+\mu^2\frac{d^2z}{dx^2}=0;$$

whence

$$\frac{\lambda}{\frac{dz}{dy}}=\frac{\mu}{-\frac{dz}{dx}},\ \text{and}\ \left(\frac{dz}{dy^2}\right)^2\frac{d^2z}{dx^2}-2\frac{dz}{dx}\frac{dz}{dy}\frac{d^2z}{dx\,dy}+\left(\frac{dz}{dx}\right)^2\frac{d^2z}{dy^2}=0.$$

Differentiating the last equation with respect to x and y successively, multiplying the results by $\frac{dz}{dy}$ and $\frac{dz}{dx}$ respectively, and adding, we obtain

$$\left(\frac{dz}{dy}\right)^3\frac{d^3z}{dx^3}-3\left(\frac{dz}{dy}\right)^2\frac{dz}{dx}\frac{d^3z}{dx^2dy}+3\frac{dz}{dy}\left(\frac{dz}{dx}\right)^2\frac{d^3z}{dx\,dy^2}-\left(\frac{dz}{dx}\right)^3\frac{d^3z}{dy^3}=0,$$

whence

$$\lambda^3 \frac{d^3z}{dx^3} + 3\lambda^2\mu \frac{d^3z}{dx^2dy} + 3\lambda\mu^2 \frac{d^3z}{dx\,dy^2} + \mu^3 \frac{d^3z}{dy^3} = 0;$$

and, proceeding similarly, the whole of the equations (A) will be seen to be satisfied.

409. *To find the general differential equation of developable surfaces.*

In this case, the two directions in which the tangent plane to the surface at any point meets it must coincide, or the values of $\lambda : \mu : \nu$ given by the equations

$$\lambda \frac{dF}{dx} + \mu \frac{dF}{dy} + \nu \frac{dF}{dz} = 0,$$

$$\left(\lambda \frac{d}{dx} + \mu \frac{d}{dy} + \nu \frac{d}{dz}\right)^2 F = 0,$$

must coincide.

This gives the equation

$$\left(\frac{dF}{dx}\right)^2 \left\{\frac{d^2F}{dy^2}\frac{d^2F}{dz^2} - \left(\frac{d^2F}{dy\,dz}\right)^2\right\} + \dots$$

$$+ 2\frac{dF}{dy}\frac{dF}{dz}\left\{\frac{d^2F}{dz\,dx}\frac{d^2F}{dx\,dy} - \frac{d^2F}{dx^2}\frac{d^2F}{dy\,dz}\right\}$$

$$+ \dots = 0.$$

If we take $F(x, y, z) \equiv z - f(x, y)$, this reduces to

$$\frac{d^2z}{dx^2}\frac{d^2z}{dy^2} - \left(\frac{d^2z}{dx\,dy}\right)^2 = 0.$$

Equation (A) may be shewn to hold as follows. The equation

$$\left(\lambda \frac{d}{dx} + \mu \frac{d}{dy}\right)^2 z = 0$$

has equal roots. We shall then have

$$\lambda \frac{d^2z}{dx^2} + \mu \frac{d^2z}{dx\,dy} = 0, \quad \lambda \frac{d^2z}{dx\,dy} + \mu \frac{d^2z}{dy^2} = 0,$$

simultaneously.

Differentiate each of these equations with respect to x and y successively, multiply the four resulting equations by λ^2, $\lambda\mu$, $\lambda\mu$, and μ^2, and add: the result will be

$$\left(\lambda \frac{d}{dx} + \mu \frac{d}{dy}\right)^3 z + \left(\lambda \frac{d\lambda}{dx} + \mu \frac{d\lambda}{dy}\right)\left(\lambda \frac{d^2z}{dx^2} + \mu \frac{d^2z}{dx\,dy}\right)$$
$$+ \left(\lambda \frac{d\mu}{dx} + \mu \frac{d\mu}{dy}\right)\left(\lambda \frac{d^2z}{dx\,dy} + \mu \frac{d^2z}{dy^2}\right) + \ldots = 0,$$

which reduces to

$$\left(\lambda \frac{d}{dx} + \mu \frac{d}{dy}\right)^3 z = 0,$$

by the former relations. Similarly, we may proceed to shew that the higher equations are also satisfied.

We may also obtain this equation from the general equation of a developable surface given in Art. 300. It is there shewn that the equation obtained by eliminating α from the equations

$$z = \alpha x + y\phi(\alpha) + \psi(\alpha), \quad 0 = x + y\phi'(\alpha) + \psi'(\alpha),$$

represents a developable surface.

Now at any point of this surface

$$\frac{dz}{dx} = \alpha + \frac{d\alpha}{dx}\{x + y\phi'(\alpha) + \psi'(\alpha)\} = \alpha,$$

$$\frac{dz}{dy} = \phi(\alpha) + \frac{d\alpha}{dy}\{x + y\phi(\alpha) + \psi'(\alpha)\} = \phi(\alpha).$$

Hence $$\frac{dz}{dx} = \phi\left(\frac{dz}{dy}\right),$$

and eliminating the function by differentiation, we obtain the equation

$$\frac{d^2z}{dx^2}\frac{d^2z}{dy^2} - \left(\frac{d^2z}{dx\,dy}\right)^2 = 0.$$

410. *To find the general differential equation of surfaces of revolution about a given axis.*

The general functional equation is

$$F(x, y, z) \equiv f_1\{(x-a)^2 + (y-b)^2 + (z-c)^2\} - f_2(lx + my + nz) = 0,$$

whence $$\frac{dF}{dx} = 2(x-a)f_1' - lf_2',$$

$$\frac{dF}{dy} = 2(y-b)f_1' - mf_2',$$

$$\frac{dF}{dz} = 2(z-c)f_1' - nf_2',$$

whence

$$\{m(z-c) - n(y-b)\}\frac{dF}{dx} + \{n(x-a) - l(z-c)\}\frac{dF}{dy} + \{l(y-b) - m(x-a)\}\frac{dF}{dz} = 0;$$

or $$\{m(z-c) - n(y-b)\}\frac{dz}{dx} + \{n(x-a) - l(z-c)\}\frac{dz}{dy} = l(y-b) - m(x-a).$$

411. *Application of the differential equations of families of surfaces to conicoids.*

Take $$F(x, y, z) \equiv ax^2 + by^2 + cz^2 + 2a'yz + 2b'zx + 2c'xy + 2a''x + 2b''y + 2c''z + d = 0;$$

then, by the equation of Art. (405), in order that this may be a cylinder, we must have at every point

$$l(ax + c'y + b'z + a'') + m(c'x + by + a'z + b'') + n(b'x + a'y + cz + c'') = 0,$$

for fixed values of $l : m : n$.

This cannot be the case at every point, unless

$$al + c'm + b'n = 0,$$
$$c'l + bm + a'n = 0,$$
$$b'l + a'm + cn = 0,$$
$$a''l + b''m + c''n = 0;$$

from the first three of these equations we obtain

$$aa'^2 + bb'^2 + cc'^2 - abc - 2a'b'c' = 0 \qquad (1),$$

and $$l(b'c' - aa') = m(c'a' - bb') = n(a'b' - cc');$$

whence by the last, we have the further condition

$$\frac{a''}{b'c'-aa'}+\frac{b''}{c'a'-bb'}+\frac{c''}{a'b'-cc'}=0 \qquad (2).$$

The conditions (1) and (2) must hold among the coefficients in order that the surface may be a cylinder. Compare Art. (233).

Similarly, if the equation $F(x, y, z)=0$ represent a cone, we must have the equation

$$(x-\alpha)(ax+c'y+b'z+a'')+(y-\beta)(c'x+by+a'z+b'') \\ +(z-\gamma)(b'x+a'y+cz+c'')=0,$$

satisfied at every point of it. This cannot be the case unless the following four equations be simultaneously true:

$$a\alpha+c'\beta+b'\gamma+a''=0,$$
$$c'\alpha+b\beta+a'\gamma+b''=0,$$
$$b'\alpha+a'\beta+c\gamma+c''=0,$$
$$a''\alpha+b''\beta+c''\gamma+d=0;$$

and, the condition that the equation of the second degree may represent a cone, is the determinant

$$\begin{vmatrix} a, & c', & b', & a'' \\ c', & b, & a', & b'' \\ b', & a', & c, & c'' \\ a'', & b'', & c'', & d \end{vmatrix}=0.$$

This is the condition of Art. (231).

The application of the condition for a conoid leads to some rather tedious work, if we take the most general form of the equation. The condition finally requires the relation among the coefficients

$$aa'^2+bb'^2+cc'^2-abc-2a'b'c'=0,$$

with the further relation for a right conoid

$$a+b+c=0.$$

We may obtain these by taking the axis of z as the axis, and the directing plane parallel to the axis of x; in which case

$$l=0,\ m=0,\ l'=0,$$

and the differential equation reduces to

$$x\frac{dF}{dx} + y\frac{dF}{dy} - \frac{m'}{n'}y\frac{dF}{dz} = 0.$$

Hence the equation

$$x(ax + c'y + b'z + a'')$$
$$+ y(c'x + by + a'z + b'') - \frac{m'}{n'}y(b'x + a'y + cz + c'') = 0,$$

or $$ax^2 + \left(b - \frac{a'm'}{n'}\right)y^2 + \left(a' - \frac{cm'}{n'}\right)yz$$
$$+ b'zx + xy\left(2c' - \frac{b'm'}{n'}\right) + a''x + \left(b'' - \frac{c''n'}{n'}\right)y = 0,$$

must be true at every point for which $F(x, y, z) = 0$.

Subtracting, we shall obtain the following equation, satisfied at every point of the surface,

$$\frac{a'm'}{n'}y^2 + cz^2 + \left(a' + \frac{cm'}{n'}\right)yz$$
$$+ b'zx + \frac{b'm'}{n'} + a''x + \left(b'' + \frac{c''m'}{n'}\right)y + 2c''z + d = 0.$$

These equations must then coincide, and we have

$$a = 0,\ b = \frac{2a'm'}{n'},\ c = 0,\ c' = \frac{b'm'}{n'},\ c'' = 0,\ d = 0.$$

and the equation of the surface will be

$$by^2 + a'yz + b'zx + c'xy + a''x + b''y = 0,$$

the coefficients being connected by the equation $bb' = 2a'c'$.

Hence the condition

$$aa'^2 + bb'^2 + cc'^2 - abc - 2a'b'c' = 0,$$

must be satisfied when these axes are taken, and therefore when any other axes are taken, since the left-hand member of the equation is unaltered by transformation of co-ordinates.

If the conoid be a right conoid, we shall have $m' = 0$, and therefore $b'' = 0$, $c' = 0$, and the equation becomes

$$a'yz + b'zx + a''x + b''y = 0,$$

in which the condition $a + b + c = 0$ is satisfied. This condition must therefore be satisfied for every right conoid of the second degree.

The only conoid of the second degree is then a hyperbolic paraboloid, and for a right conoid, the two principal sections must be equal parabolas.

The application of the condition for developable surfaces leads to the equation

$$(bc - a'^2)(ax + b'y + b'z + a'')^2 + \ldots\ldots$$
$$+2(b'c' - aa')(c'x + by + a'z + b'')(b'x + a'y + cz + c'') + \ldots = 0,$$

to be satisfied at every point of the surface.

On examination of the coefficients, this will be found to be

$$ax^2 + by^2 + cz^2 + 2a'yz + 2b'zx + 2c'xy + 2a''x + 2b''y + 2c''z - \Delta = 0,$$

Δ having the same meaning as in Art. (231). The condition for a developable surface is then

$$\Delta + d = 0,$$

which shews that the only developable surfaces of the second degree are cones, of which cylinders are a particular case.

For surfaces of revolution, we obtain the equation, assuming the co-ordinates of the center of the sphere all zero, which does not affect the generality,

$$(mz - xy)(ax + c'y + b'z + a'') + (nx - lz)(l'x + by + a'z + b'')$$
$$+ (ly - nx)(b'x + a'y + cz + c'') = 0,$$

which must either coincide with the original equation, or be identically true.

We have accordingly, using an undetermined multiplier k,

$$c'n - b'm = ka, \quad a'l - c'n = kb, \quad b'm - a'b = kc, \ldots\ldots (1),$$

$$c'm - b'n - l(b - c) = 2ka', \ldots\ldots\ldots\ldots\ldots\ldots (2),$$

$$b''n - c''m = 2kb'', \ldots\ldots\ldots\ldots\ldots\ldots\ldots\ldots (3).$$

From the equations (1) we obtain

$$k(a + b + c) = 0,$$

and taking $k=0$, we have $a'l=b'm=c'n$, and the system (2) gives us

$$\frac{c'}{b'}-\frac{b'}{c'}=\frac{bc}{a'}, \text{ \&c.}$$

or
$$c-\frac{a'b'}{c'}=b-\frac{c'a'}{b'}=a-\frac{b'c'}{a'}.$$

The third system will then give

$$a'a''=b'b''=c'c'';$$

which equations will however not be necessary in the case when the axis does not pass through the origin.

The factor $a+b+c$ may possibly be shewn to be foreign to the investigation.

XVIII.

(1) Find the general functional equation of a family of surfaces such that the tangent plane at any point (x, y, z) of one of them intercepts on the axis of z a length $\frac{k^{n+1}}{z^n}$.

Determine the arbitrary function so that the intercepts on the axes of x and y may be in the ratio $x : y$.

(2) Shew that the differential equation of all surfaces which are generated by a circle, whose plane is parallel to the plane of yz, and which passes through the axis of x, is

$$(y^2+z^2)\frac{d^2z}{dx^2}+2\left(z-y\frac{dz}{dy}\right)\left\{1+\left(\frac{dz}{dy}\right)^2\right\}=0.$$

(3) A surface is generated by a straight line always passing through the two fixed straight lines

$$y=mx,\ z=c;\quad y=-mx,\ z=-c;$$

prove that the equation to the surface generated is of the form

$$\frac{mcx-yz}{c^2-z^2}=f\left(\frac{mzx-cy}{c^2-z^2}\right);$$

also that its differential equation is

$$(cy-mzx)\frac{dF}{dx}+m(mcx-yz)\frac{dF}{dy}+(c^2-z^2)\frac{dF}{dz}=0.$$

(4) The two branches of the curve of intersection of a surface

and its tangent plane will be at right angles to each other at every point, if the equation of the surface satisfy the condition

$$\left(\frac{dF}{dx}\right)^2\left(\frac{d^2F}{dy^2}+\frac{d^2F}{dz^2}\right)+\ldots-2\frac{dF}{dy}\frac{dF}{dz}\frac{d^2F}{dy\,dz}\ldots=0.$$

(5) Shew that the only surface of revolution in which the two branches of the curve of intersection with the tangent plane are at right angles to each other at every point, is the surface generated by the revolution of a catenary about its directrix.

(6) Shew that the only conoid possessing this property is a right conoid, and that its equation may be reduced to the form

$$y = x \tan mz.$$

(7) Apply the condition of question (4), to determine at what points of the surface

$$ax^2+by^2+cz^2+2a'yz+2b'zx+2c'xy+2a''x+2b''y+2c''z+d=0,$$

the generating lines are at right angles to each other.

(8) The functional equation of surfaces generated by a straight line intersecting the axis of z, and meeting the plane of xy on the circle $x^2+y^2=a^2$, is

$$\sqrt{x^2+y^2}-a=zf\left(\frac{y}{x}\right).$$

Find also the differential equation.

(9) Find the functional equation to a family of surfaces generated by a straight line of constant length c sliding between the co-ordinate planes of yz, zx, and remaining parallel to the plane of xy.

Shew that the differential equation is

$$\left(x\frac{dz}{dx}+y\frac{dz}{dy}\right)^2\left\{\left(\frac{dz}{dx}\right)^2+\left(\frac{dz}{dy}\right)^2\right\}=c^2\left(\frac{dz}{dx}\right)^2\left(\frac{dz}{dy}\right)^2.$$

(10) Shew that a certain differential equation of the third order must be satisfied at every point of any ruled surface whatever.

(11) Shew that every right conoid of the nth degree will be cut by any plane perpendicular to the axis in a number of straight lines not exceeding $(n-1)$.

(12) In a right conoid of the third degree, in which only one generating line passes through any point of the axis, shew that the section made by any plane through the axis will consist of the axis, and two generating lines, the sum of whose distances from any fixed point on the axis is constant.

(13) The general functional equation of ruled surfaces whose generating lines pass through the given straight line

$$\frac{x-a}{l}=\frac{y-b}{m}=\frac{z-c}{n},$$

is
$$z=xf(u)+y\,\phi(u),$$

where
$$u\equiv\frac{n(y-b)-m(z-c)}{n(x-a)-l(z-c)}:$$

and the general differential equation is

$$\left(\mu+\nu\frac{dz}{dy}\right)^2\frac{d^2z}{dx^2}-2\left(\mu+\nu\frac{dz}{dy}\right)\left(\lambda+\nu\frac{dz}{dx}\right)\frac{d^2z}{dx\,dy}$$
$$+\left(\lambda+\nu\frac{dz}{dx}\right)^2\frac{d^2z}{dy^2}=0,$$

λ, μ, ν, being proportional to $m(z-c)-n(y-b)$, &c.

What do these equations become respectively when the given straight line is the axis of z?

(14) Shew that all developable surfaces of the third degree are cones or cylinders.

CHAPTER XVIII.

PROPERTIES OF CONICOIDS SATISFYING GIVEN CONDITIONS. FORMS OF THE EQUATION OF THE SPHERE.

412. *To find the form of the general equation of a conicoid passing through eight given points.*

If $u=0$, $v=0$ be the equations of two particular conicoids satisfying the given conditions, then $u+kv=0$ will be satisfied when u and v simultaneously vanish, and will therefore represent a conicoid passing through the eight given points, and since it involves one arbitrary constant, it is the *general* equation required.

413. *If a conicoid pass through eight given points, the polar plane of any other given point will pass through a fixed straight line.*

The polar plane of the point A of the fundamental tetrahedron is

$$\frac{du}{d\alpha}+k\frac{dv}{d\alpha}=0,$$

which passes through the straight line determined by the equations

$$\frac{du}{d\alpha}=0, \quad \frac{dv}{d\alpha}=0.$$

There are moreover certain points whose polar planes are altogether fixed. For if $(\alpha', \beta', \gamma', \delta')$ be a point such that

$$\frac{\frac{du'}{d\alpha'}}{\frac{dv'}{d\alpha'}}=\frac{\frac{du'}{d\beta'}}{\frac{dv'}{d\beta'}}=\frac{\frac{du'}{d\gamma'}}{\frac{dv}{d\gamma'}}=\frac{\frac{du'}{d\delta'}}{\frac{dv'}{d\delta'}},$$

its polar plane will be independent of k; or will be fixed. This number will afterwards be found to be four.

414. *If a conicoid pass through eight given points, the pole of any given plane will lie on the curve of intersection of two conicoids.*

The equations determining the pole of the plane BCD are

$$\frac{du}{d\beta}+k\frac{dv}{d\beta}=0, \quad \frac{du}{d\gamma}+k\frac{dv}{d\gamma}=0, \quad \frac{du}{d\delta}+k\frac{dv}{d\delta}=0,$$

whence the locus of the pole is the curve given by the equations

$$\frac{du}{d\beta}\frac{dv}{d\gamma}-\frac{du}{d\gamma}\frac{dv}{d\beta}=0, \quad \frac{du}{d\gamma}\frac{dv}{d\delta}-\frac{du}{d\delta}\frac{dv}{d\gamma}=0,$$

which are both of the second degree. There will be four planes whose poles are altogether fixed, namely, the polar planes of the points determined in the last article.

The locus of the center of a conicoid passing through eight fixed points is a curve of this species, since the center is the pole of the plane at infinity.

415. Reciprocating these propositions, we obtain the following.

If a conicoid touch eight given planes, the pole of any other given plane will lie on a fixed straight line. Hence the locus of its center will be a straight line.

The polar plane of any given point will envelope a developable surface of the fourth class.

In this case also there exist four polar planes whose corresponding poles are altogether fixed.

416. *Four cones of the second degree can in general be described passing through eight given points.*

The equations for determining the center of the conicoid $u+kv=0$, are

$$\frac{du}{d\alpha}+k\frac{dv}{d\alpha}=\frac{du}{d\beta}+k\frac{dv}{d\beta}=\frac{du}{d\gamma}+k\frac{dv}{d\gamma}=\frac{du}{d\delta}+k\frac{dv}{d\delta}.$$

If the conicoid be a cone, its center lies on the surface, and each member of these equations

$$\equiv\frac{\alpha\frac{du}{d\alpha}+\ldots+k\left(\alpha\frac{dv}{d\alpha}+\ldots\right)}{\alpha+\beta+\gamma+\delta}=0.$$

Now if we take four of the given points as the angular points of the fundamental tetrahedrons, we may write

$$u \equiv l\beta\gamma + m\gamma\alpha + n\alpha\beta + l'\alpha\delta + m'\beta\delta + n'\gamma\delta,$$
$$v \equiv \lambda\beta\gamma + \ldots\ldots\ldots\ldots\ldots\ldots\ldots\ldots\ldots\ldots\ldots$$

and the condition for a cone will become

$$\begin{vmatrix} 0, & n+k\nu, & m+k\mu, & l'+k\lambda' \\ n+k\nu, & 0, & l+k\lambda, & m'+k\mu' \\ m+k\mu, & l+k\lambda, & 0, & n'+k\nu' \\ l'+k\lambda', & m'+k\mu', & n'+k\nu', & 0 \end{vmatrix} = 0.$$

which, being a biquadratic in k, shews that there are in general four cones of the second degree passing through eight arbitrary points.

The vertices of these cones are the points whose polar planes are fixed with respect to any conicoid passing through the eight points. For if, in Art. (413), the value of the ratio

$$\frac{du'}{d\alpha'} : \frac{dv'}{d\alpha'} \text{ be } -k_1,$$

k_1 will be a root of this equation, and the point $(\alpha', \beta', \gamma', \delta')$ will be the vertex of the cone

$$u + k_1 v = 0.$$

Let P, Q be two of these points, then since the polar plane of P is the same for all conicoids passing through the eight points, it will be the polar plane of P with respect to the cone whose vertex is Q, it will therefore pass through Q, and similarly through the other two points. Hence of these four points, each is the pole of the plane passing through the other three, with respect to any conicoid passing through the eight points.

If we take these points as the angular points of the fundamental tetrahedron, the general equation of the conicoid will be

$$l\alpha^2 + m\beta^2 + n\gamma^2 + r\delta^2 = 0,$$

l, m, n, r each involving an arbitrary constant k in the first degree.

It follows from this result, that the equations of any two conicoids may be obtained in the form

$$l\alpha^2 + m\beta^2 + n\gamma^2 + r\delta^2 = 0, \quad l'\alpha^2 + m'\beta^2 + n'\gamma^2 + r'\delta^2 = 0.$$

417. These results will fail if the eight points lie on two planes, for in that case, taking the two planes as $\alpha = 0$, $\beta = 0$, the equation of any conicoid passing through the eight points may be written $u - 2k\,\alpha\beta = 0$, in which

$$u \equiv p\alpha^2 + q\beta^2 + r\gamma^2 + s\delta^2 + 2l\beta\gamma + \ldots + 2l'\alpha\delta + \ldots$$

and the equation giving the values of k, for which the conicoid becomes a cone, is

$$\begin{vmatrix} p, & n-k, & m, & l' \\ n-k, & q, & l, & m' \\ m, & l, & r, & n' \\ l', & m', & n', & s \end{vmatrix} = 0,$$

which is only a quadratic in k, so that only two proper cones can be described through the eight points, agreeably to Art. (343).

418. *Three conicoids can be described passing through eight given points, and touching a given plane.*

Let the equation of any conicoid passing through the eight given points be

$$(l + kl')\,\alpha^2 + (m + km')\,\beta^2 + (n + kn')\,\gamma^2 + (r + kr')\,\delta^2 = 0,$$

and let α', β', γ', δ' be the perpendiculars from the angular points on the given plane. Then if the point of contact be $(\alpha'', \beta'', \gamma'', \delta'')$, we shall have

$$\frac{(l + kl')\,\alpha''}{\alpha'} = \frac{(m + km')\,\beta''}{\beta'} = \frac{(n + kn')\,\gamma''}{\gamma'} = \frac{(r + kr')\,\delta''}{\delta'},$$

whence the condition that the given plane may touch the conicoid is

$$\frac{\alpha'^2}{l + kl'} + \frac{\beta'^2}{m + km'} + \frac{\gamma'^2}{n + kn'} + \frac{\delta'^2}{r + kr'} = 0,$$

a cubic equation for k, provided α', β', γ', δ' are all finite.

If $\alpha' = 0$, one solution is $l + kl' = 0$, which reduces the required conicoid to one of the four cones of the second degree, passing through the eight points, which is not properly a solution, as the plane will not in general be a proper tangent plane to the cone, but, passing through its vertex, of course will satisfy the analytical condition of tangency. So if the given plane pass through two of the angular points, there will be only one proper solution, and if through three, no conicoid can be described as required.

Reciprocally, we can in general find three conicoids, touching eight given planes, and passing through a given point. Corresponding cases of exception arise, when the given point lies on one, two, or three of four planes, fixed with respect to the eight given planes.

419. *To find the general form of the equation of a conicoid passing through seven given points.*

Take $u=0$, $v=0$, $w=0$, the equations of three particular conicoids, not having a common curve of intersection, and satisfying the required conditions; then the equation

$$lu+mv+nw=0$$

will be the general equation required. For it is satisfied whenever u, v, and w simultaneously vanish, and it involves two arbitrary constants, the ratios $l : m : n$, by means of which it may be made to pass through any two other given points. It can therefore be made to represent any conicoid passing through the seven points.

Since the equations $u=0$, $v=0$, $w=0$ determine eight points, we see that any conicoid which passes through seven fixed points will necessarily pass through an eighth fixed point, whose position may be determined from that of the seven.

420. *If a conicoid pass through seven given points, the polar plane of any other given point will pass through a fixed point.*

For the polar plane of the point A is

$$l\frac{du}{d\alpha}+m\frac{dv}{d\alpha}+n\frac{dw}{d\alpha}=0,$$

which passes through a fixed point determined by the equations

$$\frac{du}{d\alpha}=0,\quad \frac{dv}{d\alpha}=0,\quad \frac{dw}{d\alpha}=0.$$

If the fixed point be a vertex of any cone of the second degree passing through the seven points, the polar plane will pass through a fixed straight line. For, take $u=0$ as the equation of a cone whose vertex is A, passing through the seven points. Then $\frac{du}{d\alpha} \equiv 0$, and the polar plane of A will be

$$m\frac{dv}{d\alpha} + n\frac{dw}{d\alpha} = 0,$$

which passes through the fixed straight line $\frac{dv}{d\alpha} = 0$, $\frac{dw}{d\alpha} = 0$.

In general, no polar plane can be altogether fixed, for, in order that the polar plane of A may be independent of $l : m : n$, $\frac{du}{d\alpha} = 0$, $\frac{dv}{d\alpha} = 0$, $\frac{dw}{d\alpha} = 0$, must be equivalent to only one equation, whence each of the equations $u = 0$, $v = 0$, $w = 0$, must be of the form

$$\alpha u_1 + u_2 = 0, \quad \alpha u_1 + v_2 = 0, \quad \alpha u_1 + w_2 = 0,$$

where u_1 is of the first degree, and u_2, v_2, w_2 do not involve α. Hence through the seven points can be described two cones of the second degree $u_2 - v_2 = 0$, $v_2 - w_2 = 0$, having a common vertex A, and the seven points must lie on four fixed straight lines passing through A.

421. *When a conicoid passes through seven given points, to find the locus of the pole of a given plane.*

If the equation of the conicoid be $lu + mv + nw = 0$, and if the given plane be that of BCD, the equations determining the pole are

$$l\frac{du}{d\beta} + m\frac{dv}{d\beta} + n\frac{dw}{d\beta} = 0,$$

$$l\frac{du}{d\gamma} + m\frac{dv}{d\gamma} + n\frac{dw}{d\gamma} = 0,$$

$$l\frac{du}{d\delta} + m\frac{dv}{d\delta} + n\frac{dw}{d\delta} = 0.$$

Hence the locus of the pole is the surface of the third degree

$$\begin{vmatrix} \frac{du}{d\beta}, & \frac{du}{d\gamma}, & \frac{du}{d\delta} \\ \frac{dv}{d\beta}, & \frac{dv}{d\gamma}, & \frac{dv}{d\delta} \\ \frac{dw}{d\beta}, & \frac{dw}{d\gamma}, & \frac{dw}{d\delta} \end{vmatrix} = 0.$$

The locus of the center, which is the pole of the plane at infinity, is obtained by eliminating l, m, n from the equations

$$l\frac{du}{d\alpha}+m\frac{dv}{d\alpha}+n\frac{dw}{d\alpha}=l\frac{du}{d\beta}+\ldots=l\frac{du}{d\gamma}+\ldots=l\frac{du}{d\delta}+\ldots$$

and is therefore

$$\begin{vmatrix} \frac{du}{d\beta}-\frac{du}{d\alpha}, & \frac{dv}{d\beta}-\frac{dv}{d\alpha}, & \frac{dw}{d\beta}-\frac{dw}{d\alpha} \\ \frac{du}{d\gamma}-\frac{du}{d\alpha}, & \frac{dv}{d\gamma}-\frac{dv}{d\alpha}, & \frac{dw}{d\gamma}-\frac{dw}{d\alpha} \\ \frac{du}{d\delta}-\frac{du}{d\alpha}, & \frac{dv}{d\delta}-\frac{dv}{d\alpha}, & \frac{dw}{d\delta}-\frac{dw}{d\alpha} \end{vmatrix}=0.$$

422. *To find how many conicoids can be described passing through seven given points, and touching two given planes.*

Let the given planes be $\alpha=0$, $\beta=0$; then at the point of contact with the first we must have

$$\alpha=0,\ l\frac{du}{d\beta}+m\frac{dv}{d\beta}+n\frac{dw}{d\beta}=0,\ l\frac{du}{d\gamma}+\ldots=0,\ l\frac{du}{d\delta}+\ldots=0,$$

and eliminating $\beta:\gamma:\delta$ from these, the eliminant is of the third degree in l, m, n. Similarly the condition of touching $\beta=0$ will lead to a relation of the third degree in l, m, n, and the final equation for determining $l:m$ will be of the ninth degree. There are therefore, generally, nine conicoids satisfying the given conditions.

423. *To find a general form of the equation of a conicoid passing through seven given points, six of which lie by threes on two non-intersecting straight lines.*

The two straight lines must lie altogether on the conicoid, and if through the seventh point we draw a straight line intersecting the other two, three points on this line, and therefore the whole line, will lie on the conicoid. Take these as the edges *AB*, *BC*, *CD* of the fundamental tetrahedron, then, since they are contained by the three pairs of planes

$$\alpha\gamma=0,\quad \alpha\delta=0,\quad \beta\delta=0,$$

the general equation is

$$l\alpha\gamma + m\alpha\delta + n\beta\delta = 0.$$

If we take two such conicoids (l, m, n), and (l', m', n'), their fourth line of intersection will be found to be

$$\alpha(lm' - l'm) = \beta(mn' - m'n),$$
$$\gamma(nl' - n'l) = \delta(mn' - m'n).$$

424. *In the conicoid of the last article, the pole of a given plane lies in a fixed plane.*

Let $L\alpha + M\beta + N\gamma + R\delta = 0$ be the equation of the given plane, let $(\alpha', \beta', \gamma', \delta')$ be its pole with respect to the conicoid $l\alpha\gamma + m\alpha\delta + n\beta\delta = 0$, we shall then have

$$\frac{l\gamma' + m\delta'}{L} = \frac{n\delta'}{M} = \frac{l\alpha'}{N} = \frac{m\alpha' + n\beta'}{R}$$
$$= \frac{\alpha'(l\gamma' + m\delta') + \beta'(n\delta') - \gamma'(l\alpha') - \delta'(m\alpha' + n\beta')}{L\alpha' + M\beta' - N\gamma' - R\delta'}.$$

Hence the locus of the pole is the plane

$$L\alpha' + M\beta' = N\gamma' + R\delta'.$$

The locus of the center is therefore

$$\alpha' + \beta' = \gamma' + \delta',$$

which is a plane passing through the center of gravity of the tetrahedron, and parallel to the edges AB, CD; or the plane bisecting the edges AC, AD, BC, BD.

The polar plane of the point $(\alpha', \beta', \gamma', \delta')$ may similarly be shewn to pass through the point

$$\frac{\alpha}{\alpha'} = \frac{\beta}{\beta'} = -\frac{\gamma}{\gamma'} = -\frac{\delta}{\delta'},$$

which may be determined as follows. Take a plane through AB, and O the pole, meeting CD in P, and a plane through CD and O, meeting AB in Q. Then if R be taken on PQ such that $OPQR$ is a harmonic range, the polar plane of O will pass through R. Compare Art. (343).

425. *To find a general equation of a conicoid passing through seven given points such that four straight lines can be drawn through them and through each of four given points.*

Let the four given points be the vertices of the tetrahedron; then any seven of the eight points represented by the equations

$$\frac{\alpha^2}{l}=\frac{\beta^2}{m}=\frac{\gamma^2}{n}=\frac{\delta^2}{r}$$

will satisfy the required conditions, and the eighth point will lie on any conicoid passing through the other seven. Since the points lie on the pairs of planes

$$\frac{\alpha^2}{l}=\frac{\beta^2}{m},\quad \frac{\beta^2}{m}=\frac{\gamma^2}{n},\quad \frac{\gamma^2}{n}=\frac{\delta^2}{r},$$

the general equation required will be

$$\lambda\left(\frac{\alpha^2}{l}-\frac{\beta^2}{m}\right)+\mu\left(\frac{\beta^2}{m}-\frac{\gamma^2}{n}\right)+\nu\left(\frac{\gamma^2}{n}-\frac{\delta^2}{r}\right)=0,$$

which we may write

$$L\alpha^2+M\beta^2+N\gamma^2+R\delta^2=0,$$

L, M, N, R being connected by the equation

$$Ll+Mm+Nn+Rr=0.$$

In this conicoid each angular point of the fundamental tetrahedron is the pole of the opposite face. The polar plane of a given point $(\alpha', \beta', \gamma', \delta')$ will pass through the point

$$\frac{\alpha\alpha'}{l}=\frac{\beta\beta'}{m}=\frac{\gamma\gamma'}{n}=\frac{\delta\delta'}{r},$$

and the pole of a given plane $\alpha\alpha'+\beta\beta'+\gamma\gamma'+\delta\delta'=0$ will lie on the surface

$$\frac{l\alpha'}{\alpha}+\frac{m\beta'}{\beta}+\frac{n\gamma'}{\gamma}+\frac{r\delta'}{\delta}=0.$$

The locus of the center is therefore the surface

$$\frac{l}{\alpha}+\frac{m}{\beta}+\frac{n}{\gamma}+\frac{r}{\delta}=0.$$

The centers of the eight spheres which touch the faces of the fundamental tetrahedron are a particular case of these eight points, since they are given by the equations

$$p_1^2\alpha^2=p_2^2\beta^2=p_3^2\gamma^2=p_4^2\delta^2.$$

The locus of the center of any surface passing through them is therefore

$$\frac{1}{p_1^2\alpha}+\frac{1}{p_2^2\beta}+\frac{1}{p_3^2\gamma}+\frac{1}{p_4^2\delta}=0,$$

which equation may be interpreted geometrically to mean that the feet of the perpendiculars from any point of the surface on the faces of the fundamental tetrahedron lie in one plane.

426. *To find the number of conicoids passing through six given points, and touching three given planes.*

If we take four of the given points as the angular points of the fundamental tetrahedron, we may write the equation of the conicoid

$$l\beta\gamma+m\gamma\alpha+n\alpha\beta+l'\alpha\delta+m'\beta\delta+n'\gamma\delta=0,$$

and l, m, n, l', m', n' will be connected by two linear equations expressing the conditions of passing through the two other given points.

The condition that this conicoid may touch a fixed plane

$$L_1\alpha+M_1\beta+N_1\gamma+R_1\delta=0,$$

may be found by eliminating $(\alpha, \beta, \gamma, \delta)$ from the equation

$$\frac{l'\alpha+n\beta+m\gamma}{L_1}=\frac{n\alpha+m'\beta+l\gamma}{M_1}=\frac{m\alpha+l\beta+n'\gamma}{N_1}=\frac{l'\alpha+m'\beta+n'\gamma}{R_1},$$

and the eliminant is of the third degree. Hence l, m, n, l', m', n' will be connected by two homogeneous equations of the first degree, and three of the third, and the number of solutions will therefore be twenty-seven.

427. *To find the number of conicoids passing through five given points, and touching four given planes.*

Proceeding as in the last article, we shall have l, m, n, l', m', n' connected by one homogeneous equation of the first degree, and four of the third degree, and the number of solutions will be eighty-one.

428. Hence by reciprocating the results of Arts. (428), (427), and (423), we see that eighty-one conicoids can be described pass-

ing through four given points, and touching five given planes; twenty-seven conicoids through three given points, and touching six given planes; and nine through two given points, and touching seven given planes.

429. *To find the general equation of a conicoid containing two non-intersecting straight lines.*

If these straight lines be taken for the edges $\alpha=0$, $\delta=0$; $\beta=0$, $\gamma=0$; of the fundamental tetrahedron, we see that the terms in the general equation involving α^2, β^2, γ^2, δ^2, $\alpha\delta$, $\beta\gamma$, must vanish, and the resulting equation may be written

$$m\gamma\alpha+n\alpha\beta+m'\beta\delta+n'\gamma\delta=0.$$

430. *To find the condition that the conicoid of the last article may be a paraboloid.*

In order that this conicoid may be a paraboloid, it must have contact with the plane at infinity; we must therefore have, at the point of contact,

$$m\gamma+n\beta=n\alpha+m'\delta=m\alpha+n'\delta=m'\beta+n'\gamma;\ \alpha+\beta+\gamma+\delta=0.$$

These give

$$\frac{\alpha}{m'-n'}=\frac{\beta}{m-n'}=\frac{\gamma}{m'-n}=\frac{\delta}{m-n},$$

whence the required condition is

$$m+m'=n+n';$$

which will give at the point of contact

$$\alpha+\delta=0,\ \ \beta+\gamma=0.$$

Hence the axis of the paraboloid and the two generating lines are parallel to the same plane.

The polar plane of any point $(\alpha', \beta', \gamma', \delta')$ may be readily shewn to pass through the fixed point

$$\frac{\alpha}{\alpha'}=-\frac{\beta}{\beta'}=-\frac{\gamma}{\gamma'}=\frac{\delta}{\delta'},$$

which may be determined geometrically as before (Art. 425).

431. *If a conicoid contain two given non-intersecting straight lines, the pole of a given plane will lie on a fixed plane.*

Let the given plane be

$$L\alpha + M\beta + N\gamma + R\delta = 0;$$

its pole will be given by the equations

$$\frac{m\gamma + n\beta}{L} = \frac{n\alpha + m'\delta}{M} = \frac{n\alpha + n'\delta}{N} = \frac{m'\beta + n'\gamma}{R},$$

each member of which is therefore equal either to

$$\frac{\alpha(m\gamma + n\beta) - \gamma(m\alpha + n'\delta)}{L\alpha - N\gamma} \text{ or to } \frac{\beta(n\alpha + m'\delta) - \delta(m'\beta + n'\gamma)}{M\beta - R\delta};$$

that is, to $$\frac{n\alpha\beta - n'\gamma\delta}{L\alpha - N\gamma}, \text{ or } \frac{n\alpha\beta - n'\gamma\delta}{M\beta - R\delta}.$$

Hence the locus of the pole is

$$L\alpha - M\beta - N\gamma + R\delta = 0.$$

The locus of the center is therefore

$$\alpha - \beta - \gamma + \delta = 0,$$

or the plane parallel to the two straight lines, and equidistant from them.

432. *If three conicoids have a common plane section, the other planes in which, taken two and two, they intersect will meet in one straight line.*

Let $u = 0$ be the equation of one of the conicoids, and let $\alpha = 0$ be the plane section, the equations of the other two conicoids may then be taken to be

$$u + k\alpha\beta = 0, \quad u + k'\alpha\gamma = 0,$$

and the equations of the second planes of intersection will be

$$\beta = 0, \quad \gamma = 0, \quad k\beta = k'\gamma,$$

which all intersect in the straight line $\beta = 0$, $\gamma = 0$.

Reciprocating, it follows that if three conicoids have one common enveloping cone, the vertices of the other enveloping cones common to them, taken two and two, will lie on one straight line. This is the interpretation of the above equations if α, β, γ, δ be taken as Four Point Co-ordinates of the tangent planes, instead of Tetrahedral Co-ordinates of the points on the surfaces.

433. *If a series of conicoids have two common plane sections, the polar plane of a given point passes through a given straight line, and the pole of a given plane lies on a plane conic.*

Let the planes of section intersect in the edge CD of the tetrahedron, and let C, D be the points in which one of the plane conics meets the line of intersection: the other must meet it in the same points, or the sections could not lie on the same conicoid, since a straight line would meet them in four points. Also, let A, B be the vertices of the two cones which can be drawn containing the two plane conics, then the equation of the planes of section may be taken to be, since they divide AB harmonically, $\alpha^2 - k^2\beta^2 = 0$; and if we assume the equation of any conicoid containing these conics to be

$$p\alpha^2 + q\beta^2 + r\gamma^2 + s\delta^2 + 2l\beta\gamma + 2m\gamma\alpha + 2n\alpha\beta + 2l'\alpha\delta + 2m'\beta\delta + 2n'\gamma\delta = 0,$$

we shall have $r = 0$, $s = 0$, because it passes through C and D, and $l = 0$, $m = 0$, $n = 0$, $l' = 0$, $m' = 0$, because the substitutions $\alpha = k\beta$, $\alpha = -k\beta$ must lead to the same equation, namely, that of the cone whose vertex is A, and which contains the two conics; and similarly for $\beta = \dfrac{\alpha}{k}$, $\beta = -\dfrac{\alpha}{k}$.

The general equation of any conicoid containing these two given conics is therefore

$$\lambda\,(p\alpha^2 + q\beta^2 + 2n'\gamma\delta) + \mu\,(\alpha^2 - k^2\beta^2) = 0.$$

The polar plane of any point $(\alpha', \beta', \gamma', \delta')$ therefore contains the straight line

$$p\alpha\alpha' + q\beta\beta' + n'\gamma\gamma' + n'\delta\delta' = 0, \quad \alpha\alpha' - k^2\beta\beta' = 0;$$

a straight line intersecting CD in the point

$$\alpha = 0, \quad \beta = 0, \quad \gamma\delta' + \delta\gamma' = 0,$$

so that CD is divided harmonically by the plane passing through AB and the given point, and by the polar plane of the given point, as might be foreseen from the fact that AB, CD are reciprocal straight lines for any one of the series of conicoids.

The pole of any given plane $\alpha'\alpha + \beta'\beta + \gamma'\gamma + \delta'\delta = 0$ is given by the equations

$$(\lambda p + \mu)\,\frac{\alpha}{\alpha'} = (\lambda q - \mu k^2)\,\frac{\beta}{\beta'} = n'\,\frac{\lambda\delta}{\gamma'} = n'\,\frac{\lambda\gamma}{\delta'},$$

and therefore lies on the fixed conic section

$$\gamma\gamma' = \delta\delta', \quad (pk^2 + q)\,\alpha\beta\gamma' = n'\beta'\alpha\delta + n'k^2\alpha'\beta\delta,$$

a conic passing through A, B, and a point on CD which with the given plane divides CD harmonically.

The locus of the center is therefore the conic

$$\gamma = \delta, \quad (pk^2 + q)\,\alpha\beta = n'\alpha\delta + n'k^2\beta\delta,$$

passing through A, B, and the middle point of CD.

434. *The tangential equation of any surface is the equation, referred to tetrahedral co-ordinates, of its reciprocal with respect to the surface* $\alpha^2 + \beta^2 + \gamma^2 + \delta^2 = 0$.

Let $(\alpha_1, \beta_1, \gamma_1, \delta_1)$ be the Four-point Co-ordinates of a tangent plane to any surface whose equation, referred to Tetrahedral Co-ordinates, is $F(\alpha, \beta, \gamma, \delta) = 0$. Also let α', β', γ', δ' be the Tetrahedral Co-ordinates of the pole of this plane with respect to the surface $\alpha^2 + \beta^2 + \gamma^2 + \delta^2 = 0$, α'', β'', γ'', δ'' the co-ordinates of the point of contact.

Then the equation of the plane will be

$$\alpha_1\alpha + \beta_1\beta + \gamma_1\gamma + \delta_1\delta = 0, \qquad \text{(Art. 81)},$$

and we shall have, to determine $(\alpha_1, \beta_1, \gamma_1, \delta_1)$, the equations

$$\frac{\alpha_1}{\dfrac{dF}{d\alpha''}} = \frac{\beta_1}{\dfrac{dF}{d\beta''}} = \frac{\gamma_1}{\dfrac{dF}{d\gamma''}} = \frac{\delta_1}{\dfrac{dF}{d\delta''}};$$

from which, determining the ratios $\alpha'' : \beta'' : \gamma'' : \delta''$ in terms of α_1, β_1, γ_1, δ_1, and substituting in the equation

$$F(\alpha'', \beta'', \gamma'', \delta'') = 0,$$

we obtain the relation between α_1, β_1, γ_1, δ_1, which is the tangential equation of the surface.

But since $(\alpha', \beta', \gamma', \delta')$ is the pole of this plane with respect to the surface $\alpha^2 + \beta^2 + \gamma^2 + \delta^2 = 0$, the equation of the plane must also be

$$\alpha'\alpha + \beta'\beta + \gamma'\gamma + \delta'\delta = 0,$$

which shews that the equations for determining $(\alpha', \beta', \gamma', \delta')$ are the same as those for $(\alpha_1, \beta_1, \gamma_1, \delta_1)$, and the resulting homogeneous equation must be the same in both cases.

Hence, if any proposition with respect to surfaces be proved by the use of Tetrahedral Co-ordinates, the reciprocal proposition may be deduced from a different interpretation of the same equations, namely, by considering them throughout as relations between the Point-Co-ordinates of the tangent planes to surfaces, *i. e.* as the tangential equations of surfaces.

435. *To find the equation of the sphere circumscribing the fundamental tetrahedron.*

The equation will be of the form

$$l\beta\gamma + m\gamma\alpha + n\alpha\beta + l'\alpha\delta + m'\beta\delta + n'\gamma\delta = 0,$$

since the surface passes through the angular points of the tetrahedron.

For points in the section of this surface by one of the fundamental planes, $\delta = 0$, we must have

$$l\beta\gamma + m\gamma\alpha + n\alpha\beta = 0 \;\ldots\ldots\ldots\ldots\ldots\ldots\; (1).$$

But this section is the circle circumscribing the triangle ABC, and therefore for every point on this circle

$$a'^2\beta'\gamma' + b'^2\gamma'\alpha' + c'^2\alpha'\beta' = 0,$$

$(\alpha', \beta', \gamma')$ being triangular co-ordinates measured from the triangle ABC.

Now P being any point in the plane ABC,

$$\alpha' = \frac{\Delta PBC}{\Delta ABC} = \frac{\text{vol. } DPBC}{\text{vol. } DABC} = \alpha,$$

and, similarly, $\beta' = \beta$, $\gamma' = \gamma$. Hence, for points on the circle circumscribing the triangle ABC,

$$a'^2\beta\gamma + b'^2\gamma\alpha + c'^2\alpha\beta = 0,$$

whence, comparing this equation with (1), we have

$$\frac{l}{a'^2} = \frac{m}{b'^2} = \frac{n}{c'^2},$$

and proceeding similarly, we shall obtain

$$\frac{l}{a'^2} = \frac{m}{b'^2} = \frac{n}{c'^2} = \frac{l'}{a^2} = \frac{m'}{b^2} = \frac{n'}{c^2}.$$

The equation of the circumscribing sphere is therefore

$$a'^2\beta\gamma + b'^2\gamma\alpha + c'^2\alpha\beta + a^2\alpha\delta + b^2\beta\delta + c^2\gamma\delta = 0.$$

436. *To find the general equation of a sphere in tetrahedral co-ordinates.*

Taking the general equation of a conicoid

$$p\alpha^2 + q\beta^2 + r\gamma^2 + s\delta^2 + l\beta\gamma + \dots\dots + l'\alpha\delta + \dots\dots = 0,$$

let us consider the segments made by this surface of the edges DA, DB, DC of the tetrahedron.

Let P_1, P_2 be the two points where the surface meets DA. Then, at P_1, P_2, we shall have $\beta = 0$, $\gamma = 0$, and therefore $p\alpha^2 + s\delta^2 + l'\alpha\delta = 0$, $\alpha + \delta = 1$; and the equation for determining α is $p\alpha^2 + s(1-\alpha)^2 + l'\alpha(1-\alpha) = 0$. Now if α_1, α_2 be the values of α at P_1, P_2 respectively,

$$\alpha_1 = \frac{\text{vol. } P_1BCD}{\text{vol. } ABCD} = \frac{DP_1}{DA}.$$

Hence $$DP_1 \,.\, DP_2 = a^2\alpha_1\alpha_2 = \frac{sa^2}{p+s-l'}.$$

If the surface meet DB, DC in Q_1, Q_2; R_1, R_2; respectively, we shall also have

$$DQ_1 \,.\, DQ_2 = \frac{sb^2}{q+s-m'}, \quad DR_1 \,.\, DR_2 = \frac{sc^2}{r+s-n'},$$

if the surface be a sphere, we shall therefore have the conditions

$$\frac{p+s-l'}{a^2} = \frac{q+s-m'}{b^2} = \frac{r+s-n'}{c^2},$$

each of which may similarly be shewn to be equal to

$$\frac{q+r-l}{a'^2} = \frac{r+p-m}{b'^2} = \frac{p+q-n}{c'^2}.$$

These are necessary conditions in order that the surface may represent a sphere, and, since they are five in number, they are also sufficient.

If we eliminate l, m, n, l', m', n' by means of these relations, the equation of the sphere takes the form

$$(p\alpha + q\beta + r\gamma + s\delta)(\alpha + \beta + \gamma + \delta)$$
$$- (a'^2\beta\gamma + b'^2\gamma\alpha + c'^2\alpha\beta + a^2\alpha\delta + b^2\beta\delta + c^2\gamma\delta) = 0,$$

a form which shews that all spheres have a common impossible section on the plane at infinity.

When the equation is reduced to this form, the coefficients p, q, r, s are the rectangles of the segments of chords of the sphere passing through A, B, C, D respectively. Hence, since the left-hand member of the equation is always proportional to the rectangle of the segments of any chord drawn through the point $(\alpha, \beta, \gamma, \delta)$, Art. (137), it must be equal to that rectangle.

437. *To find the radius of the circumscribing sphere in terms of the lengths of the edges.*

Its equation being

$$a'^2\beta\gamma + b'^2\gamma\alpha + c'^2\alpha\beta + a^2\alpha\delta + b^2\beta\delta + c^2\gamma\delta \equiv f(\alpha, \beta, \gamma, \delta) = 0,$$

the equations of its center are

$$f'(\alpha) = f'(\beta) = f'(\gamma) = f'(\delta) \equiv \lambda.$$

Hence $\alpha f'(\alpha) + \beta f'(\beta) + \gamma f'(\gamma) + \delta f'(\delta) = (\alpha + \beta + \gamma + \delta)\lambda = \lambda$,

or

$$2f(\alpha, \beta, \gamma, \delta) = \lambda.$$

But $-f(\alpha, \beta, \gamma, \delta)$ is the rectangle of the segments of any chord through $(\alpha, \beta, \gamma, \delta)$, and therefore is equal to $-R^2$, if R be the radius: or $\lambda = 2R^2$.

The equations determining the center are then

$$\begin{aligned}
c'^2\beta + b'^2\gamma + a^2\delta &= 2R^2,\\
c'^2\alpha + a'^2\gamma + b^2\delta &= 2R^2,\\
b'^2\alpha + a'^2\beta + c^2\delta &= 2R^2,\\
a^2\alpha + b^2\beta + c^2\gamma &= 2R^2,\\
\alpha + \beta + \gamma + \delta &= 1,
\end{aligned}$$

which give, on eliminating $\alpha, \beta, \gamma, \delta$ between these equations,

$$4R^2\{b^2c^2a'^2 + c^2a^2b'^2 + a^2b^2c'^2 + a'^2b'^2c'^2 - a^2a'^2(b^2 + b'^2 + c^2 + c'^2 - a^2 - a'^2) - \ldots\}$$
$$= a^4a'^4 + b^4b'^4 + c^4c'^4 - 2b^2b'^2c^2c'^2 - 2c^2c'^2a^2a'^2 - 2a^2a'^2b^2b'^2.$$

Now, if V be the volume of the tetrahedron, $6V$ will be the volume of a parallelopiped whose edges are DA, DB, DC, or

$6V = abc \times$
$\sqrt{1 - \cos^2 BDC - \cos^2 CDA - \cos^2 ADB - 2\cos BDC \cos CDA \cos DAB}$,

$$144V^2 = 4a^2b^2c^2 - a^2(b^2 + c^2 - a'^2)^2 - b^2(c^2 + a^2 - b'^2)^2 - c^2(a^2 + b^2 - c'^2)^2$$
$$- (b^2 + c^2 - a'^2)(c^2 + a^2 - b'^2)(a^2 + b^2 - c'^2)$$

$-144V^2$

$$= b^2c^2a'^2 + c^2a^2b'^2 + a^2b^2c'^2 + a'^2b'^2c'^2 - a^2a'^2(b^2 + b'^2 + c^2 + c'^2 - a^2 - a'^2) - \ldots$$

whence

$4R^2 \times 144\, V^2 = 16s\,(s-aa')\,(s-bb')\,(s-cc')$, where $2s \equiv aa'+bb'+cc'$,

we have then, finally,

$$R = \frac{\sqrt{s\,(s-aa')\,(s-bb')\,(s-cc')}}{6V}.$$

438.. *To find the general tangential equation of a sphere.*

Let $(\alpha', \beta', \gamma', \delta')$ be the tetrahedral co-ordinates of the center of the sphere, ρ its radius, then we shall have for any tangent plane, whose co-ordinates are $(\alpha, \beta, \gamma, \delta)$,

$$\alpha'\alpha + \beta'\beta + \gamma'\gamma + \delta'\delta = \pm\,\rho, \text{ Art. (109)},$$

and the homogeneous equation of the sphere will therefore be

$$(\alpha'\alpha+\beta'\beta+\gamma'\gamma+\delta'\delta)^2 = \rho^2\left\{\frac{\alpha^2}{p_1^{\,2}} + \frac{\beta^2}{p_2^{\,2}} + \frac{\gamma^2}{p_3^{\,2}} + \frac{\delta^2}{p_4^{\,2}} - 2\,\frac{\beta\gamma}{p_2p_3}\cos(AD) - \dots\right\}.$$

439. *To find the tangential equations of the spheres touching the faces of the fundamental tetrahedron.*

In order that the surface may be touched by the plane BCD, the tangential equation must be satisfied by the plane $\beta = 0$, $\gamma = 0$, $\delta = 0$; hence the coefficient of α^2 must vanish, and hence, for the spheres touching all the faces of the tetrahedron, we must have

$$p_1^{\,2}\alpha'^2 = p_2^{\,2}\beta'^2 = p_3^{\,2}\gamma'^2 = p_4^{\,2}\delta'^2 = \rho^2.$$

The tangential equation of the inscribed sphere will therefore be

$$\frac{\beta\gamma}{p_2p_3}\cos^2\frac{AD}{2} + \dots = 0,$$

and of the remaining seven spheres touching the faces,

$$\frac{\beta\gamma}{p_2p_3}\cos^2\frac{AD}{2} + \frac{\gamma\alpha}{p_3p_1}\cos^2\frac{BD}{2} + \frac{\alpha\beta}{p_1p_2}\cos^2\frac{CD}{2}$$
$$- \frac{\alpha\delta}{p_1p_4}\sin^2\frac{BC}{2} - \frac{\beta\delta}{p_2p_4}\sin^2\frac{CA}{2} - \frac{\gamma\delta}{p_3p_4}\sin^2\frac{AB}{2} = 0,$$

with three similar equations, and

$$\frac{\beta\gamma}{p_2p_3}\cos^2\frac{AD}{2} + \frac{\alpha\delta}{p_1p_4}\cos^2\frac{BC}{2} - \frac{\gamma\alpha}{p_3p_1}\sin^2\frac{BD}{2}$$
$$- \frac{\alpha\beta}{p_1p_2}\sin^2\frac{CD}{2} - \frac{\beta\delta}{p_2p_4}\sin^2\frac{CA}{2} - \frac{\gamma\delta}{p_3p_4}\sin^2\frac{AB}{2} = 0,$$

with two similar equations.

If $\alpha, \beta, \gamma, \delta$ be the co-ordinates of any tangent plane to one of these spheres, we shall have

$$\pm\frac{\alpha}{p_1}\pm\frac{\beta}{p_2}\pm\frac{\gamma}{p_3}\pm\frac{\delta}{p_4}=1,$$

the signs being all positive for the inscribed sphere, three positive and one negative for the second set of spheres; and two positive, two negative, for the last set.

440. *To find the centers of similarity of the inscribed, and the four escribed, spheres of the fundamental tetrahedron.*

If any point be taken on the line joining the center of the inscribed sphere, and that of the escribed sphere opposite A, dividing the distance between them in the ratio $\mu : \lambda$, the equation of this point is

$$\lambda\frac{\left(\frac{\alpha}{p_1}+\frac{\beta}{p_2}+\frac{\gamma}{p_3}+\frac{\delta}{p_4}\right)}{\frac{1}{p_1}+\frac{1}{p_2}+\frac{1}{p_3}+\frac{1}{p_4}}\pm\mu\frac{-\frac{\alpha}{p_1}+\frac{\beta}{p_2}+\frac{\gamma}{p_3}+\frac{\delta}{p_4}}{-\frac{1}{p_1}+\frac{1}{p_2}+\frac{1}{p_3}+\frac{1}{p_4}}=0,$$

but if this point be a center of similarity, the ratio $\mu : \lambda$ must be the ratio of the radii of the two spheres, or

$$\lambda\left(-\frac{1}{p_1}+\frac{1}{p_2}+\frac{1}{p_3}+\frac{1}{p_4}\right)\pm\mu\left(\frac{1}{p_1}+\frac{1}{p_2}+\frac{1}{p_3}+\frac{1}{p_4}\right)=0,$$

and the equations of the two centers of similarity are

$$\alpha=0,\quad \frac{\beta}{p_2}+\frac{\gamma}{p_3}+\frac{\delta}{p_4}=0,$$

with similar equations for the centers of similarity of the inscribed sphere, and any other of the escribed spheres.

In a similar manner for the two escribed spheres opposite A and D, we shall find the centers of similarity to be

$$\frac{\alpha}{p_1}=\frac{\delta}{p_4};\quad \frac{\beta}{p_2}+\frac{\gamma}{p_4}=0;$$

the former being external, the latter internal.

The centers of similarity of the escribed spheres are therefore the points in which the planes, bisecting the internal and external angles between the faces, meet the opposite edges.

In tetrahedral co-ordinates these results are, (1) the center of similarity of the inscribed sphere, and the escribed sphere opposite A, are given by the equations

$$\beta = 0,\ \gamma = 0,\ \delta = 0\,;\ \alpha = 0,\ p_2\beta = p_3\gamma = p_4\delta\,;$$

(2) the centers of similarity of the inscribed spheres opposite A and D are given by the equations

$$\beta = 0,\ \gamma = 0,\ p_1\alpha + p_4\delta = 0\,;\ \alpha = 0,\ \delta = 0,\ p_2\beta = p_3\gamma.$$

The twelve centers of similarity of the inscribed sphere and the three escribed spheres opposite A, B, and C, will lie by sixes on the eight planes

$$\delta = 0\,;\quad \alpha = 0,\ \beta = 0,\ \gamma = 0,\ p_1\alpha + p_2\beta + p_3\gamma - 2p_4\delta = 0\,;$$

$$p_2\beta + p_3\gamma - p_4\delta = 0,\ p_3\gamma + p_1\alpha - p_4\delta = 0,\ p_1\alpha + p_2\beta + p_4\delta = 0\,;$$

the first containing the six external centers; each of the next four containing three external and three internal; and each of the last three containing two external and four internal poles.

The twelve centers of similarity of the four escribed spheres will lie in the same manner on the eight planes

$$p_1\alpha + p_2\beta + p_3\gamma + p_4\delta = 0\,;\ \alpha = 0,\ \beta = 0,\ \gamma = 0,\ \delta = 0\,;$$

$$p_1\alpha + p_2\beta - p_3\gamma - p_4\delta = 0,\ p_1\alpha - p_2\beta - p_3\gamma + p_4\delta = 0,$$

$$p_1\alpha - p_1\beta + p_3\gamma - p_4\delta = 0.$$

441. *To find the distance between the centers of the inscribed and circumscribed spheres.*

The tetrahedral co-ordinates of the center of the inscribed sphere are $\dfrac{r}{p_1}, \dfrac{r}{p_2}, \dfrac{r}{p_3}, \dfrac{r}{p_4}$, if r be its radius; hence the rectangle of the segments of any chord of the circumscribed sphere passing through this point

$$= -r^2\left(\frac{a'^2}{p_2p_3} + \frac{a^2}{p_1p_4} + \frac{b'^2}{p_3p_1} + \frac{b^2}{p_2p_4} + \frac{c'^2}{p_1p_2} + \frac{c^2}{p_3p_4}\right);$$

or, if Δ be the distance between the centers,

$$\Delta^2 = R^2 - r^2\left(\frac{a'^2}{p_2p_3} + \frac{a^2}{p_1p_4} + \frac{b'^2}{p_3p_1} + \frac{b^2}{p_2p_4} + \frac{c'^2}{p_1p_2} + \frac{c^2}{p_3p_4}\right):$$

a result due to Mr Salmon. *Quarterly Journal*, No. 15.

XIX.

1. If A, B, C, D be the vertices of the four cones of the second degree, which can be described through the curve of intersection of two conicoids, the triangle BCD will be a conjugate triad of the section made by its plane of the cone whose vertex is A.

2. If of eight given points six lie by threes on two non-intersecting straight lines, shew that no cones can be described through the eight points; but that there is an infinite number of points, lying on two straight lines, which have their polar planes, with respect to any conicoid containing the eight points, fixed.

3. If of seven given points six lie by threes on two non-intersecting straight lines, shew that the remaining line of intersection of any two paraboloids passing through the seven points will be a fixed straight line at infinity.

4. If a conicoid be described containing the edges AB, BC, CD of a tetrahedron, the pole of the plane bisecting the edges AB, CD, AC, BD will lie on the plane bisecting the edges AB, CD, AD, BC.

5. Shew that only one conicoid can be described containing two given non-intersecting straight lines, and touching three given planes.

6. Shew that eight conicoids can in general be described passing through four given points, touching three given planes, and having the intersections of the tangent planes at the four given points, with the corresponding planes containing the points, lying in one plane.

7. Shew that four conicoids can in general be described passing through five points in one plane, two other given points not in that plane, and touching two given planes.

8. Shew that a sphere can be described touching the edges of a tetrahedron, whose lengths are a, a'; b, b'; c, c'; if

$$a \pm a' = b \pm b' = c \pm c',$$

the ambiguities being independent.

9. If a tetrahedron be self-conjugate with respect to a sphere, shew that the opposite edges are, two and two, at right angles; and that all the plane angles containing one of the solid angles must be obtuse. Shew that this angular point will lie within the sphere, and the three others without, and determine the radius of the sphere.

10. The number of paraboloids, which can be drawn through eight given points, is, in general, three.

11. Prove that the equations of two conicoids cannot both be obtained in the form of Art. (416), if they have a common generating line.

12. Four cones of the second degree can be drawn, each containing the locus of the center of a conicoid passing through eight given points; and having their vertices coincident with the vertices of the cones on which lie the eight given points.

13. Prove that, in general, one conicoid can be described, which is self-conjugate with respect to one given tetrahedron, and also with respect to another of which three angular points are given. Shew also that the fourth angular point of the second tetrahedron will be fixed.

14. If two conicoids be so related that a tetrahedron can be drawn, whose faces touch one of the conicoids, and two pairs of whose opposite edges lie on the other, an infinite number of tetrahedrons can be so drawn.

15. If the conicoids

$$l\alpha^2 + m\beta^2 + n\gamma^2 + r\delta^2 = 0, \qquad l'\alpha^2 + m'\beta^2 + n'\gamma^2 + r'\delta^2 = 0,$$

be so related, prove that a similar relation will exist between the conicoids

$$l'\alpha^2 + m'\beta^2 + n'\gamma^2 + r'\delta^2 = 0 \text{ and } \frac{l'^2}{l}\alpha^2 + \frac{m'^2}{m}\beta^2 + \frac{n'^2}{n}\gamma^2 + \frac{r'^2}{r}\delta^2 = 0.$$

16. Prove that two tetrahedrons may be inscribed in the conicoid

$$l\alpha^2 + m\beta^2 + n\gamma^2 + r\delta^2 = 0,$$

having their faces parallel to the faces of the fundamental tetrahedron, provided that

$$(l + m + n + r)\left(\frac{1}{l} + \frac{1}{m} + \frac{1}{n} + \frac{1}{r}\right) < 4.$$

17. Prove that, in order that the two conicoids in (15) may be so related, one of the following conditions must hold:

$$\frac{l'}{l} + \frac{m'}{m} = \frac{n'}{n} = \frac{r'}{r},$$

$$\frac{l'}{l} + \frac{n'}{n} = \frac{m'}{m} + \frac{r'}{r},$$

$$\frac{l'}{l} + \frac{r'}{r} = \frac{m'}{m} + \frac{n'}{n}.$$

CHAPTER XIX.

CURVES.

442. We have already considered curves as the complete or partial intersection of surfaces; but in the investigation of the properties of curves we may also have to consider curves as the locus of points each of which satisfies given laws, the algebraical statement of which assumes the form of equations between the co-ordinates of any point in the curve, and variable parameters, the number of equations being two more than the number of parameters.

Instances of this mode of representation of a curve occur in dynamical problems in which the curve is defined by equations between the co-ordinates of the position of a particle and the time of its arrival at that position.

If from such equations the parameters were eliminated, the result would be two final equations, that is, the equations of two surfaces whose complete or partial intersections are the curve in question.

As an example of a curve in space considered in this point of view, we may take the Helix, which is generated by the uniform motion of a point along a generating line of a right cylinder as the generating line revolves with uniform angular velocity about the axis of the cylinder.

If we take the axis for the axis of z, and the axis of x through the generating point at any initial time, θ the angle through which the generating line has revolved when the point has moved through a space z on the generating line, we have, for the co-ordinates of the point, a being the radius of the cylinder,

$$x = a\cos\theta,\ y = a\sin\theta,\ z = na\theta;$$

here θ is the variable parameter, and the curve is the intersection of the surfaces

$$x^2 + y^2 = a^2, \text{ and } y = x\tan\frac{z}{na}.$$

Tangent line to a Curve.

443. *To find the equations of the tangent line at a given point in a curve.*

Let (x, y, z) be the given point P, $(x+\Delta x,\ y+\Delta y,\ z+\Delta z)$ an adjacent point Q, the equations of the line PQ are

$$\frac{\xi - x}{\Delta x} = \frac{\eta - y}{\Delta y} = \frac{\zeta - z}{\Delta z}.$$

If Q move up to and ultimately coincides with P, the limiting position of PQ is the tangent at P, if, therefore, $dx : dy : dz$ be the ultimate value of $\Delta x : \Delta y : \Delta z$, the equations of the tangent at P are

$$\frac{\xi - x}{dx} = \frac{\eta - y}{dy} = \frac{\zeta - z}{dz}.$$

(1) Let the equations of the curve be given in terms of a variable parameter θ, in the form

$$\xi = \phi(\theta), \quad \eta = \psi(\theta), \quad \zeta = \chi(\theta),$$
$$dx : dy : dz = \phi'(\theta) : \psi'(\theta) : \chi'(\theta),$$

and the equations of the tangent at a point corresponding to θ are

$$\frac{\xi - x}{\phi'(\theta)} = \frac{\eta - y}{\psi'(\theta)} = \frac{\xi - z}{\chi'(\theta)}.$$

(2) Let the equations be those of surfaces containing the curve, $F(\xi, \eta, \zeta) = 0$, and $G(\xi, \eta, \zeta) = 0$.

Then on any point P of the curve,

$$F'(x)\,dx + F'(y)\,dy + F'(z)\,dz = 0,$$

and $$G'(x)\,dx + G'(y)\,dy + G'(z)\,dz = 0;$$

whence the equations of the tangent PQ may be written

$$F'(x)(\xi - x) + F'(y)(\eta - y) + F'(z)(\zeta - z) = 0;$$

and $$G'(x)(\xi - x) + G'(y)(\eta - y) + G'(z)(\zeta - z) = 0,$$

which equations represent analytically the fact that the tangent to the curve lies in each of the tangent planes to the surfaces at the common point P.

(3) If the surfaces, of which the intersection gives the curve, be cylindrical surfaces whose sides are parallel to the two

axes of z and y, and their equations be $\eta = f(\xi)$, $\zeta = \phi(\xi)$, the equations of the tangent will be

$$\eta - y = f'(x)\,(\xi - x),$$
$$\zeta - z = \phi'(x)\,(\xi - x).$$

Those equations are the analytical representation of the fact that the projections of the tangent to the curve are the tangents to the projections of the curve on the co-ordinate planes of xy, zx; which is obviously true, since the projections of P and Q have their ultimate coincidence simultaneously with that of P and Q.

444. If the point P be a multiple point on the curve, there being more than one tangent line, the equations given in (2) of the last article will be satisfied by more than one system of values of $dx : dy : dz$; therefore,

either $$F'(x) = F'(y) = F'(z) = 0 \qquad (1),$$
or $$G'(x) = G'(y) = G'(z) = 0 \qquad (2),$$

or both sets of equations hold simultaneously, or else

$$\frac{F'(x)}{G'(x)} = \frac{F'(y)}{G'(y)} = \frac{F'(z)}{G'(z)} \qquad (3).$$

The case (1) occurs when the first surface has a multiple point at P, in which case the tangent lines at the multiple point of the curve are the intersections of the conical tangent to the first surface with the tangent plane to the second.

The case of (1) and (2) holding simultaneously is that of both surfaces having multiple points at P, and the tangent lines are the intersections of the conical tangents.

The case (3) occurs when the surfaces have a common tangent plane at P.

445. *To find the directions of the branches of the curve of intersection of two surfaces, at a multiple point.*

The equations of the surfaces being

$$F(\xi, \eta, \zeta) = 0, \quad \text{and} \quad G(\xi, \eta, \zeta) = 0,$$

and (x, y, z) being a multiple point P on the curve, let

$$\frac{\xi - x}{l} = \frac{y - \eta}{m} = \frac{z - \zeta}{n} = r \qquad (1)$$

be the equations of a line through P, the points in which this line meets the surfaces are given by the equations

$$\left.\begin{array}{r} F(x+lr,\ y+mr,\ z+nr)=0 \\ \text{and}\ \ G(x+lr,\ y+mr,\ z+nr)=0 \end{array}\right\} \qquad (2),$$

there are an infinite number of directions which give two values of r equal to zero, since the curve has a multiple point at P: therefore the two equations

$$\left(l\frac{d}{dx}+m\frac{d}{dy}+n\frac{d}{dz}\right)F(x,y,z)=0 \qquad (3),$$

$$\text{and}\ \ \left(l\frac{d}{dx}+m\frac{d}{dy}+n\frac{d}{dz}\right)G(x,y,z)=0 \qquad (4),$$

must be one or both identically satisfied, or else they must not be independent equations.

1. If (3) be identically satisfied, then of the values of r common to the equations, two become zero for the directions given by equation (4), and

$$\left(l\frac{d}{dx}+m\frac{d}{dy}+n\frac{d}{dz}\right)^2 F(x,y,z)=0,$$

and these determine the directions of the tangent lines at the multiple point.

2. If (3) and (4) be both identically satisfied, the directions of the tangent lines are given by the two equations

$$\left.\begin{array}{r} \left(l\frac{d}{dx}+m\frac{d}{dy}+n\frac{d}{dz}\right)^2 F(x,y,z)=0 \\ \text{and}\ \left(l\frac{d}{dx}+m\frac{d}{dy}+n\frac{d}{dz}\right)^2 G(x,y,z)=0 \end{array}\right\}.$$

3. If neither be identically satisfied, let $(3)\times\lambda+(4)\times\mu=0$ be an identical equation, and by (2),

$$\lambda F(x+lr,\ y+mr,\ z+nr)+\mu G(x+lr,\ y+mr,\ z+nr)=0;$$

therefore the directions of the line (1) which give two values of r common to (2), each equal to zero, are given by the equations (3) *or* (4), and

$$\left(l\frac{d}{dx}+m\frac{d}{dy}+n\frac{d}{dz}\right)^2\{\lambda F(x,y,z)+\mu G(x,y,z)\}=0 \ldots\ldots (5).$$

If (5) combined with (3) or (4) give equal values of $l:m:n$, the two tangents coincide and the point P is in this case a cusp on the curve.

We can proceed in a similar manner for higher orders of multiplicity.

446. *To find the differential coefficient of the arc of a curve.*

If x, y, z and $x+\Delta x$, $y+\Delta y$, $z+\Delta z$ be the co-ordinates of P, and an adjacent point on the curve, Δs the length of the arc between them,

$$PQ^2 = \overline{\Delta x}|^2 + \overline{\Delta y}|^2 + \overline{\Delta z}|^2;$$

therefore, since limit $\dfrac{PQ}{\Delta s} = 1$, dividing by $(\Delta s)^2$ and proceeding to the limit we obtain

$$\left(\frac{dx}{ds}\right)^2 + \left(\frac{dy}{ds}\right)^2 + \left(\frac{dz}{ds}\right)^2 = 1.$$

If the position of P be determined by the variable t,

$$\left(\frac{ds}{dt}\right)^2 = \left(\frac{dx}{dt}\right)^2 + \left(\frac{dy}{dt}\right)^2 + \left(\frac{dz}{dt}\right)^2;$$

and $\dfrac{dx}{ds}, \dfrac{dy}{ds}, \dfrac{dz}{ds}$ are the direction-cosines of the tangent.

447. *To find the differential coefficients of the arc referred to polar co-ordinates.*

Transforming to polar co-ordinates

$$x = r\sin\theta\cos\phi = \rho\cos\phi,$$
$$y = r\sin\theta\sin\phi = \rho\sin\phi,$$
$$z = r\cos\theta,$$
$$\rho = r\sin\theta,$$

$$\left(\frac{dx}{d\theta}\right)^2 + \left(\frac{dy}{d\theta}\right)^2 = \left(\frac{d\rho}{d\theta}\right)^2 + \rho^2\left(\frac{d\phi}{d\theta}\right)^2,$$

$$\left(\frac{dz}{d\theta}\right)^2 + \left(\frac{d\rho}{d\theta}\right)^2 = \left(\frac{dr}{d\theta}\right)^2 + r^2;$$

$$\therefore \left(\frac{ds}{d\theta}\right)^2 = \left(\frac{dr}{d\theta}\right)^2 + r^2 + r^2\sin^2\theta\left(\frac{d\phi}{d\theta}\right)^2.$$

The equation is easily obtained geometrically by observing that ultimately

$$\overline{\Delta s}|^2 = \overline{\Delta r}|^2 + \overline{r\Delta\theta}|^2 + (r\sin\theta\Delta\phi)^2.$$

Also, if p be the perpendicular from the pole upon the tangent, and ψ the angle between r and the tangent,

$$\frac{\Delta s}{\Delta r} = \sec \psi \text{ ultimately,}$$

$$\text{and } \sin \psi = \frac{p}{r};$$

$$\therefore \left(\frac{ds}{dr}\right)^2 = \frac{r^2}{r^2 - p^2}.$$

Normal Plane.

448. *To find the equation of the normal plane at any point of a curve.*

Since the tangent line is perpendicular to the normal plane, the direction-cosines of the normal plane are in the ratio

$$dx : dy : dz.$$

The equation is therefore

$$(\zeta - x)\, dx + (\eta - y)\, dy + (\xi - z)\, dz = 0.$$

449. *To find the edge of regression of the developable surface formed by the normal planes to a curve.*

The equation of a normal plane at (x, y, z) is

$$(\xi - x)\, dx + (\eta - y)\, dy + (\zeta - z)\, dz = 0.$$

The line of intersection of this normal plane with the normal plane at the point $(x + dx,\ y + dy,\ z + dz)$ is given by the above equation and the equation

$$(\xi - x)\, d^2x + (\eta - y)\, d^2y + (\zeta - z)\, d^2z - \overline{dx}|^2 - \overline{dy}|^2 - \overline{dz}|^2 = 0,$$

and any point in the edge of regression is therefore given by these equations and the equation

$$(\xi - x)\, d^3x + (\eta - y)\, d^3y + (\zeta - z)\, d^3z$$
$$- 3\,(dx\, d^2x + dy\, d^2y + dz\, d^2z) = 0.$$

The two equations of the edge of regression are therefore obtained by eliminating x, y, z between these equations and the equations of the curve.

Osculating Plane.

450. Def. The *osculating plane* is the position which a plane passing through three adjacent points assumes as a limiting position, when the three points are ultimately coincident.

451. *To find the equation of the osculating plane at any point of a curve.*

Let the co-ordinates of any point P in the curve be considered as functions of a variable t, and let $t-\tau$ and $t+\tau$ be the values of t corresponding to points Q and R on opposite sides of P.

The direction-cosines of QP and PR are proportional to the differences of the co-ordinates of their extremities, and if λ, μ, ν be the direction-cosines of the normal to the plane QPR,

$$\lambda\left\{-\frac{dx}{dt}\tau+\left(\frac{d^2x}{dt^2}+\epsilon\right)\frac{\tau^2}{2}\right\}+\mu\left\{-\frac{dy}{dt}\tau+\left(\frac{d^2y}{dt^2}+\epsilon'\right)\frac{\tau^2}{2}\right\}$$
$$+\ldots\ldots\ldots\ldots\ldots=0,$$

and
$$\lambda\left\{\frac{dx}{dt}\tau+\left(\frac{d^2x}{dt^2}+\epsilon_1\right)\frac{\tau^2}{2}\right\}+\mu\left\{-\frac{dy}{dt}\tau+\left(\frac{d^2y}{dt^2}+\epsilon_1'\right)\frac{\tau^2}{2}\right\}$$
$$+\ldots\ldots\ldots\ldots\ldots=0,$$

where ϵ, ϵ', ... vanish when τ is indefinitely diminished, therefore ultimately

$$\lambda\frac{dx}{dt}+\mu\frac{dy}{dt}+\nu\frac{dz}{dt}=0,$$

and
$$\lambda\frac{d^2x}{dt^2}+\mu\frac{d^2y}{dt^2}+\nu\frac{d^2z}{dt^2}=0,$$

whence
$$\frac{\lambda}{\dfrac{dy}{dt}\dfrac{d^2z}{dt^2}-\dfrac{dz}{dt}\dfrac{d^2y}{dt^2}}=\frac{\mu}{\dfrac{dz}{dt}\dfrac{d^2x}{dt^2}-\dfrac{dx}{dt}\dfrac{d^2z}{dt^2}}=\ldots\ldots$$

therefore the equation of the osculating plane is

$$\left(\frac{dy}{dt}\frac{d^2z}{dt^2}-\frac{dz}{dt}\frac{d^2y}{dt^2}\right)(\xi-x)+\left(\frac{dz}{dt}\frac{d^2x}{dt^2}-\frac{dx}{dt}\frac{d^2z}{dt^2}\right)(\eta-y)$$
$$+\left(\frac{dx}{dt}\frac{d^2y}{dt^2}-\frac{dy}{dt}\frac{d^2x}{dt^2}\right)(\zeta-z)=0.$$

452. The osculating plane may be considered also as the plane which has the closest contact with the curve at the given point.

Thus, if $L(\xi - x) + M(\eta - y) + N(\zeta - z) = 0$ be the equation of the osculating plane at (x, y, z) and $x + \Delta x$, $y + \Delta y$, $z + \Delta z$ be the co-ordinates of a point near (x, y, z), the perpendicular distance from this point on the plane is

$$\frac{L\Delta x + M\Delta y + N\Delta z}{(\overline{\Delta x}|^2 + \overline{\Delta y}|^2 + \overline{\Delta z}|^2)^{\frac{1}{2}}}.$$

If x, y, z be functions of t,

$$\Delta x = \frac{dx}{dt}\Delta t + \frac{1}{2}\frac{d^2x}{dt^2}\overline{\Delta t}|^2 (1 + \epsilon),$$

where ϵ vanishes in the limit; therefore L, M, N must be chosen so as to make $L\Delta x + M\Delta y + N\Delta z$ the least possible when Δt is made as small as we please; therefore

$$Ldx + Mdy + Ndz = 0,$$

and $$Ld^2x + Mdy + Nd^2z = 0,$$

whence the equation of the osculating plane is found as before.

453. The osculating plane is also the plane containing two tangent lines at consecutive points.

Their equations will be $\dfrac{\xi - x}{dx} = \dfrac{\eta - y}{dy} = \dfrac{\zeta - z}{dz}$,

and $$\frac{\xi - x - dx}{dx + d^2x} = \frac{\eta - y - dy}{dy + d^2y} = \frac{\zeta - z - dz}{dz + d^2z}.$$

If therefore the equation of such a plane be

$$l(\xi - x) + m(\eta - y) + n(\zeta - z) = 0,$$

we obtain the equations

$$l\,dx + m\,dy + n\,dz = 0,$$

and $$l\,d^2x + m\,d^2y + n\,d^2z = 0.$$

454. *To find the direction-cosines of the osculating plane.*

If l, m, n be the direction cosines,

$$\frac{l}{dy\,d^2z - dz\,d^2y} = \frac{m}{dz\,d^2x - dx\,d^2z} = \frac{n}{dx\,d^2y - dy\,d^2x} = \frac{1}{u},$$

where $u^2 = (dy\, d^2z - dz\, d^2y)^2 + \ldots\ldots$

$$= (\overline{dx}|^2 + \overline{dy}|^2 + \overline{dz}|^2)(\overline{d^2x}|^2 + \overline{d^2x}| + \overline{d^2z}|^2)$$

$$- (dx\, d^2x + dy\, d^2y + dz\, d^2z)^2,$$

and
$$\overline{dx}|^2 + \overline{dy}|^2 + \overline{dz}|^2 = \overline{ds}|^2;$$

$$\therefore\ dx\, d^2x + dy\, d^2y + dz\, d^2z = ds\, d^2s;$$

$$\therefore\ u^2 = ds^2\,(\overline{d^2x}|^2 + \overline{d^2y}|^2 + \overline{d^2z}|^2 - \overline{d^2s}|^2;$$

$$\therefore\ l = \frac{dy\, d^2z - dz\, d^2y}{ds\sqrt{\overline{d^2x}|^2 + \overline{d^2y}|^2 + \overline{d^2z}|^2 - \overline{d^2s}|^2}}, \ \&c.,$$

in which $d^2s = 0$ if s be the independent variable.

455. *The osculating plane is perpendicular to the line of intersection of consecutive normal planes.*

This is obvious geometrically, since consecutive normal planes are the limiting positions of planes perpendicular to two adjacent sides of the polygon whose limit is the curve, and their line of intersection is perpendicular to the plane containing the sides which is ultimately the osculating plane.

The equation of the normal plane being

$$(\xi - x)\, dx + (\eta - y)\, dy + (\zeta - z)\, dz = 0,$$

and of the consecutive normal plane

$$(\xi^2 - x - dx)\,(dx + d^2x) + (\eta - y - dy)\,(dy + d^2y) + (\zeta - z - dz)\,(dz + d^2z) = 0,$$

at the line of intersection we have

$$(\xi - x)\, d^2x + (\eta - y)\, d^2y + (\zeta - z)\, d^2z - \overline{dx}|^2 - \overline{dy}|^2 - \overline{dz}|^2 = 0.$$

Hence, the direction-cosines of the line of intersection are proportional to $dy\, d^2z - dz\, d^2y$, &c.

The line is therefore perpendicular to the osculating plane.

Principal Normal.

456. DEF. The *principal normal* is that normal which lies in the plane of two consecutive tangents or in the osculating plane.

457. *To find the equations of the principal normal at any point of a curve.*

The direction-cosines of the tangent at any point P are

$$\frac{dx}{ds},\ \frac{dy}{ds},\ \frac{dz}{ds},$$

and if x, y, z be given functions of t, let $t+\tau$ correspond to a point Q, the direction-cosines of the tangent at Q will be

$$\frac{dx}{ds}+\left\{\frac{d}{dt}\left(\frac{dx}{ds}\right)+\epsilon\right\}\tau,\quad \frac{dy}{ds}+\ldots\ldots,\quad \frac{dz}{ds}+\ldots\ldots$$

If λ, μ, ν be the direction-cosines of a line perpendicular to the principal normal, and the tangents at P and Q, l, m, n those of the principal normal,

$$l\lambda+m\mu+n\nu=0,$$

$$\lambda\frac{dx}{ds}+\mu\frac{dy}{ds}+\nu\frac{dz}{ds}=0,$$

$$\lambda\left[\frac{dx}{ds}+\left\{\frac{d}{dt}\left(\frac{dx}{ds}\right)+\epsilon\right\}\tau\right]+\ldots\ldots=0;$$

therefore, ultimately, $\lambda\dfrac{d}{dt}\left(\dfrac{dx}{ds}\right)+\mu\dfrac{d}{dt}\left(\dfrac{dy}{ds}\right)+\nu\dfrac{d}{dt}\left(\dfrac{dz}{ds}\right)=0.$

If $$l=A\frac{d}{dt}\left(\frac{dx}{ds}\right)+B\frac{dx}{ds},$$

and $$m=A\frac{d}{dt}\left(\frac{dy}{ds}\right)+B\frac{dy}{ds},$$

then $$n=A\frac{d}{dt}\left(\frac{dz}{ds}\right)+B\frac{dz}{ds},$$

and $$\left(\frac{dx}{ds}\right)^2+\left(\frac{dy}{ds}\right)^2+\left(\frac{dz}{ds}\right)^2=1;$$

$$\therefore\ \frac{dx}{ds}\frac{d}{dt}\left(\frac{dx}{ds}\right)+\ldots\ldots=0.$$

Also $$l\frac{dx}{ds}+m\frac{dy}{ds}+n\frac{dz}{ds}=0;$$

$$\therefore\ B=0;$$

$$\therefore\quad \frac{l}{\dfrac{d}{dt}\left(\dfrac{dx}{ds}\right)}=\frac{m}{\dfrac{d}{dt}\left(\dfrac{dy}{ds}\right)}=\frac{n}{\dfrac{d}{dt}\left(\dfrac{dz}{ds}\right)}.$$

458. *If from any point in a curve equal distances be measured along the curve and its tangent, the limiting position of the line joining the extremities of these distances is the principal normal.*

From the point (x, y, z) let equal distances σ be measured along the curve and the tangent to the points Q, T. The co-ordinates of Q are

$$x + \frac{dx}{ds}\sigma + \left(\frac{d^2x}{ds^2} + \epsilon\right)\frac{\sigma^2}{2}, \text{ \&c.}$$

and those of T

$$x + \frac{dx}{ds}\sigma, \text{ ....,}$$

ϵ vanishing in the limit.

The equations of the line QT are

$$\frac{\xi - x}{\frac{d^2x}{ds^2} + \epsilon} = \frac{\eta - y}{\frac{d^2y}{ds^2} + \epsilon'} = \frac{\zeta - z}{\frac{d^2z}{ds^2} + \epsilon''};$$

therefore the limiting position of QT is the principal normal, being perpendicular to the tangent, since

$$\frac{dx}{ds}\frac{d^2x}{ds^2} + \frac{dy}{ds}\frac{d^2y}{ds^2} + \frac{dz}{ds}\frac{d^2z}{ds^2} = 0.$$

Cauchy proposed, as a definition of the Principal Normal at any point, the limiting position of the line joining the points on the curve and tangent, whose distances from the point of contact measured along the curve and tangent respectively are equal; by which means the definition was made independent of the osculating and normal planes.

Four-point System.

459. *Equation of a curve of double curvature.*

If $\alpha, \beta, \gamma, \delta$ be the four-point co-ordinates of any plane, the equation of a point is

$$l\alpha + m\beta + n\gamma + r\delta = 0 \qquad (1),$$

and we have seen (Art. 112) that if l, m, n, r involve one variable in the first degree, the locus of all the points, which can be obtained by giving to the variable values from $-\infty$ to $+\infty$, is a straight line.

Let l, m, n, r be any functions of a single variable x, we can show that the locus of points corresponding to all values of the variable is a curve line; for, if the locus be cut by any plane, and the co-ordinates of the plane be substituted for α, β, γ, δ in the equation of the point, the resulting equation determines a series of values of the variable x, which correspond to the point in which the locus is intersected by the plane; and, by shifting the plane, we obtain a continuous series of such points which form the different portions of the curve line of which (1) may therefore be considered to be the equation.

If l, m, n, r be rational and integral functions of x, not having a common factor, and one at least being of the nth degree, any plane determines n values of x, real or imaginary, and therefore meets the curve in n points, hence the curve is of the nth degree.

We may observe that, in order to be sure that the curve is of the nth degree, it must not be possible to make any substitution of a new variable so as to diminish the degree, while the functions remain rational.

460. If the curve which is the locus of $m\beta + n\gamma + r\delta = 0$ be traced on the fundamental plane BCD, every point in the curve which is the locus of (1) lies on a line joining A with a point of this curve, that is, on the surface of a cone whose vertex is A and guiding curve $m\beta + n\gamma + r\delta = 0$.

461. If
$$l = a_0 + a_1x + a_2x^2 + \dots\dots$$
$$m = b_0 + b_1x + b_2x^2 + \dots\dots$$
$$\dots\dots\dots\dots\dots\dots\dots$$
and
$$a_0\alpha + b_0\beta + c_0\gamma + d_0\delta = a'\alpha',$$
$$a_1\alpha + b_1\beta + c_1\gamma + d_1\delta = b'\beta',$$
$$\dots\dots\dots\dots\dots\dots\dots$$
the equation of the curve may be written
$$a'\alpha' + b'x\beta' + c'x^2\gamma' + d'x^3\delta' + \dots = 0.$$

This reduced equation exhibits that a curve of the first degree is the straight line joining the points
$$\alpha' = 0, \quad \beta' = 0.$$

Also that a curve of the second degree is a plane curve, the plane containing the three points

$$\alpha' = 0, \quad \beta' = 0, \quad \gamma' = 0.$$

Again, by the preceding article, a curve of the third degree, which is not necessarily a plane curve, lies on two cones whose vertices are A' and D', and whose guiding curves are the conics traced on $B'C'D'$ and $A'B'C'$ whose equations are

$$b'\beta' + c'x\gamma' + d'x^2\delta' = 0, \text{ and } a'\alpha' + b'x\beta' + c'x^2\gamma' = 0,$$

which have a common generating line $A'D'$.

462. *To find the equation of the tangent to a curve.*

Let $f(x) \equiv l\alpha + m\beta + n\gamma + r\delta = 0$ be the equation of the curve, and let x_0 determine any point P in the curve, $x_0 + \lambda$ a point Q adjacent to it, whose equation will be

$$f(x_0 + \lambda) \equiv f(x_0) + \lambda f'(x_0) + \frac{\lambda^2}{2} f''(x_0) + \ldots = 0.$$

The straight line whose equation is

$$f(x_0) + \mu f'(x_0) = 0$$

is the equation of the tangent at P, since, when Q moves up to P and ultimately coincides with it, the straight line ultimately passes through Q.

The distance between adjacent points in the tangent and curve is evidently of the order λ^2, generally.

If $f''(x_0) = 0$, the distance is of the order λ^3, and the curve, which in ordinary cases lies on the same side of the tangent on each side of the point of contact, in this case lies on opposite sides, or there is a point of inflexion in the osculating plane.

The equation $f(x_0) + \mu f'(x_0) + \frac{\mu^2}{2} f''(x_0) = 0$ is the equation of a conic which has a contact of the second order.

A double point occurs when l, m, n, r retain the same ratio for two values of x.

463. *To find the equation of the osculating plane at any point of a curve.*

The plane whose equation is

$$f(x_0) + \mu f'(x_0) + \nu f''(x_0) = 0$$

is the plane which passes through the points $f(x_0)=0$, $f'(x_0)=0$, and $f''(x_0)=0$, and therefore coincides with the limiting position of the plane which passes through three contiguous points of the curve.

The equation is therefore the equation required.

The distance of a point $f(x_0+\lambda)=0$ from the osculating plane is ultimately $c\lambda^3 f'''(x_0)$, or the curve generally lies on opposite sides of the osculating plane in passing through the point of contact.

Singularities of Curves and Developables.

464. A curve of double curvature and the developable surface of which it is the edge of regression, may be considered in connexion with one another; and we may expect that singularities in one will have corresponding singularities in the other. Cayley in Liouville's *Journal*, Tom. x., and in the *Cambridge and Dublin Mathematical Journal*, Vol. v., and Salmon, in the same place in the latter Journal, have investigated equations among the number of such singularities; and Salmon has proceeded to shew how curves of double curvature may be classified by the consideration of the number of apparent double points in the curves.

We shall introduce the student to some of the methods employed, and leave him to consult the papers referred to, if he desire to enter more fully into the subject.

465. A curve of double curvature may be considered as the locus of a system of points, or as the envelope of a system of straight lines, and the corresponding developable surface as the locus of the system of lines, or as the envelope of a system of planes.

We may consider the three systems of points, lines, and planes as connected in the following ways.

1. If the system of points be supposed given, each line of the second system joins two consecutive points of the given system, and each plane contains three consecutive points of the same system.

2. If the system of planes be supposed given, each line of the second system is the intersection of two consecutive planes of

the given system, and each point of the first system is the intersection of three consecutive planes of the given system.

3. If the system of lines be supposed given, each point of the first system is the intersection of two consecutive lines of the given system, and each plane contains two consecutive lines of the same system.

466. The following terms will be employed:

A line through two points denotes a line joining *any* two arbitrary points of the system of points.

A line in two planes denotes the line of intersection of *any* two planes of the system of planes.

A point in two lines is the intersection of *any* two lines of the system of lines which intersect.

A plane through two lines is the plane containing *any* two lines which intersect.

A stationary plane is a singular plane which contains four consecutive points, or three consecutive lines, and occurs when two consecutive planes coincide.

A stationary point is a singular point which lies in four consecutive planes, or in three consecutive lines, and occurs when two consecutive points coincide.

Any plane not belonging to the system contains a certain number of *lines in two planes*.

Any point not belonging to the system lies in a certain number of *lines through two points*.

Any plane contains a certain number of *points in two lines*.

Any point lies in a certain number of *planes through two lines*.

These numbers will be denoted by l, λ, p, and ϖ, respectively, and the numbers of stationary planes and points by s and σ.

467. The *degree* of a curve is the number of points in which it intersects an arbitrary plane.

The *class* of a curve is the number of osculating planes which contain an arbitrary point.

The *degree* of the developable surface of which the curve is the edge of regression is the number of points in which it meets an arbitrary straight line.

The *class* of the surface is the number of tangent planes which can be drawn through an arbitrary point.

The *rank* of the system is the number of planes which can be drawn through an arbitrary straight line so as to contain lines of the system.

Hence the rank of the system is the same as the degree of the surface.

The class of the curve is the same as the class of the surface.

468. Singularities in curves of double curvature are by these considerations made to depend upon the singularities of plane curves.

Let a given plane intersect the surface, the plane curve is thus connected with the lines and planes of the system as follows.

Every point of the plane curve is in a line of the system, every tangent to the plane curve is the intersection of the cutting plane with a plane of the system.

A straight line in the cutting plane meets the surface in r points, if r be the degree of the surface; therefore the degree of the curve of intersection is r.

A point in the cutting plane lies in n tangent planes to the surface if n be the class of the surface, hence, n tangent lines to the curve of intersection can be drawn through the point; therefore the class of the curve is n.

When the cutting plane has a *point in two lines* the plane curve has a double point, since the curve of intersection may be supposed generated by the intersection of lines of the system with the cutting plane, and the generating line in this twice passes through the same point.

When the cutting plane has a *line in two planes*, it may be seen similarly that the plane curve has a double tangent.

At the points in which the cutting plane meets the curve of double curvature, two lines of the system meet the plane curve in the same point, this point is therefore a cusp in the plane curve.

For every stationary plane, two consecutive planes coincide, and therefore two tangents to the plane curve coincide, or there is a point of inflexion.

If m be the degree of the curve of double curvature, the plane curve is of the degree r, and of the class n; it has l double points, p double tangents, m cusps, and s points of inflexion.

Hence, the formulæ for plane curves give three independent equations among these numbers.

These formulæ are given by Salmon in the following forms.

If μ be the degree of a plane curve, r its class, δ the number of double points, κ of cusps, τ of double tangents, ι of points of inflexion,

$$\nu = \mu\,\overline{\mu - 1} - 2\delta - 3\kappa$$

$$\iota - \kappa = 3(\nu - \mu),$$

and $$2(\tau - \delta) = (\nu - \mu)(\nu + \mu - 9);$$

whence $$n = r(r - 1) - 2l - 3m,$$

$$s - m = 3(n - r),$$

and $$2(p - l) = (n - r)(n + r - 9).$$

469. If, instead of considering the system in connexion with a plane, which intersects the developable surface in a curve, we consider it in connexion with a point, which is made the vertex of a conical surface whose guiding curve is the curve of double curvature, we shall obtain other relations among the number of singularities, which can be connected with those of a plane section of the conical surface.

Every plane through the vertex cuts the curve in m points, corresponding to which are m generating lines of the cone; also a plane which cuts the cone meets the curve in m points; therefore the line of intersection contains m points on the curve, the plane section of the curve is therefore of the mth degree.

Again, r tangent lines meet the straight line joining any point in the cutting plane with the vertex of the cone, and hence r tangent planes to the cone can be drawn through the point in the cutting plane; therefore there are r tangents to the plane section, which can be drawn through the point, that is, the class of the plane section is r.

The vertex lies in ϖ *lines through two points*, hence there are ϖ double generating lines of the cone, or ϖ double points on the plane section.

The vertex lies in λ *planes through two lines*, hence there are λ double tangent planes to the cone, and therefore λ double tangent lines to the plane section.

The r planes of the system which pass through the vertex correspond to three generating lines, therefore there are n points of inflexion of the plane section.

For every stationary point, two consecutive points coincide; therefore the cone has a cuspidal edge, and therefore the number of cusps of the plane section is σ.

Hence, the plane section of the cone is of the degree m and of the class r; and it has ϖ double points, λ double tangents, σ cusps, and n points of inflexion.

Thus, three more independent equations are obtained,

$$r = m(m-1) - 2\varpi - 3\sigma,$$

$$n - \sigma = 3(r - m),$$

and

$$2(\lambda - \varpi) = (r - m)(r + m - 9).$$

470. As an exercise, the student may calculate the number of such points, and the order and class of a section made by a tangent plane of the developable surface, and of a conical surface in which the vertex is on the curve of double curvature.

He will find that

$\mu = r - 2$ for the first, and $m - 1$ for the second,
$\nu = n - 1$ .... $r - 2$ ....
$\delta = l - 2r + 8$ .... $\varpi - m + 2$ ....
$\kappa = m - 3$ .... σ ....
$\tau = p - n + 2$ .... $\lambda - 2r + 8$ ....
$\iota = s$ .... $n - 3$ ....

and the results of substitution in the three equations for plane curves lead to the same six equations among the number of singularities.

471. If a cone be described whose guiding curve is a given curve of double curvature, λ *lines through two points* pass through the vertex and determine a double side of the cone.

The two points through which any *line through two points* passes may be either distinct or coincident, as in the case of a multiple point of the curve; to an eye placed at the vertex of the cone two different branches will in both cases *appear* to intersect, but will actually intersect only in the latter case; and in the case of actual intersection the intersection will take place for all positions of the vertex. The sum of the *apparent* and *actual* double points is λ.

Salmon has employed the number of these double points to construct a classification of the curves which are the complete or partial intersections of two surfaces of given degrees; in which the distinctions are made according to the number of points in which the surfaces touch, and the nature of the constants where they do touch. The student is referred to the article in the *Cambridge and Dublin Journal*, Vol. v.

XX.

(1) The equations of the tangent to the curve of intersection of the surfaces

$$ax^2 + by^2 + cz^2 = 1$$

and

$$bx^2 + cy^2 + az^2 = 1,$$

are

$$\frac{x(\xi - x)}{ab - c^2} = \frac{y(\eta - y)}{bc - a^2} = \frac{z(\zeta - z)}{ac - b^2}.$$

The tangent line at the point $x = y = z$ lies in the plane

$$(a - b)x + (b - c)y + (c - a)z = 0.$$

(2) If $ac = b^2$ in the curve of the last problem, the tangent lines trace on the plane of xy the two straight lines whose equation is

$$\frac{cx^2}{c^3 - b^3} = \frac{ay^2}{a^3 - b^3}.$$

(3) The equations of a sphere and cylinder being

$$x^2 + y^2 + z^2 = 4a^2$$

and

$$x^2 + z^2 = 2ax,$$

prove that the equations of the tangent to the curve of intersection at the point (α, β, γ) are

$$(\alpha - a)x + \gamma z = 0$$

and

$$\beta y + ax = a(4a - \alpha),$$

and that the equation of the normal plane is

$$\frac{x}{a}-\frac{y}{\beta}=\left(1-\frac{a}{a}\right)\left(\frac{z}{\gamma}-\frac{y}{\beta}\right).$$

(4) The paraboloid whose equation is $ax^2+by^2=4z$ has traced upon it a curve, every point of which is the extremity of the latus rectum of the parabolic section through the axis of z; shew that the tangent to the curve traces upon the plane of xy the curves whose equations are

$$r\sin 2\theta=\pm(a\sim b).$$

(5) A curve is described on the surface of a cylinder, such that it cuts each generating line at an angle proportional to the angular distance of the principal plane through the point of intersection from a fixed principal plane; prove that the equations of the curve are

$$x^2+y^2=a^2, \qquad e^{-\mu\frac{z}{a}}=\cos\left(\mu\tan^{-1}\frac{y}{x}\right),$$

the axis of z being in the fixed principal plane.

(6) Find the projection on yz of any curve of greatest inclination to that plane, traced on the surface

$$ax^2+by^2+cz^2=1.$$

(7) Prove that the edge of regression of the developable surface generated by the normal planes to a helix, is another helix.

(8) If two surfaces touch, the point of contact is a double point on their curve of intersection.

(9) Prove that in the common tangent plane of two surfaces which touch there are two directions, any planes through which meet both surfaces in three coincident points.

(10) Find the condition that two surfaces which touch may have a cusp in the curve of intersection at the point of contact.

(11) Investigate a construction for the apparent double points in the curve of intersection of two conicoids.

(12) The axes of two elliptic cylinders intersect at right angles, shew that the locus of points for which the apparent double points coincide is a hyperbolic cylinder.

(13) If an eye be placed in either of the axes of x, y or z at a distance c from the origin, the two apparent double points of the curve of intersection of the surfaces

$$xyz=a^3 \text{ and } x^2+y^2+z^2=3a^2$$

are distant $\sqrt{9a^2-c^2}$, and the tangent line which passes through the axis of x meets the first surface in another point whose distance from yz is $4a$.

(14) Prove that the edge of regression of the developable whose tangent planes are given by

$$\alpha + 3\beta t + 3\gamma t^2 + \delta t^3 = 0,$$

is the intersection of two cones

$$\gamma^2 = \beta\delta \text{ and } \beta^2 = \alpha\gamma.$$

Prove also that the equation of the developable is

$$(\beta\gamma - \alpha\delta)^2 = 4(\alpha\gamma - \beta^2)(\beta\delta - \gamma^2).$$

(15) Prove that in the curve of the last problem

$$m = 3, \quad n = 3, \quad r = 4,$$
$$s = 0, \quad l = 0, \quad p = 1,$$
$$\sigma = 0, \quad \lambda = 0, \quad \varpi = 1.$$

(16) If A, B, C, D be fixed points and P, Q be drawn in AB, CD dividing them in any the same ratio, and R divides PR in the duplicate ratio, shew that the locus of R is a curve of the third degree, and shew that if SRT meet AC, BD, R divides ST in the ratio of $AP : BP$, and S, T divide AC, BD in the ratio of $PR : QR$.

CHAPTER XX.

ENVELOPES.

472. The problem of finding the locus of the ultimate intersection, or the *envelope*, of a series of surfaces, will separate into two distinct classes of cases, in which the general equation of the series involves one arbitrary parameter, and those in which it involves two. If more than two are involved, there can exist no *locus* of ultimate intersections, for by making small variations in each of the parameters, which are by hypothesis independent, we shall obtain more than three equations for determining the points in which any given surface intersects the consecutive surfaces, from which the current co-ordinates may be eliminated, and one or more equations obtained involving only the parameters, and constants, which are the same for the whole series. Unless the equation, or equations, so obtained, subsist, there can be no locus of ultimate intersection; and if such equations do hold, the case is reduced to that of not more than two independent parameters.

We have already had occasion to notice the difference of the two classes of envelopes here mentioned, a developable surface being the envelope of a series of planes whose equations involve only one parameter, and other surfaces being the envelope of the corresponding tangent planes whose equations involve the co-ordinates of the point of contact, which are equivalent to only two *independent* parameters. The distinction, in the nature of the contact, which exists between the tangent plane to a developable, and to a non-developable surface, will be found to be general; the envelope of a series of surfaces whose equation involves only one parameter touches each of the series in an infinite number of points lying on a curve, and that of a series whose equation involves two parameters, touches each separate surface in one, two, or more points, but in some *fixed* number.

473. *To find the locus of ultimate intersection of a series of surfaces, the general equation of which involves one arbitrary parameter.*

Let
$$F(x, y, z, a) = 0$$
be the equation of one of the series, a being the parameter; and let a consecutive surface be obtained by taking a new value $a+h$ of the parameter.

The equations giving the points of intersection are
$$F(x, y, z, a) = 0, \quad F(x, y, z, a+h) = 0;$$
the second of which may be written
$$F(x, y, z, a) + hF'(a + \theta h) = 0.$$

The points in question then lie on the surfaces
$$F(x, y, z, a) = 0, \quad F'(a + \theta h) = 0,$$
and the ultimate intersection is the curve
$$F(x, y, z, a) = 0, \quad \frac{d}{da} F(x, y, z, a) = 0,$$
from which we may, by the elimination of a, obtain the equation of a surface on which all these curves lie, or the locus of ultimate intersections of the surfaces.

Since this locus is generated by the series of curves whose equations are
$$F = 0, \quad \frac{dF}{da} = 0,$$
such a curve is called the *characteristic* of the envelope.

474. *To find the equations of the edge of the envelope.*

It will generally happen that any characteristic will intersect the consecutive characteristic in a definite number of points. The locus of these points of ultimate intersection is called the edge of the envelope.

If $u \equiv F(x, y, z, a) = 0$ be the equation of one of the series of enveloped surfaces, the characteristic corresponding to this surface is given by the equations
$$u = 0, \quad \frac{du}{da} = 0;$$

the characteristic corresponding to a consecutive surface is given by the equations

$$u+\left(\frac{du}{da}+\epsilon\right)\delta a=0, \text{ and } \frac{du}{da}+\left(\frac{d^2u}{da^2}+\eta\right)\delta a=0,$$

ϵ and η vanishing when $\delta a=0$.

The corresponding point in the edge of the envelope will therefore satisfy the equations

$$u=0, \quad \frac{du}{da}=0, \text{ and } \frac{d^2u}{da^2}=0,$$

which, by the elimination of a, give the two equations of the edge of the envelope.

Thus, in the case of a developable surface, which is the envelope of a plane whose equation involves one parameter, the characteristic is a straight line and the edge is the edge of regression.

475. *The envelope of a series of surfaces, whose equation involves one parameter, will in general touch each of the surfaces along a curve.*

If from the equation

$$\frac{dF}{da}=0,$$

we obtain the equation $a=\phi(x, y, z)$, the equation of the envelope is

$$F\{x, y, z, \phi(x, y, z)\}=0,$$

and hence the envelope meets any particular surface

$$F(x, y, z, a)=0,$$

in all the points in which the surface $\phi(x, y, z)=a$ meets it, or in all the points lying on the curve

$$F(x, y, z, a)=0, \quad \frac{d}{da}F(x, y, z, a)=0.$$

It will also touch the surface at all such points, for the values of $\frac{dz}{dx}, \frac{dz}{dy}$ in the two surfaces are given at such a point by the equations

$$\frac{dF}{dx}+\frac{dF}{dz}\frac{dz}{dx}=0, \quad \frac{dF}{dy}+\frac{dF}{dz}\frac{dz}{dy}=0,$$

and
$$\frac{dF}{dx}+\frac{dF}{dz}\frac{dz}{dx}+\frac{dF}{d\phi}\left\{\frac{d\phi}{dx}+\frac{d\phi}{dz}\frac{dz}{dx}\right\}=0,$$
$$\frac{dF}{dy}+\frac{dF}{dz}\frac{dz}{dy}+\frac{dF}{d\phi}\left\{\frac{d\phi}{dy}+\frac{d\phi}{dz}\frac{dz}{dy}\right\}=0,$$
respectively.

But, at all the points in question, $\phi(x, y, z)=0$,

or
$$\frac{d}{da}F(x, y, z, a)=0,$$
whence we have $\frac{dF}{d\phi}=0$, and the two pairs of equations coincide. The tangent planes to the two surfaces will consequently be coincident, or the envelope will touch any one of the surfaces along the curve in which that surface meets its consecutive. The contact will be real or unreal, according as the characteristic is real or unreal; and may be real for one portion of a series of surfaces, and unreal for the remainder.

It is this property which gives rise to the name envelope for the locus of the ultimate intersections.

476. That the locus of the ultimate intersections of the surfaces which involve one parameter touches, in general, each of the surfaces, or is the envelope of the surfaces, may be seen by geometrical considerations.

For, if U_1, U_2, U_3 be three consecutive surfaces, the two curves of intersection of U_1, U_2 and U_2, U_3 are, ultimately, two consecutive curves which generate the locus of ultimate intersections, and which ultimately coincide; these consecutive curves lie both on the locus and on the surface U_2, therefore, at every point common to the surface U_2 and the locus of ultimate intersections, there will be a common tangent plane; and the same is true of every other surface of the series; hence the locus of ultimate intersections envelopes every surface of the series along a curve line.

As an example of this class of envelopes, we may take a series of spheres having for diametral planes one series of circular sections of an ellipsoid.

Let the equation of the ellipsoid be
$$\frac{x^2}{a^2}+\frac{y^2}{b^2}+\frac{z^2}{c^2}=1,$$

and let the equation of a circular section be

$$\frac{x}{a}\sqrt{a^2-b^2}+\frac{z}{c}\sqrt{b^2-c^2}=\varpi. \qquad \text{(Art. 184.)}$$

If we take a circular section of the other system

$$\frac{x}{a}\sqrt{a^2-b^2}-\frac{z}{c}\sqrt{b^2-c^2}=\varpi',$$

the equation of the sphere containing the two will be

$$x^2+y^2+z^2-b^2-(\varpi+\varpi')\frac{x}{a}\sqrt{a^2-b^2}$$
$$+(\varpi-\varpi')\frac{z}{c}\sqrt{b^2-c^2}+\varpi\varpi'=0; \quad \text{(Art. 186.)}$$

and the condition that the former section may be a diametral plane of the sphere supplies the condition

$$\varpi'=\varpi\frac{a^2+c^2}{a^2-c^2},$$

whence the equation of the sphere may be written

$$x^2+y^2+z^2-b^2-\frac{2\varpi}{a^2-c^2}\{ax\sqrt{a^2-b^2}+cz\sqrt{b^2-c^2}\}+\varpi^2\frac{a^2+c^2}{a^2-c^2}=0.$$

This gives, as the characteristic, the plane section made by

$$ax\sqrt{a^2-b^2}+cz\sqrt{b^2-c^2}=\varpi(a^2+c^2),$$

and for the envelope, the surface whose equation is

$$(a^4-c^4)(x^2+y^2+z^2-b^2)=(ax\sqrt{a^2-b^2}+cz\sqrt{b^2-c^2})^2.$$

This equation may be written in the form

$$\frac{x^2}{a^2}+\frac{y^2}{b^2}+\frac{z^2}{c^2}-1=\frac{1}{a^2b^2c^2}\,\frac{(c^3x\sqrt{a^2-b^2}+a^3z\sqrt{b^2-c^2})^2}{a^4-c^4},$$

which proves that the envelope required is a prolate spheroid, concentric with the ellipsoid, and touching it along a central section.

The equations of the axis of this spheroid are

$$\frac{x}{a\sqrt{a^2-b^2}}=\frac{y}{0}=\frac{z}{c\sqrt{b^2-c^2}},$$

the locus of the centres of the circular sections; and the squares of its semi-axes may readily be found to be a^2+c^2, and b^2. The foci will therefore be two of the umbilici of the ellipsoid.

The characteristic in this case is given by the equations

$$ax\sqrt{a^2-b^2}+cz\sqrt{b^2-c^2}=\varpi(a^2+c^2),$$

$$x^2+y^2+z^2=b^2+\varpi^2\frac{a^2+c^2}{a^2-c^2},$$

and the curve will therefore only be real for values of ϖ satisfying the condition

$$\frac{\varpi^2(a^2+c^2)^2}{a^2(a^2-b^2)+c^2(b^2-c^2)} \not> b^2+\varpi^2\frac{a^2+c^2}{a^2-c^2},$$

or,
$$\varpi^2\frac{a^2+c^2}{a^2-c^2}\left\{\frac{a^2+c^2}{a^2+c^2-b^2}-1\right\} \not> b^2,$$

or,
$$\varpi^2 \not> \frac{a^2-c^2}{a^2+c^2}(a^2+c^2-b^2).$$

Also, since the characteristics are circles on a series of parallel planes, their ultimate intersections will be impossible and at infinity; and the edge of the envelope is in this case the two impossible circular points at infinity, which lie in the plane

$$ax\sqrt{a^2-b^2}+cz\sqrt{b^2-c^2}=0.$$

477. *To find the locus of ultimate intersection of a series of surfaces, the general equation of which involves two arbitrary parameters.*

Let the equation of any such surface be

$$F(x, y, z, a, b)=0,$$

in which a, b are independent arbitrary parameters. The equation of any consecutive surface may be taken to be

$$F(x, y, z, \ a+\delta a, \ b+\delta b)=0,$$

and the points of ultimate intersection of the two surfaces will be given by the equations

$$F=0, \quad \delta a\frac{dF}{da}+\delta b\frac{dF}{db}=0;$$

or, since δa, δb are independent, on the surfaces

$$F=0, \quad \frac{dF}{da}=0, \quad \frac{dF}{db}=0;$$

and, by the elimination of a and b from these equations, the equation of the locus of all such ultimate intersections will be determined.

478. *To shew that the locus of ultimate intersections is the envelope of a series of surfaces, whose general equation involves two parameters.*

The locus of ultimate intersections being obtained by the elimination of a and b from the equations

$$F=0, \quad \frac{dF}{da}=0, \quad \frac{dF}{db}=0,$$

we may assume that a and b are found from the two latter equations in the forms

$$a=\phi_1(x, y, z), \quad b=\phi_2(x, y, z),$$

and that, consequently, the equation of the locus is

$$F(x, y, z, \phi_1, \phi_2)=0.$$

Now, let x, y, z, be the co-ordinates of any point in which the envelope meets the surface

$$F(x, y, z, a, b)=0,$$

we shall have, to determine $\frac{dz}{dx}$, $\frac{dz}{dy}$ at the point, the following pairs of equations in the two surfaces, respectively,

$$\left.\begin{array}{l} \dfrac{dF}{dx}+\dfrac{dF}{dz}\dfrac{dz}{dx}+\dfrac{dF}{d\phi_1}\left\{\dfrac{d\phi_1}{dx}+\dfrac{d\phi_1}{dz}\dfrac{dz}{dx}\right\}+\dfrac{dF}{d\phi_2}\left\{\dfrac{d\phi_2}{dx}+\dfrac{d\phi_2}{dz}\dfrac{dz}{dx}\right\}=0 \\ \dfrac{dF}{dy}+\dfrac{dF}{dz}\dfrac{dz}{dy}+\dfrac{dF}{d\phi_1}\left\{\dfrac{d\phi_1}{dy}+\dfrac{d\phi_1}{dz}\dfrac{dz}{dy}\right\}+\dfrac{dF}{d\phi_2}\left\{\dfrac{d\phi_2}{dy}+\dfrac{d\phi_2}{dz}\dfrac{dz}{dy}\right\}=0 \end{array}\right\},$$

and

$$\left.\begin{array}{l} \dfrac{dF}{dx}+\dfrac{dF}{dz}\dfrac{dz}{dx}=0 \\ \dfrac{dF}{dy}+\dfrac{dF}{dz}\dfrac{dz}{dy}=0 \end{array}\right\}.$$

But, at the point in question, we have

$$a=\phi_1(x, y, z), \quad b=\phi_2(x, y, z), \quad \frac{dF}{da}=0, \quad \frac{dF}{db}=0;$$

whence, also, at this point,

$$\frac{dF}{d\phi_1}=0,\quad \frac{dF}{d\phi_2}=0\,;$$

from which equations it follows that the values of $\frac{dz}{dx}$, $\frac{dz}{dy}$ are the same in the two surfaces, or that the locus of ultimate intersections touches each of the series of surfaces in the points in which that surface meets the consecutive surfaces. The locus of ultimate intersections is therefore the envelope.

This contact may, as in the former case, be either real or unreal.

479. *To find the envelope of a series of surfaces whose equation involves* n *parameters, connected by either* n − 1 *or* n − 2 *equations.*

We have already remarked that no envelope can exist for a series of surfaces, if the general equation involve more than two independent arbitrary parameters. If therefore the equation involve n parameters, we must have either $n-1$ or $n-2$ equations of condition, by means of which we might eliminate all the parameters but one, or two, respectively, and the envelope might then be obtained as before. A more convenient method, however, may be deduced from the consideration that the equation of the envelope of the surfaces

$$F(x, y, z, a, b)=0$$

is

$$\phi(x, y, z)=0,$$

where $\phi(x, y, z)$ is the maximum or minimum value of

$$F(x, y, z, a, b)$$

obtained by variation of a and b.

The envelope of the surfaces, whose general equation is

$$u\equiv F(x, y, z, a_1, a_2 \ldots\ldots a_n)=0,$$

in which $a_1, a_2, \ldots\ldots a_n$ are connected by the equations

$$\phi_1=0,\ \phi_2=0\ \ldots\ldots\ \phi_{n-2}=0,$$

will be $v=0$, if v be the maximum or minimum value of u obtained by variation of $a_1, a_2, \ldots\ldots a_n$ subject to the $n-1$ or $n-2$ equations of condition.

The ordinary method of proceeding by undetermined multipliers is described and explained in all treatises on the Differential Calculus. We will only exemplify it in the following case.

480. *To find the envelope of a series of planes passing through the centre of an ellipsoid and intersecting it in sections of constant area.*

The equation of the plane may be taken to be

$$u \equiv lx + my + nz = 0,$$

the parameters l, m, n being connected by the equations

$$l^2 + m^2 + n^2 = 1,$$

$$l^2a^2 + m^2b^2 + n^2c^2 = d^2;$$

whence, differentiating, and using undetermined multipliers λ, μ, we obtain

$$\lambda x + \mu l + a^2 l = 0,$$

$$\lambda y + \mu m + b^2 m = 0,$$

$$\lambda z + \mu n + c^2 n = 0,$$

and multiplying by l, m, n, respectively, and adding

$$\lambda u + \mu + d^2 = 0,$$

whence we may obtain the equation

$$\frac{x^2}{a^2 - d^2 - \lambda u} + \frac{y^2}{b^2 - d^2 - \lambda u} + \frac{z^2}{c^2 - d^2 - \lambda u} = 0;$$

or, since for the envelope, $u = 0$,

$$\frac{x^2}{a^2 - d^2} + \frac{y^2}{b^2 - d^2} + \frac{z^2}{c^2 - d^2} = 0,$$

the equation of the envelope, which is a cone whose focal lines are the asymptotes of the focal hyperbola of the ellipsoid.

We may observe that since the maximum or minimum value of u is zero for all points of the envelope, we may put it zero at any stage of the operation, which would have given above the equation

$$\mu + d^2 = 0,$$

and the subsequent work would have been correspondingly simplified.

481. *To find the envelope of a series of spheres, having for diameters a series of parallel chords of an ellipsoid.*

Take the diametral plane of the ellipsoid bisecting the chords for the plane of xy, its principal axes for those of x and y, and the axis of z perpendicular to this plane. Then, if $2a$, $2b$ be the principal axes of this section, $2c$ the diameter parallel to the chords, the radius of a sphere, whose center is $(x_0, y_0, 0)$, will be given by the equation

$$r^2 = c^2\left(1 - \frac{x_0^2}{a^2} - \frac{y_0^2}{b^2}\right),$$

and the equation of the corresponding sphere will be

$$(x - x_0)^2 + (y - y_0)^2 + z^2 + \frac{c^2}{a^2}x_0^2 + \frac{c^2}{b^2}y_0^2 = c^2.$$

Hence, for the envelope

$$x - x_0 - \frac{c^2}{a^2}x_0 = 0, \quad y - y_0 - \frac{c^2}{b^2}y_0 = 0,$$

and the equation of the envelope is

$$\frac{c^4x^2}{(a^2+c^2)^2} + \frac{c^4y^2}{(b^2+c^2)^2} + z^2 + \frac{a^2c^2x^2}{(a^2+c^2)^2} + \frac{b^2c^2y^2}{(b^2+c^2)^2} = c^2,$$

or,

$$\frac{x^2}{a^2+c^2} + \frac{y^2}{b^2+c^2} + \frac{z^2}{c^2} = 1.$$

The envelope is therefore an ellipsoid, whose focal ellipse is the section of the given ellipsoid made by a plane diametral to the given chords. Also, we see that if α, β, γ be the semi-axes of the envelope, 2γ being the axis perpendicular to the diametral plane,

$$\alpha^2 + \beta^2 - \gamma^2 = a^2 + b^2 + c^2,$$

or, is equal to the sum of the squares of the semi-axes of the given ellipsoid.

It may easily be shewn that the envelope has double contact with the given ellipsoid at the points where a normal coincides with one of the given system of chords, and that the contact of the envelope will be real only when the center of the sphere lies within the ellipse whose equation is

$$\frac{x^2}{a^4}(a^2+c^2) + \frac{y^2}{b^4}(b^2+c^2) = 1, \quad z = 0.$$

Differential Equations of Envelopes.

482. *To find the differential equation of the envelope of surfaces containing two parameters, one of which is any arbitrary function of the other.*

Let $u \equiv F\{x, y, z, a, \phi(a)\} = 0$ be the equation of one of the surfaces.

The equation of the envelope is given by the elimination of a from the equations

$$u = 0, \quad \frac{du}{da} = 0;$$

therefore the equations

$$u = 0, \quad \frac{du}{dx} + p\frac{du}{dz} = 0, \quad \frac{du}{dy} + q\frac{du}{dz} = 0,$$

are satisfied whether a be considered constant or a function of x, y, z; hence, the differential equation in p and q obtained by eliminating a and $\phi(a)$ between these equations, will belong either to the envelope or to any one of the family of surfaces.

483. *To find the differential equation of the envelope of a series of surfaces involving three parameters, two of which are arbitrary functions of the third.*

If the equation of one of the surfaces be

$$u \equiv F\{x, y, z, a, \phi(a), \psi(a)\} = 0,$$

we have the equations

$$\frac{du}{dx} + p\frac{du}{dz} = 0, \text{ and } \frac{du}{dy} + q\frac{du}{dz} = 0.$$

Differentiating these equations with respect to x and y, considering a as a function of x and y,

$$\frac{d^2u}{dx^2} + 2\frac{d^2u}{dx\,dz}p + \frac{d^2u}{dz^2}p^2 + \frac{du}{dz}r + \left(\frac{d^2u}{dx\,da} + p\frac{d^2u}{dz\,da}\right)\frac{da}{dx} = 0,$$

$$\frac{d^2u}{dx\,dy} + \frac{d^2u}{dx\,dz}q + \frac{d^2u}{dy\,dz}p + \frac{d^2u}{dz^2}pq + \frac{du}{dz}s$$
$$+ \left(\frac{d^2u}{dx\,da} + p\frac{d^2u}{dz\,du}\right)\frac{da}{dy} = 0,$$

$$\frac{d^2u}{dx\,dy} + \frac{d^2u}{dx\,dz}q + \frac{d^2u}{dy\,dz}p + \frac{d^2u}{dz^2}pq + \frac{du}{dz}s$$
$$+ \left(\frac{d^2u}{dy\,da} + q\frac{d^2u}{dz\,da}\right)\frac{du}{dx} = 0,$$

$$\frac{d^2u}{dy^2} + 2\frac{d^2u}{dy\,dz}q + \frac{d^2u}{dz^2}q^2 + \frac{du}{dz}t + \left(\frac{d^2u}{dy\,da} + q\frac{d^2u}{dx\,da}\right)\frac{du}{dy} = 0,$$

or writing these equations in the form

$$P + r = -A\frac{da}{dx},$$

$$Q + t = -B\frac{da}{dy},$$

$$R + s = -A\frac{da}{dy},$$

$$R + s = -B\frac{da}{dx},$$

$$(P+r)(Q+t) = AB\frac{da}{dx}\frac{da}{dy} = (R+s)^2;$$

$$\therefore\ PQ - R^2 + Qr + Pt - 2Rs + rt - s^2 = 0,$$

and if a, $\phi(a)$, $\psi(a)$ be eliminated between this equation and the equations

$$u = 0, \quad \frac{dy}{dx} + p\frac{du}{dz} = 0, \quad \frac{du}{dy} + q\frac{du}{dz} = 0,$$

the differential equation of the envelope or of any of the surfaces will be obtained, which is linear in r, s, t and $rt - s^2$.

484. *When a family of surfaces depends on two parameters, one of which is an arbitrary function of the other, to find the form of the function in order that the envelope may contain a given directing curve.*

Let $u = 0$ be the equation of one of the surfaces, a, $\phi(a)$ being the parameters; and $v = 0$, $w = 0$ be the equations of the given curve.

Since this curve must be a tangent to each of the enveloped surfaces, the same values of $dx : dy : dz$ will satisfy each of the equations $u = 0$, $v = 0$, and $w = 0$.

Hence,
$$\begin{vmatrix} \frac{du}{dx}, & \frac{du}{dy}, & \frac{du}{dz} \\ \frac{dv}{dx}, & \frac{dv}{dy}, & \frac{dv}{dz} \\ \frac{dw}{dx}, & \frac{dw}{dy}, & \frac{dw}{dz} \end{vmatrix} = 0,$$

and eliminating x, y, z between these four equations, we obtain an equation involving $\phi(a)$ and a, which will completely determine the form of $\phi(a)$, and substituting this in $u=0$, we can obtain the envelope.

485. *To find the form of the arbitrary function of the parameter, when the envelope touches a given surface.*

Let $u=0$, $v=0$ be the equations of one of the enveloped surfaces and of the given surface.

Then, since the two surfaces touch one another, they must have a common normal, therefore besides the two equations, we have

$$\frac{\frac{du}{dx}}{\frac{dv}{dx}} = \frac{\frac{du}{dy}}{\frac{dv}{dy}} = \frac{\frac{du}{dz}}{\frac{dv}{dz}},$$

and, eliminating x, y, and z, we have an equation for determining $\phi(a)$.

486. If the surfaces depend upon two arbitrary functions of the parameter, we can obtain the form of the functions, in order that the envelope may pass through two directing curves or touch two surfaces.

487. *To find the edge of a tubular surface.*

A *tubular surface* is the envelope of a sphere whose center moves along any curve.

Let the equations of the curve be

$$y = \phi(x), \quad z = \chi(x);$$

the equation of the sphere will be in the form

$$(x-a)^2 + \{y - \phi(a)\}^2 + \{z - \chi(a)\}^2 = a^2.$$

The corresponding characteristic has for its equations this equation and

$$(x-a)+\phi'(a)\{y-\phi(a)\}+\chi'(a)\{z-\chi(a)\}=0,$$

which being the normal plane to the curve on which the center lies, shews that the characteristic is the circle which is the intersection of the normal plane through the center of the sphere with the sphere.

The edge is given by those equations and the equation

$$1+\overline{\phi'(a)}\,\big|^2+\overline{\chi'(a)}\,\big|^2=\phi''(a)\{y-\phi(a)\}+\chi''(a)\{z-\chi(a)\}.$$

If we eliminate a between the last two of the equations, we obtain the developable surface touching the normal planes of the curve of the centers, which joins the edge of the tubular surface whenever it is met by the developable surface.

488. To find the envelope of a plane, whose equation is

$$lx+my+nz=v,$$

the parameters l, m, n, v being connected by the equations

$$l^2+m^2+n^2=1,$$

$$\frac{l^2}{v^2-a^2}+\frac{m^2}{v^2-b^2}+\frac{n^2}{v^2-c^2}=0.$$

We have, for the point of contact, the equations

$$xdl+ydm+zdn=dv \qquad (1),$$

$$ldl+mdm+ndn=0 \qquad (2),$$

$$\frac{ldl}{v^2-a^2}+\frac{mdm}{v^2-b^2}+\frac{ndn}{v^2-c^2}=vdv\left\{\frac{l^2}{(v^2-a^2)^2}+\dots\right\} \qquad (3).$$

Using undetermined multipliers, λ, μ, we obtain

$$\lambda x+\mu l+\frac{l}{v^2-a^2}=0, \qquad (4),$$

$$\lambda y+\mu m+\frac{m}{v^2-b^2}=0, \qquad (5),$$

$$\lambda z+\mu n+\frac{n}{v^2-c^2}=0, \qquad (6),$$

$$\lambda+v\left\{\frac{l^2}{(v^2-a^2)^2}+\dots\right\}=0, \qquad (7).$$

Multiplying (4), (5), (6) respectively by l, m, n, and adding, we obtain

$$\lambda v + \mu = 0, \qquad (8),$$

and, multiplying them by x, y, z, adding, and putting

$$x^2 + y^2 + z^2 \equiv r^2,$$

$$\lambda r^2 + \mu v + \frac{lx}{v^2 - a^2} + \frac{my}{v^2 - b^2} + \frac{nz}{v^2 - c^2} = 0,$$

or

$$\lambda (r^2 - v^2) + \frac{lx}{v^2 - a^2} + \ldots = 0, \qquad (9).$$

Again, from (4), (5), (6) we obtain

$$\lambda^2 r^2 = \mu^2 + \frac{l^2}{(v^2 - a^2)^2} + \frac{m^2}{(v^2 - b^2)^2} + \frac{n^2}{(v^2 - c^2)^2}$$

$$= \lambda^2 v^2 - \frac{\lambda}{v} \text{ by (7) and (8).}$$

Therefore $\lambda = \dfrac{1}{v(v^2 - r^2)}$, $\mu = -\dfrac{1}{v^2 - r^2}$,

and substituting in (4), (5), (6),

$$\frac{x}{r^2 - a^2} = \frac{vl}{v^2 - a^2}, \quad \frac{y}{r^2 - b^2} = \frac{vm}{v^2 - b^2}, \quad \frac{z}{r^2 - c^2} = \frac{vn}{v^2 - c^2};$$

and, multiplying these by x, y, z, and adding, we obtain the equation of the envelope

$$\frac{x^2}{r^2 - a^2} + \frac{y^2}{r^2 - b^2} + \frac{z^2}{r^2 - c^2} = 1.$$

This is the equation of the Wave Surface, and was first obtained in this manner by Mr A. Smith. See *Cambridge Transactions*, Vol. VI.

XXI.

1. A series of similar ellipsoids are described, having a series of sections of a paraboloid, perpendicular to the axis, as principal sections; prove that their envelope will be a paraboloid, similar to the former.

2. Find the surface always touched by a plane which cuts off a pyramid of constant volume from three given planes.

3. The envelope of the plane

$$lx + my + nz = a,$$

l, m, n being connected by the equations

$$l^2 + m^2 + n^2 = 1, \quad \lambda l + \mu m + \nu n = 0,$$

is a right circular cylinder, whose equation is

$$(x^2 + y^2 + z^2 - a^2)(\lambda^2 + \mu^2 + \nu^2) = (\lambda x + \mu y + \nu z)^2.$$

4. Find the envelope of planes cutting off a constant volume from the cone

$$\frac{x^2}{a^2} + \frac{y^2}{b^2} = \frac{z^2}{c^2}.$$

5. Find the envelope of the surface

$$(\alpha x + \beta y + \gamma z)(ax + by + cz) = m,$$

α, β, γ being parameters, satisfying the equation

$$\alpha^2 + \beta^2 + \gamma^2 = 1.$$

6. If an enveloping cone of an ellipsoid be a cone of revolution, the plane of contact will touch a hyperbolic cylinder.

7. If a cone be described with any point of a central conicoid as vertex, and the conjugate central section as base, this cone will envelope a similar, concentric, and similarly situated conicoid.

8. Two generating lines of a hyperboloid, of the same system, are fixed, and a third is equally inclined to the two former, and at a constant distance from the middle point of their shortest distance, prove that all the hyperboloids will touch a hyperboloid of revolution of one sheet.

Prove that, in this case, the characteristic is a straight line, and the edge is two straight lines.

9. Find the envelope of a series of spheres described on parallel chords of a hyperbolic paraboloid as diameters.

10. Find the envelope of the locus of a point, the rectangle of whose distances from two planes is constant, these planes being at right angles respectively to two fixed planes.

11. Find the envelopes of the surfaces

$$(1)\quad \begin{vmatrix} a, & b, & c \\ \alpha, & \beta, & \gamma \\ x, & y, & z \end{vmatrix} \times \begin{vmatrix} a', & b' & c' \\ \alpha, & \beta, & \gamma \\ x, & y, & z \end{vmatrix} = m,$$

$$(2)\quad (\alpha x + \beta y + \gamma z) \times \begin{vmatrix} a, & b, & c \\ \alpha, & \beta, & \gamma \\ x, & y, & z \end{vmatrix} = m,$$

$$(3)\quad (\alpha x + \beta y + \gamma z)^2 + 2 \begin{vmatrix} a, & b, & c \\ \alpha, & \beta, & \gamma \\ x, & y, & z \end{vmatrix} = m,$$

α, β, γ in each case satisfying the condition

$$\alpha^2 + \beta^2 + \gamma^2 = 1.$$

12. The envelope of the plane, whose equation is

$$\alpha \cos(\theta + \phi) + \beta \cos(\theta - \phi) + \gamma \sin(\theta + \phi) + \delta \sin(\theta - \phi) = 0,$$

θ, ϕ being parameters, is the surface

$$\alpha^2 - \beta^2 + \gamma^2 - \delta^2 = 0.$$

CHAPTER XXI.

VOLUMES, AREAS OF SURFACES, &c.

489. *To find the differential coefficients of the solid contained between a surface, given in rectangular co-ordinates, the co-ordinate planes, and planes parallel to them, drawn through any point of the surface.*

Let x, y, z and $x+\Delta x$, $y+\Delta y$, $z+\Delta z$ be the co-ordinates of two points P and Q upon the surface.

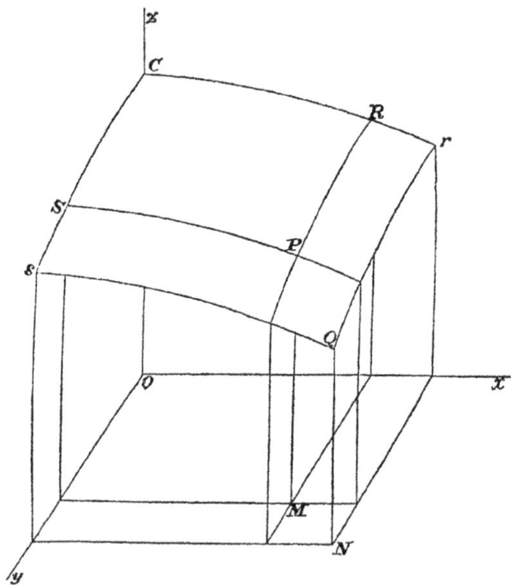

Draw planes through P and Q parallel to the planes of yz, zx, and let V be the volume $CRPSOM$ cut off by these planes from the given solid. If $\Delta_x V$ be the increment of V, when x is changed to $x+\Delta x$, while y remains constant, and a similar interpretation be given to the operation Δ_y, the volume $PrM = \Delta_x V$; also the volume $PQNM$, which is the increment of $\Delta_x V$ when y changes to $y + \Delta y = \Delta_y (\Delta_x V)$, which is easily seen to be the same as $\Delta_x (\Delta_y V)$.

Let z_1, z_2 be the greatest and least values of z within the portion of the surface PQ, therefore $PQNM$ lies between $z_1 \Delta x \Delta y$ and $z_2 \Delta x \Delta y$;

$$\therefore \frac{\Delta_y \left(\dfrac{\Delta_x V}{\Delta x}\right)}{\Delta y} \text{ or } \frac{\Delta_x \left(\dfrac{\Delta_y V}{\Delta y}\right)}{\Delta x} \text{ lies between } z_1 \text{ and } z_2.$$

If we proceed to the limit, in which $z_1 = z_2 = z$, we obtain

$$\frac{d^2 V}{dy\, dx} \text{ or } \frac{d^2 V}{dx\, dy} = z.$$

Since the volume PrM is ultimately equal to the area $RM \times \Delta x$, the partial differential coefficient $\dfrac{dV}{dx}$ represents the area RM, and similarly $\dfrac{dV}{dy}$ the area SM.

490. The differential coefficient of the volume of a wedge of the solid contained between the planes of zx, xy, a plane through the axis of z, and a plane parallel to yOz may be obtained as follows.

If V be the volume included between the planes zOx, xOy, the surface, the plane whose equation is $y = tx$, and a plane parallel to yOz through any point (x, y, z), $\Delta_t V$ is the increment of V when t changes to $t + \Delta t$, x remaining constant; $\Delta_x (\Delta_t V)$ is the increment of $\Delta_t V$ when t changes to $x + \Delta x$; and this is the volume of the prism standing on a base between $x \Delta t\, \Delta x$, and $(x + \Delta x)\, \Delta t \Delta x$.

$$\Delta_x (\Delta_t V) \text{ is between } z_1 x \Delta t \Delta x \text{ and } z_2 (x + \Delta x)\, \Delta t \Delta x;$$

$$\therefore \frac{\Delta_x \dfrac{\Delta_t V}{\Delta t}}{\Delta x} \text{ is between } z_1 x \text{ and } z_2 (x + \Delta x),$$

and proceeding to the limit

$$\frac{d^2 V}{dx\, dt} = zx.$$

491. *To find the differential coefficient of the portion of a surface given in rectangular co-ordinates, cut off by the co-ordinate planes, and planes parallel to them drawn through any point of the surface.*

Let P, Q be the points (x, y, z) and $(x+\Delta x, y+\Delta y, z+\Delta z)$, S the surface $PRCS$, cut off by the planes through P. $\Delta_x S$ is the surface Pr, which is the increment of S when x is changed to $x+\Delta x$.

$\Delta_y(\Delta_x S)$ is the surface PQ, which is the increment of $\Delta_x S$ when y is changed to $y+\Delta y$, and is evidently the same as $\Delta_x(\Delta_y S)$.

Let γ_1, γ_2 be the greatest and least inclinations of the tangent plane to the plane of xy for any points within the surface PQ.

Therefore PQ is intermediate between $\Delta x \Delta y \sec \gamma_1$ and $\Delta x \Delta y \sec \gamma_2$.

Hence $\dfrac{\Delta_y\left(\dfrac{\Delta_x S}{\Delta x}\right)}{\Delta y}$ or $\dfrac{\Delta_x\left(\dfrac{\Delta_y S}{\Delta y}\right)}{\Delta x}$ is intermediate between $\sec \gamma_1$ and $\sec \gamma_2$, which are, in the limit, each equal to $\sec \gamma$.

Therefore, $$\frac{d^2 S}{dy\,dx} \text{ or } \frac{d^2 S}{dx\,dy} = \sec \gamma = \sqrt{1 + \left(\frac{dz}{dx}\right)^2 + \left(\frac{dz}{dy}\right)^2}.$$

492. If S be the surface contained between the plane zOx, and a plane whose equation is $y = tx$; we can shew, by proceeding as in Art. (491), that

$$\frac{d^2 S}{dx\,dt} = \sqrt{\left\{1 + \left(\frac{dz}{dt}\right)^2\right\} x^2 + \left(\frac{dx}{dt}\right)^2}.$$

493. *To find the differential coefficients of the volume of a surface referred to polar co-ordinates.*

Let r, θ, ϕ be the polar co-ordinates of a point P in the surface, θ being measured from Oz, and ϕ from the plane zOx, and let V be the volume of the wedge of a cone contained between the planes zOx and zOP, and the given surface, the axis of the cone being Oz, and θ the semi-vertical angle.

$OPRrS$ is the increase of the volume when θ increases by $\Delta\theta$, ϕ remaining constant, therefore $OPRrS = \Delta_\theta V$.

$OPSQT$ is the increase of $\Delta_\theta V$ when ϕ becomes $\phi + \Delta\phi$, and therefore $= \Delta_\phi (\Delta_\theta V)$, and similarly $= \Delta_\theta (\Delta_\phi V)$.

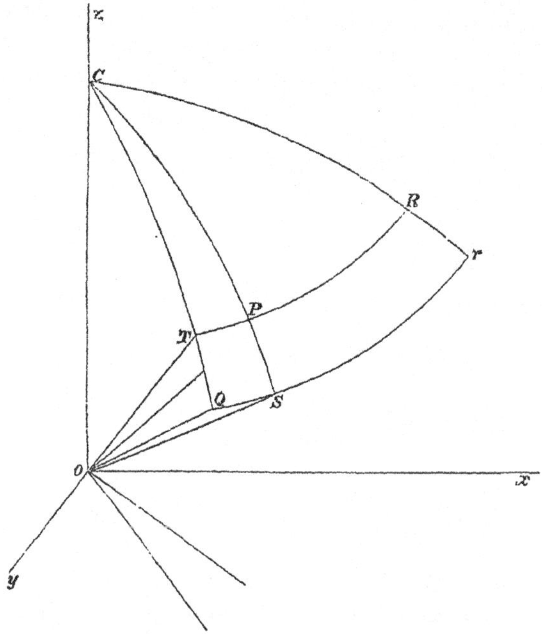

Let s_1, s_2 be the greatest and least values of $r^3 \sin\theta$ for the portion $PSQT$ of the surface, then

$$\Delta_\phi \Delta_\theta V \text{ lies between } \frac{1}{3} s_1 \,\Delta\phi\,\Delta\theta, \text{ and } \frac{1}{3} s_2 \,\Delta\phi\,\Delta\theta,$$

and proceeding to the limit, when $s_1 = s_2 = r^3 \sin\theta$ we obtain

$$\frac{d^2 V}{d\phi\, d\theta} = \frac{1}{3} r^3 \sin\theta.$$

494. *To find the differential coefficient of a surface referred to polar co-ordinates.*

Let r, θ, ϕ be the polar co-ordinates of P, and let S be the surface CPR,

$\Delta_\theta S$ is the increment Pr when θ changes to $\theta + \Delta\theta$,

$\Delta_\phi (\Delta_\theta S)$ is the increment PQ when ϕ changes to $\phi + \Delta\phi$.

Let ψ_1, ψ_2 be the greatest and least inclinations of the tangent planes at points taken within PQ to the corresponding tangent planes of the sphere whose radius is r and center O.

Therefore PQ is intermediate between $r_1^2 \sin\theta_1 \Delta\theta\,\Delta\phi \sec\psi_1$,

and $r_2^2 \sin\theta_2 \Delta\theta\,\Delta\phi \sec\psi_2$,

$r_1^2 \sin\theta_1$ and $r_2^2 \sin\theta_2$ being the greatest and least values of $r^2 \sin\theta$.

Hence $\dfrac{\Delta_\phi\left(\dfrac{\Delta_\theta S}{\Delta\theta}\right)}{\Delta\phi}$ is intermediate between $r_1^2 \sin\theta_1 \sec\psi_1$ and $r_2^2 \sin\theta_2 \sec\psi_2$, which are ultimately equal to $r^2 \sin\theta \sec\psi$;

$$\therefore \frac{d^2 S}{d\phi\, d\theta} = r^2 \sin\theta \sec\psi = \frac{r^3 \sin\theta}{p},$$

$$\text{and } \frac{1}{p^2} = \frac{1}{r^2} + \frac{1}{r^4}\left(\frac{dr}{d\theta}\right)^2 + \frac{1}{r^4 \sin^2\theta}\left(\frac{dr}{d\phi}\right)^2;$$

$$\therefore \frac{d^2 S}{d\phi\, d\theta} = r\sqrt{\left\{r^2 + \left(\frac{dr}{d\theta}\right)^2\right\}\sin^2\theta + \left(\frac{dr}{d\phi}\right)^2}.$$

495. *To find the differential coefficient of a wedge of a volume or surface contained between a cylindrical surface, whose generating lines are parallel to the edge of the wedge, and a given surface.*

Let Oz be the edge of the wedge, zOx one of the faces, and let the base of the cylinder on the plane of xy be referred to polar co-ordinates ρ, ϕ, the equation of the given surface being $F(\rho, \phi, z) = 0$, V the volume corresponding to any point (ρ, ϕ, z).

$\Delta_\phi (\Delta_\rho V)$ is the prism whose base is $\rho\Delta\phi\Delta\rho$, which lies between $z_1\rho_1\Delta\phi\Delta\rho$ and $z_2\rho_2\Delta\phi\Delta\rho$,

therefore $\dfrac{\Delta_\phi\left(\dfrac{\Delta_\rho V}{\Delta\rho}\right)}{\Delta\phi}$ lies between $z_1\rho_1$, and $z_2\rho_2$; hence, proceeding to the limit,

$$\frac{d^2 V}{d\phi\, d\rho} = z\rho.$$

496. If S be the surface corresponding to the point (ρ, ϕ, z), proceeding in the same way, we obtain, as before, that

$$\frac{\Delta_\phi\left(\dfrac{\Delta_\rho S}{\Delta\rho}\right)}{\Delta\phi} \text{ lies between } \rho_1\Delta\phi\,\Delta\rho\sec\gamma_1 \text{ and } \rho_2\Delta\phi\,\Delta\rho\sec\gamma_2,$$

$$\text{and } \frac{d^2S}{d\phi\,d\rho} = \rho\sec\gamma$$

$$= \sqrt{\rho^2 + \left(\frac{dz}{d\phi}\right)^2 + \rho^2\left(\frac{dz}{d\rho}\right)^2}.$$

497. *To find the volume of a solid contained between a cylindrical surface, and two surfaces or two portions of the same surface, whose equations are given.*

Let the cylindrical surface have its generating lines parallel to the axis of z, and a guiding curve traced on the plane of xy.

Suppose the volume to be divided by a series of planes into slices parallel to the plane of yz, one of which (P) is contained between the planes whose distances from yz are x, $x+\Delta x$; the length of this slice measured on the trace of the first plane will be $y_2 - y_1$ in which $y_1 = f_1(x)$, $y_2 = f_2(x)$, the forms of f_1 and f_2 being obtained from the equation of the trace of the cylindrical surface on xy.

Let the slice (P) be subdivided, by planes parallel to the plane of zx, into slips, of which (Q) is between the planes whose distances from zx are y, $y+\Delta y$, the length of this slip measured along an edge will be $z_2 - z_1$, where $z_1 = \phi_1(x, y)$, $z_2 = \phi_2(x, y)$, the forms of ϕ_1 and ϕ_2 being obtained from the equations of the two surfaces, or of the portions of the same surface.

Let the slip (Q) be subdivided by planes parallel to the plane of xy into elementary portions (R), one of which is between planes at distances z and $z+\Delta z$ from the plane of xy,

$$(Q) = \Sigma(R) = \Delta x\,\Delta y\,\Sigma(\Delta z) + \epsilon\,\Delta x\,\Delta y,$$

$\epsilon\,\Delta x\,\Delta y$ being the two portions of (Q) contained between the curved surfaces and the complete parallelopiped, therefore ϵ vanishes when Δx, Δy are indefinitely diminished;

$$\therefore\ (Q) = \Delta x\,\Delta y\,(z_2 - z_1 + \epsilon).$$

Again, if ϵ_1, ϵ_2 be the greatest and least values of ϵ through the slice P, $(P) = \Sigma(Q)$ lies between

$$\Delta x\,\Sigma\,\{\Delta y\,(z_2 - z_1 + \epsilon_1)\} \text{ and } \Delta x\,\Sigma\,\{\Delta y\,(z_2 - z_1 + \epsilon_2)\},$$

or between

$$\Delta x\left\{\int_{y_1}^{y_2}\int_{z_1}^{z_2} dz\,dy + \epsilon_1\,(y_2 - y_1)\right\} \text{ and } \Delta x\left\{\int_{y_1}^{y_2}\int_{z_1}^{z_2} dz\,dy + \epsilon_2\,(y_2 - y_1)\right\};$$

$$\therefore\ (P) = \Delta x\left(\int_{y_1}^{y_2}\int_{z_1}^{z_2} dz\,dy + \eta\right),$$

where η vanishes with Δx.

Again, the whole volume, $= \Sigma\,(P)$, lies between

$$\Sigma\left\{\Delta x\left(\int_{y_1}^{y_2}\int_{z_1}^{z_2} dz\,dy + \eta_1\right)\right\} \text{ and } \Sigma\left\{\Delta x\left(\int_{y_1}^{y_2}\int_{z_1}^{z_2} dz\,dy + \eta_2\right)\right\},$$

where η_1, η_2 are the greatest and least values of η and ultimately vanish;

$$\therefore\ \Sigma\,(P) = \int_{x_1}^{x_2}\int_{y_1}^{y_2}\int_{z_1}^{z_2} dz\,dy\,dx, \text{ which is the volume required,}$$

the integrations being performed with respect to z first, considering x and y constant, next with respect to y, considering x constant, and lastly with respect to x.

The following examples will be useful to explain the method of determining the limits of integration.

498. *To find the volume contained between the ellipsoid whose equation is* $\frac{x^2}{a^2} + \frac{y^2}{b^2} + \frac{z^2}{c^2} = 1$, *and the cylinder whose equation is* $x^2 + y^2 = 2rx$, $2r$ *being less than* a.

Here, using the above notation,

$$z_1 = -c\sqrt{1 - \frac{x^2}{a^2} - \frac{y^2}{b^2}},\quad z_2 = c\sqrt{1 - \frac{x^2}{a^2} - \frac{y^2}{b^2}},\quad y_1 = -\sqrt{2rx - x^2},$$

$$y_2 = +\sqrt{2rx - x^2},\quad x_1 = 0,\quad x_2 = 2r,$$

and the volume is

$$\int_0^{2r}\int_{y_1}^{y_2} 2c\sqrt{1 - \frac{x^2}{a^2} - \frac{y^2}{b^2}}\,dy\,dx$$

$$= 4c\int_0^{2r}\int_0^{\sqrt{2rx - x^2}}\sqrt{1 - \frac{x^2}{a^2} - \frac{y^2}{b^2}}\,dy\,dx.$$

499. *To find the volume contained between the surface whose equation is $(x+y)^2 = 4az$, the tangent plane at a point (α, β, γ), and the planes of zx and yz.*

The equation of the tangent plane at (α, β, γ) is

$$(\alpha+\beta)(x-\alpha)+(\alpha+\beta)(y-\beta)-2a(z-\gamma)=0,$$

or

$$x+y=\sqrt{\frac{a}{\gamma}}(z+\gamma).$$

In this case

$$z_1=-\gamma+\sqrt{\frac{\gamma}{a}}(x+y), \quad z_2=\frac{(x+y)^2}{4a},$$

and for a given value of x the tangent plane meets the surface when $(x+y)^2=4\sqrt{a\gamma}(x+y)-4a\gamma$, or $y=2\sqrt{a\gamma}-x$;

$$\therefore y_1=0, \quad y_2=2\sqrt{a\gamma}-x;$$

$$\therefore x_1=0, \quad x_2=2\sqrt{a\gamma};$$

therefore the volume $=\int_{x_1}^{x_2}\int_{y_1}^{y_2}\int_{z_1}^{z_2} dz\,dy\,dx$

$$=\int_{x_1}^{x_2}\int_{y_1}^{y_2}\frac{1}{4a}(x+y-2\sqrt{a\gamma})^2\,dy\,dx$$

$$=\int_0^{2\sqrt{a\gamma}}-\frac{1}{12a}(x-2\sqrt{a\gamma})^3\,dx$$

$$=\frac{(2\sqrt{a\gamma})^4}{4\,.\,12a}=\frac{1}{3}a\gamma^2.$$

This result may be verified thus. Let AOB be the surface, ACB the tangent plane along the line AB, ADB parallel to xOy, adb a section parallel to xOy of the surface of thickness dz,

area adb : area ADB :: $ad^2 : AD^2$

:: $Od : OD$;

$$\therefore \text{volume } AOBD=\int_0^{\gamma} 2a\gamma\,.\,\frac{z}{\gamma}\,dz=a\gamma^2;$$

$$\text{............} \quad ACDB=\frac{1}{3}2a\gamma\,.\,2\gamma=\frac{4}{3}a\gamma^2;$$

$$\therefore \text{volume required}=\frac{a\gamma^2}{3}.$$

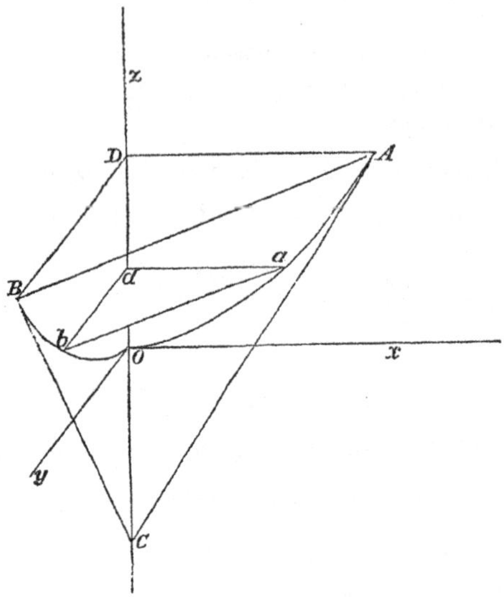

500. *To find the volume of the elliptic paraboloid* $\frac{y^2}{b} + \frac{z^2}{c} = 2x$, *cut off by the plane* $lx + my + nz = p$.

Perform the integration in the order x, y, z,

$$x_1 = \frac{y^2}{2b} + \frac{z^2}{2c}, \quad x_2 = \frac{p - my - nz}{l}.$$

For a given value of z, the values of y at the curve of intersection are given by the equation $x_1 = x_2$,

or $$y^2 + \frac{2bm}{l} y + \frac{b}{c} z^2 - \frac{2b}{l}(p - nz) = 0 \qquad (1),$$

of which y_1, y_2 are the roots, and z must be taken between the limits which correspond to $y_1 = y_2$, or z_1, z_2 on the roots of the equation

$$\frac{b}{c} z^2 - \frac{2b}{l}(p - nz) = \frac{b^2 m^2}{l^2} \qquad (2).$$

$$\text{The volume} = \int_{z_1}^{z_2}\int_{y_1}^{y_2} \left(\frac{p - my - nz}{l} - \frac{y^2}{2b} - \frac{z^2}{2c}\right) dy\, dz$$

$$= \frac{1}{2b}\int_{z_1}^{z_2}\int_{y_1}^{y_2} (y - y_1)(y_2 - y)\, dy\, dz \text{ by (1)},$$

$$=\frac{1}{2b}\int_{z_1}^{z_2}\int_{y_1}^{y_2}\{(y-y_1)(y_2-y_1)-(y-y_1)^2\}\,dy\,dz$$

$$=\frac{1}{2b}\int_{z_1}^{z_2}\frac{1}{6}(y_2-y_1)^3\,dz,$$

$$(y_2-y_1)^2=(y_2+y_1)-4y_2y_1=\frac{4b}{c}(z-z_1)(z_2-z)\quad\text{by (2)},$$

therefore the volume

$$=\frac{1}{12b}\int_{z_1}^{z_2}8\left(\frac{b}{c}\right)^{\frac{3}{2}}\{(z_2-z_1)(z-z_1)-(z-z_1)^2\}^{\frac{3}{2}}\,dz$$

$$=\frac{2}{3}\frac{b^{\frac{1}{2}}}{c^{\frac{3}{2}}}\int_{-\gamma}^{\gamma}(\gamma^2-u^2)^{\frac{3}{2}}\,du,\ \text{where } 2\gamma=z_2-z_1,$$

$$=\frac{4}{3}\frac{b^{\frac{1}{2}}}{c^{\frac{3}{2}}}\gamma^4\int_0^{\frac{\pi}{2}}\cos^4\theta\,d\theta,\quad\text{putting } u=\gamma\sin\theta,$$

$$=\frac{\pi}{4}\cdot\frac{b^{\frac{1}{2}}}{c^{\frac{3}{2}}}\frac{(z_2-z_1)^4}{2^4},$$

and
$$\frac{1}{4}(z_2-z_1)^2=\frac{c^2n^2}{l^2}+\frac{2cp}{l}+\frac{bcm^2}{l^2};$$

$$\therefore\ \text{volume}=\frac{\pi}{4}\sqrt{bc}\,\frac{(2pl+bm^2+cn^2)^2}{l^4}.$$

The student may verify this result by the summation of elementary slices bounded by planes parallel to the given plane.

501. *To find the volume contained between surfaces given by polar co-ordinates.*

The volume of an elementary parallelopiped is

$$r\sin\theta\,dr\,d\theta\,d\phi.$$

If we integrate this expression from $r=r_1$ to $r=r_2$, r_1, r_2 being the radii of the bounding surfaces, corresponding to θ, ϕ, we obtain a frustum of a pyramid the angular breadths of whose faces are $d\theta$, $d\phi$, intercepted between the two surfaces or the two sheets of the same surface, $=\frac{1}{3}\sin\theta\,d\theta\,d\phi\,(r_2^3-r_1^3)$, the radii being given in terms of θ and ϕ.

If now we integrate, considering ϕ and $\phi + d\phi$ constant, $\theta = \theta_1$ to $\theta = \theta_2$, θ_1, θ_2 being given in terms of ϕ by the boundaries of the volume considered, we obtain the portion included between the planes inclined to zOx at angles ϕ and $\phi + d\phi$,

$$= d\phi \int_{\theta_1}^{\theta_2} \frac{1}{3}(r_2^3 - r_1^3) \sin\theta.$$

The whole volume is found by integrating from $\phi = \phi_1$ to $\phi = \phi_2$, the extreme planes between which the volume is included.

The volume is therefore $\int_{\phi_1}^{\phi_2}\int_{\theta_1}^{\theta_2} \frac{1}{3}(r_2^3 - r_1^3) \sin\theta \, d\theta \, d\phi$.

502. *To find the volume of a sphere cut off by three planes through the center.*

Let a = radius of the sphere, ABC the spherical triangle cut off, OC the axis of z, OCA the plane of zx.

The equation of the plane OAB is

$$\cos \overline{\phi - \beta} = \tan \alpha \cot \theta.$$

The limits of integration are

$$r = 0 \text{ to } r = a,$$

$$\theta = 0 \text{ to } \theta = \cot^{-1}(\cot\alpha \cos \overline{\phi - \beta}),$$

$$\phi = 0 \text{ to } \phi = C,$$

$$\text{volume} = \iint \frac{a^3}{3} \sin\theta \, d\theta \, d\phi$$

$$= \int_0^C \frac{a^3}{3} \frac{d\phi}{\sqrt{\sin^2\alpha + \cos^2\alpha \cos^2(\phi - \beta)}}$$

$$= \int_{-\beta}^{C-\beta} \frac{a^3}{3} \frac{d\phi}{\sqrt{1 - \cos^2\alpha \sin^2\phi}}.$$

503. *To find the volume of a wedge of a sphere cut off by a right circular cylinder, a diameter of whose base is a radius of the sphere.*

Let the equation of the sphere be $\rho^2 + z^2 = a^2$, and that of the cylinder $\rho = a \cos\phi$.

The volume is $$\iint 2\rho\sqrt{a^2-\rho^2}\,d\rho\,d\phi,$$

from $$\rho = 0 \text{ to } \rho = a\cos\phi,$$
$$\phi = 0 \text{ to } \phi = \alpha,$$
$$= \int_0^\alpha \frac{2}{3}(a^3 - a^3\sin^3\phi)\,d\phi$$
$$= \frac{2}{3}a^3\left\{\alpha - \frac{1}{4}\int_0^\alpha (3\sin\phi - \sin 3\phi)\,d\phi\right\}$$
$$= \frac{2}{3}a^3\left\{\alpha - \frac{3}{4}(1-\cos\alpha) + \frac{1}{12}(1-\cos 3\alpha)\right\}.$$

If $\alpha = \frac{\pi}{2}$, the volume $= \frac{\pi a^3}{3} - \frac{4a^3}{9}$.

The surface $$= \iint\sqrt{\rho^2 + \left(\frac{dz}{d\phi}\right)^2 + \rho^2\left(\frac{dz}{d\rho}\right)^2}\,d\rho\,d\phi,$$

between the same limits,
$$= \iint \rho\sqrt{1 + \frac{\rho^2}{a^2-\rho^2}}\,d\rho\,d\phi$$
$$= \iint \frac{a\rho\,d\rho}{\sqrt{a^2-\rho^2}}\,d\phi$$
$$= a^2\int_0^\alpha (1-\sin\phi)\,d\phi$$
$$= a^2(\alpha - 1 + \cos\alpha).$$

If $\alpha = \frac{\pi}{2}$, the surface is $a^2\left(\frac{\pi}{2} - 1\right)$.

504. *To find the volume of a solid whose bounding surfaces are given by Tetrahedral Co-ordinates.*

If x, y, z be co-ordinates referred to rectangular axes of a point whose tetrahedral co-ordinates are $\alpha, \beta, \gamma, \delta$.

Since α, β, γ are linear functions of x, y, z and
$$\alpha + \beta + \gamma + \delta = 1,$$
$$\iiint dx\,dy\,dz = C\iiint d\alpha\,d\beta\,d\gamma,$$

and if V be the volume of the tetrahedron of reference,

$$\iiint dx\,dy\,dz = V.$$

If the limits of x, y, z correspond to the boundaries of the tetrahedron, and we evaluate $\iiint d\alpha\,d\beta\,d\gamma$ for the tetrahedron, the limits of γ are from $\gamma = 0$ to $\delta = 0$ or $\gamma = 1 - \alpha - \beta$, β from 0 to $1 - \alpha$, α from 0 to 1;

$$\therefore \iiint d\alpha\,d\beta\,d\gamma = \frac{1}{6}; \quad \therefore C = 6V.$$

Hence if $F(\alpha, \beta, \gamma, \delta) = 0$ be the equation of any closed surface, the volume is $6V\iiint d\alpha\,d\beta\,d\gamma$, the limits of integration being obtained from

$$F(\alpha, \beta, \gamma, 1 - \alpha - \beta - \gamma) = 0.^*$$

XXII.

1. Find the volume of the surface $xy + yz + zx + a^2 = 0$, cut off by a plane $x + y + z = c$.

2. State limits which can be used to find the volume of a closed conicoid whose equation is

$$ax^2 + by^2 + cz^2 + 2a'yz + 2b'zx + 2c'xy = 1.$$

3. Find the portion of the cylinder, $x^2 + y^2 - 2rx = 0$, intercepted between the planes,

$$ax + by + cz = 0 \text{ and } a'x + by + cz = 0.$$

4. State between what limits the summation of $dx\,dy\,dz$ must be taken in order to obtain the volume of the cone, whose equation is $x^2 + y^2 = (a - z)^2$, cut off by the planes $x = 0$ and $x = z$.

5. Find the volume contained between the surfaces

$$y^2 + z^2 = 4ax, \text{ and } x - z = a.$$

6. Find the volume included between the surfaces $r = a$, $z = 0$, $\theta = 0$, $z = mr\cos\theta$, r and θ being polar co-ordinates in the plane xy.

* From an article by Professor Slesser, *Journal of Mathematics*, Vol. II.

7. Find the volume enclosed by the surfaces $x^2+y^2=az$ $x^2+y^2=ax$, and $z=0$, and draw a figure representing the progress of summation.

8. A cavity is just large enough to allow of the complete revolution of a circular disk of radius c, whose center describes a circle of the same radius c, while the plane of the disk is constantly parallel to a fixed plane, and perpendicular to that of the circle in which the center moves. Shew that the volume of the cavity is

$$\frac{2c^2}{3}(3\pi+8).$$

9. If ΔS be an element of the surface of an ellipsoid at any point, and A the area of a section by a plane parallel to the tangent plane at that point, drawn through the center, prove that the limit of $\Sigma\frac{\Delta S}{A}=4$, the summation being taken over the whole surface.

Find ΔS in terms of α, β if $x=a\cos\alpha$, $y=b\sin\alpha\cos\beta$, and $z=c\sin\alpha\sin\beta$.

10. Prove that the volume cut off by the plane $y=k$ from the surface $a^2x^2+b^2z^2=2(ax+bz)y^2$ is $\frac{\pi k^5}{10ab}$.

11. Two cones have a common vertex in the center of an ellipsoid and bases curves in which the surface is intersected by planes parallel to the same principal plane, prove that the volume of the ellipsoid contained between the cones varies as the distance between the planes.

12. Prove that the volume contained between the surface $xz^2+(x-c)r^2=0$, and the plane $z=(c-x)\tan\alpha$ is

$$\frac{\pi c^3}{96}\left(6\cot\alpha\operatorname{cosec}\alpha-4\cos^3\alpha-3l_e\cot\frac{\alpha}{2}\right).$$

13. Prove that the volume, intercepted between the surfaces whose equations are $xyz=a^3$, $x^2=by$, $y^2=bx$ and $x=c$, is trisected by the planes $y=c$ and $y=x$.

14. If S be a closed surface, dS an element about P, at a distance r from a fixed point O, ϕ the angle, which the normal drawn inwards makes with OP, shew that the volume contained by the surface $=\frac{1}{3}\iint r\cos\phi dS$, the summation being extended over the whole surface.

O being the center of an ellipsoid, apply the formula to find its volume, interpreting geometrically the steps of the integration.

15. If θ, ϕ be the vectorial angles of a line OP and $f(\theta, \phi)$ represent the sum of the projections of the elements of any curved surface S, all taken positively, upon a plane perpendicular to OP, prove that

$$\int_{-\pi}^{+\pi} d\phi \int_{0}^{\pi} d\theta f(\theta, \phi) \sin\theta = 2\pi S.$$

Adapt this formula to the case of an ellipsoid, testing its accuracy by any independent process.

16. Prove that the area of a closed surface, no plane section of which has singular points, may be expressed by the definite integral

$$\int_{-\pi}^{+\pi}\int_{0}^{\pi} \frac{r^3 \sin\phi \, d\phi \, d\theta}{p},$$

where p is the perpendicular from the origin upon the tangent plane.

17. Find $\iint \frac{dS}{p}$, where dS is an element of the surface of an ellipsoid, p being the perpendicular from the center upon the tangent plane of the element, the integral being extended over the whole surface.

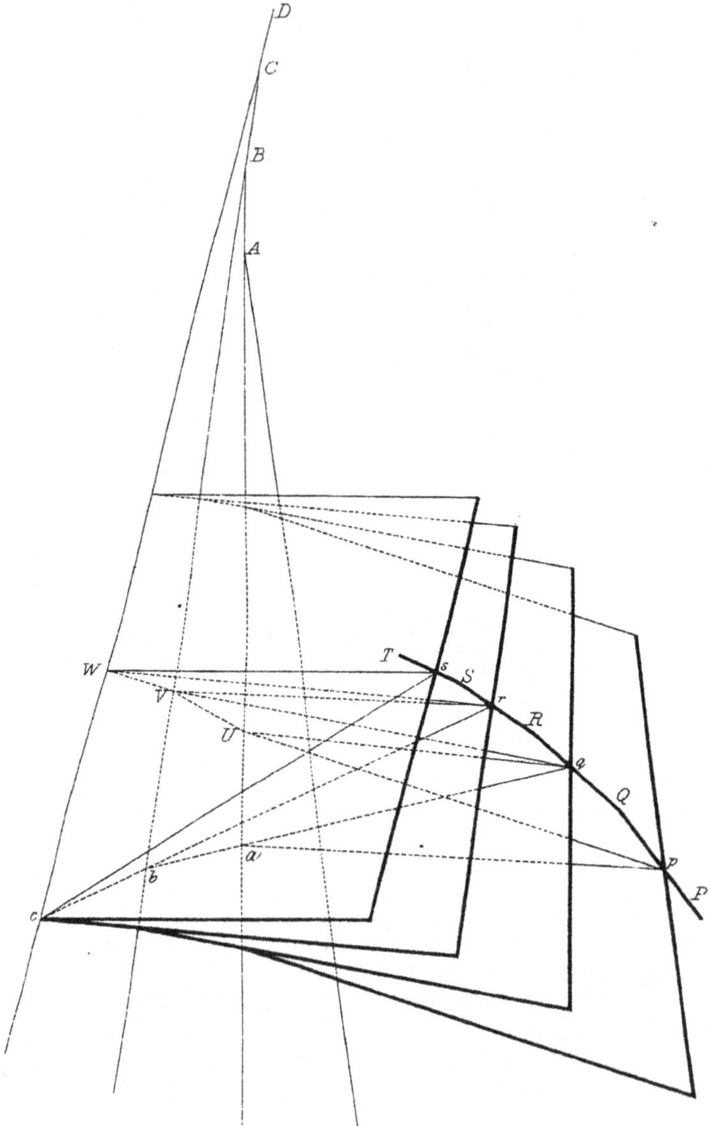

FROST AND WOLSTENHOLME'S GEOMETRY

Macmillan & Co. Cambridge.

W. Dickes lith.

CHAPTER XXII.

CURVATURE OF CURVES IN SPACE.

505. In this Chapter we shall exhibit some of the methods by which the degree of curvature of curves of double curvature has been estimated; this curvature is of two kinds, one having reference to the rapidity with which at different points the curve deflects from its tangent in the osculating plane, and the other having reference to the rapidity with which the planes containing consecutive elements change their position. The first is of the same nature as the curvature of plane curves, the second is called the curvature of torsion, and is peculiar to curves in space.

506. Let an equilateral polygon be inscribed in a curve, of which consecutive sides are PQ, QR, RS, ST, and let p, q, r, s be the middle points of these sides.

Let Aap, Bbq, Ccr be planes perpendicular to these sides forming the polygon $ABCD$ by their intersections.

If the sides PQ, QR,...... be diminished indefinitely, their directions are ultimately those of tangents to the curve, the planes Aap, Bbq,...... are ultimately normal planes to the curve, the planes PQR, QRS,...... are osculating planes, the surface generated by the plane elements Aab, Bbc, Ccd,...... is ultimately the developable surface enveloped by the normal planes of the curve, of which $ABCD$...... is ultimately the edge of regression.

The developable enveloped by the normal planes is called by Monge the *Polar Developable*.

507. *Osculating Circle.* A circle can be described containing the points P, Q, R; this circle, therefore, lies in the osculating plane, when the sides are indefinitely diminished, and its curvature may be taken as the measure of curvature of the curve

in the osculating plane. Let the plane PQR meet Aa in U and pU, qU be joined, since AU is the intersection of planes perpendicular to PQ, QR, it is perpendicular to the plane PQR and therefore to pU, qU; hence, since pU, qU are perpendicular to PQ, QR, U is the center of the circle.

Therefore the center of the osculating circle is the point of intersection of two consecutive normal planes and the osculating plane.

508. *Angle of Contingence.* The angle pUq which is equal to the angle between the two consecutive sides PQ, QR of the polygon, is ultimately equal to the angle between two consecutive tangents, and is called the angle of contingence, this angle is also the angle between two consecutive principal normals.

509. *Spherical Curvature.* If pa, qa be drawn to any point in Aa, since $Pp = Qp$, a is equally distant from P and Q, and similarly from Q and R, hence any point in Aa is equally distant from P, Q and R: similarly, any point in Bb is equally distant from Q, R, and S, therefore A their point of intersection is equally distant from the four points P, Q, R, S.

Hence, it follows that a sphere can be described whose center is A and which contains the four points P, Q, R, S, this sphere is ultimately the sphere which has the closest possible contact with the curve, since no sphere can be made to pass through more than four consecutive points, and is therefore called the sphere of curvature; the locus of the center of spherical curvature is therefore the edge of regression of the polar developable.

The line of intersection of two consecutive normal planes, a property of which is that *all points are equidistant from the osculating circle*, Monge calls the *Polar line.*

510. *Curvature of Torsion.* The plane pUq perpendicular to AUa contains the sides PQ, QR, and the plane qVr perpendicular to BVb contains the sides QR, RS, and since qU, qV are perpendicular to the line of intersection QR of the two planes, the angle UqV is their angle of inclination.

This angle which is ultimately the angle between consecutive osculating planes is called the *angle of torsion.*

Also, since the angles UqV and UBV are equal, it is evident that the angle of torsion of the curve PQR,...... is equal to the angle of contingence of the edge of regression of the polar developable.

511. *Osculating Cone.* The osculating cone at any point is a circular cone which touches three consecutive osculating planes, having its vertex at the point of the curve in which these planes intersect.

512. *Evolutes.* If a be *any* point in the intersection of the planes normal to PQ, QR, at their middle points p, q, it has been shewn that $ap = aq$ and they make equal angles with Aa. Produce qa to meet Bb in b, a string, placed in the position bap, would remain in that position if subject to tension, since the tensions of the portions ab, ap resolved parallel to Aa would be equal, and if fixed at a might pass through q without shifting the position of a. Similarly, if rb be produced to c in Cc, and if Sc be produced to d in Dd.

If we proceed to the limit, it follows that a string may be stretched upon the polar developable in such a manner that the free end, passing through any point in the curve, would describe the curve, if the string were unwrapped from the surface so that the part in contact with the surface remained stationary. The portion in contact lies on a curve called an evolute.

Also, since the position of the point a is arbitrary, the curve which is the limit of a, b, c, d, ... will change its position according to the position of a, hence the number of evolutes is infinite.

All the evolutes of a curve are geodesic lines of the polar developable.

513. *Locus of Centers of Circular Curvature.* Since qU will not, if produced, pass through V, because qU and qV include an angle in the same normal plane, the locus of the centers of circular curvature is not one of the evolutes.

514. *The Rectifying Developable.*

If through every point of a curve a plane be drawn perpendicular to the corresponding principal normal, these planes will envelope a surface on which the curve will be a geodesic line, since its osculating plane contains the normal to the surface at every point. This surface is called the *Rectifying Developable*, since if it be developed into a plane, the curve will be developed into a straight line.

The line of intersection of two consecutive planes is called the *rectifying line* for any point of the curve, being the line about which the surface is turned in order to rectify the element of the curve on that point.

It may be observed that the rectifying line is not generally coincident with the *binormal*, which is the normal perpendicular to the osculating plane.

In the figure at page (393) the surface whose edge of regression is ultimately $ABC\ldots$ is the rectifying surface to the curve which is the limit of $abc\ldots$ Aa is the rectifying line at a, and the binormal does not coincide with the rectifying line unless pa is perpendicular to Aa, or a be the center of circular curvature of the involute of $abc\ldots$

515. If the polygon $PQRS\ldots$ were transformed into a plane polygon by turning the portion $QRST\ldots$ through the angle of torsion VqU about QR, and the portion $RST\ldots$ about RS through the corresponding angle of torsion, the inclination of any side ST in the new position in the plane of PQR would be inclined to PQ, at an angle equal to the sum of the inclinations of the sides taken in order, and estimated in the same direction.

Proceeding to the limit, we see that if, as a point moves along a curve of double curvature, the curve be turned about the tangent line at every position which the point assumes through the angle of torsion, the curve will be replaced by a plane curve, such that the inclination of the tangents at the starting point, and any other point, will be the sum of all the angles of contingence; if, therefore, ϵ be taken for the angle between the tangents in the plane curve, $d\epsilon$ will be the angle of contingence corresponding to the extremity of the arc traversed by the moving point.

516. The rate at which the osculating plane twists about

the tangent line at any point, called the rate of torsion, is measured by the limit of the ratio of the angle of torsion to the arc at the extremities of which the osculating planes are taken.

If, as we pass from PQ to QR, see figure, page (393), QR be turned in the plane PQR so that PQR is a straight line, and the plane QRS be then turned through the angle VqU, and the same process be repeated, the perimeter becomes rectified, and the inclination of the last to the first position of the plane containing two elements is the sum of all angles such as VqU.

Proceeding to the limit, it follows that, if osculating planes be taken along the curve, and the elements of the arc be rectified in each osculating plane in order, the angle between the first and final positions of the osculating plane when the curve is so rectified, is the sum of all the angles of torsion.

If therefore τ be this angle, $d\tau$ is angle of torsion, corresponding to the point at which the last osculating plane is drawn.

517. *To find the angle between consecutive radii of curvature of a curve.*

Let PQ, rQR, sRS be directions of sides of a polygon which are ultimately tangents to a curve.

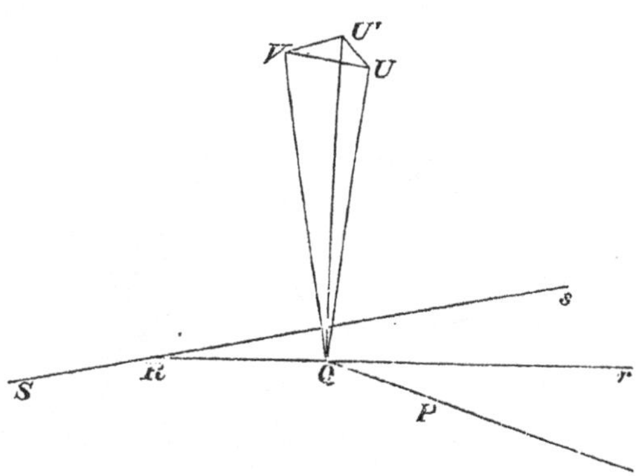

In the planes, PQr, rRs, respectively draw QU, QV perpendicular to rQR, sRS, these are ultimately directions of consecutive radii of curvature.

Draw QU' in the plane QRS perpendicular to QR.

Therefore $\angle U'QU = d\tau$ ultimately,

$\angle VQU' = d\epsilon$ ....

and $\angle VQU = d\psi$ the angle between consecutive radii of curvature.

Also, if $VQ = VU = VU'$,

$$VU^2 = VU'^2 + UU'^2 \text{ ultimately;}$$

$$\therefore \overline{d\psi}|^2 = \overline{d\epsilon}|^2 + \overline{d\tau}|^2,$$

and, if $\dfrac{ds}{d\psi} \equiv R$ be called the radius of complex curvature, ρ, σ the radii of circular curvature and of torsion,

$$\frac{1}{R^2} = \frac{1}{\rho^2} + \frac{1}{\sigma^2}.$$

518. *To find the vertical angle of the osculating cone of a curve.*

Let pOo, qPp, rQq be three consecutive planes which become ultimately the osculating planes of a curve. These planes intersect in P.

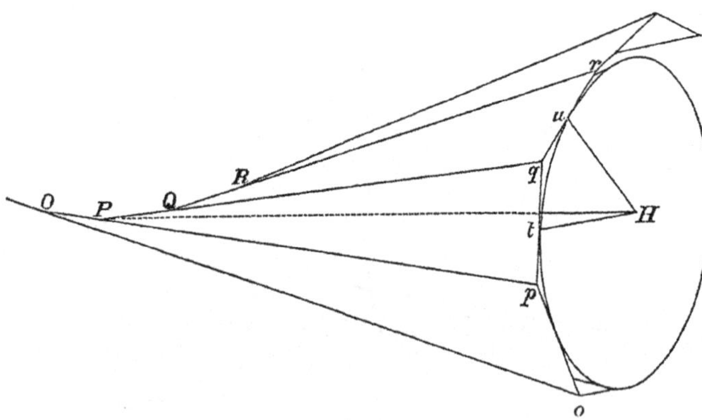

Take P as the vertex of a circular cone which touches each of the planes, and let PH be the axis, op, pq, qr the sections of the planes made by a plane perpendicular to the axis, t, u the points of contact with pq and qr.

Draw tH, uH perpendicular to the planes pPq, qQr; therefore if ψ be the semi-vertical angle of the cone,

$$\tan\psi = \frac{tH}{Pt},$$

and
$$d\epsilon = \frac{2qt}{Pt} \text{ ultimately},$$

$$d\tau = \frac{2qt}{tH} \ldots\ldots\ldots\ldots;$$

$$\therefore \tan\psi = \frac{d\epsilon}{d\tau},$$

which gives the vertical angle required.

519. *The rectifying line is the axis of the osculating cone at any point of a curve.*

For, in the figure in the last article, each of the planes through the tangent lines PQ, QR perpendicular o the osculating planes pPq, qQr, ultimately contains the axis PH.

520. *The rectifying surface is the locus of the centers of principal curvature of the developable of a curve.*

It will be shewn in the following chapter that the normal sections of least and greatest curvature in any surface are perpendicular to one a other, and the section of least curvature in a developable surface is that through a generating line, the normal section perpen.icular to this line is therefore the section of greatest curvature.

Now the plane uHt is ultimately the normal section perpendicular to PQq, and H is therefore the center of principal curvature, every point of the rectifying line is also such a center, and the rectifying surface is the locus of all the centers of pr ncipal curvature of the developable of the original curve.

Also the radius of principal curvature of a point in the developable whose distance measured along a tangent to the curve is c, will therefore be

$$c\tan\psi = c\frac{d\epsilon}{d\tau}.$$

521. *To find the angle of contingence of the locus of the centers of circular curvature.*

Let BV be the intersection of the planes perpendicular to QR, RS the sides of a polygon which are ultimately tangents to a curve; rV, qV perpendicular to BV; qU, rW perpendicular to

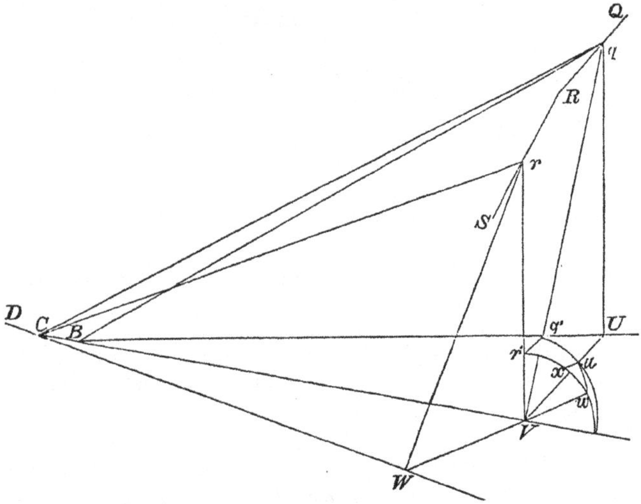

BU, CW, the polar lines preceding and succeeding BV. UV, WV will ultimately be tangents to the locus of the centers of curvature, and if WV be produced to w, UVw will be the angle of contingence required. Since qVB, qUB are right angles, U, V lie in a semi-circle on Bq, and $qVU = qBU = \phi$ suppose; also

$$wVr = rCW = rCV + VCW = \phi + d\phi + d\tau;$$

if therefore a sphere, center V and radius unity, meet Vq, Vr, VU, and Vw in q', r', u, w and ux be parallel to $q'r'$, $uw^2 = ux^2 + xw^2$ ultimately, uw = angle required, $uq' = \phi$, $q'r' = d\epsilon$, $xw = d\phi + d\tau$; therefore square of angle of contingence $= (\cos\phi\, d\epsilon)^2 + (d\phi + d\tau)^2$, where $\sin\phi$ = ratio of the radius of circular to that of spherical curvature.

522. *To find the element of the arc of the locus of centers of circular curvature.*

Employing the figure of the last article, we see that UV is ultimately the element required, and $VqU = d\tau$; hence

$$UV^2 = (qV - qU)^2 + (qUd\tau)^2;$$

$$\therefore \overline{d\sigma}\rceil^2 = \overline{d\rho}\rceil^2 + \overline{\rho d\tau}\rceil^2.$$

Also $$\frac{UV}{BV} = \frac{\sin UBV}{\sin BUV} = \frac{\sin UqV}{\sin BqV};$$

$$\therefore UV = Bq \sin UqV;$$

$$\therefore d\sigma = Rd\tau,$$

R being the radius of the spherical curvature.

523. *To find the radius of spherical curvature.*

Since $UBV = VqU = d\tau$, and Bq is ultimately the radius of spherical curvature, $= R$ suppose,

$$qU = \rho, \ qV = rV = \rho + d\rho,$$

let $qBU = \phi$; $\therefore$ $R \sin \phi = \rho$, and $R \sin (\phi + d\tau) = \rho + d\rho$;

$$\therefore R \cos \phi = \frac{d\rho}{d\tau};$$

$$\therefore R^2 = \rho^2 + \left(\frac{d\rho}{d\tau}\right)^2;$$

whence also $\frac{d\rho}{d\tau} =$ distance between the centers of spherical and circular curvature.

524. *To find the radius and the position of the center of the circle of circular curvature at any point of a curve in space.*

Let x, y, z, and $x + \Delta x$, $y + \Delta y$, $z + \Delta z$ be the co-ordinates of any point P, and an adjacent point Q; l, m, n, and $l + \Delta l$, $m + \Delta m$, $n + \Delta n$, the direction cosines of the tangents at P and Q, and let $\Delta\epsilon$ be the angle between the tangents;

then $$\cos \Delta\epsilon = l(l + \Delta l) + m(m + \Delta m) + n(n + \Delta n)$$

$$= 1 + l\Delta l + m\Delta m + n\Delta n,$$

but $$(l + \Delta l)^2 + (m + \Delta m)^2 + (n + \Delta n)^2 = 1 = l^2 + m^2 + n^2;$$

$$\therefore 2(l\Delta l + m\Delta m + n\Delta n) + \overline{\Delta l}\rceil^2 + \overline{\Delta m}\rceil^2 + \overline{\Delta n}\rceil^2 = 0;$$

$$\therefore\ 4\sin^2\tfrac{1}{2}\Delta\epsilon = \overline{\Delta l}|^2 + \overline{\Delta m}|^2 + \overline{\Delta n}|^2,$$

and $$\left(\frac{\Delta\epsilon}{\Delta s}\right)^2\cdot\left\{\frac{\sin(\frac{1}{2}\Delta\epsilon)}{\frac{1}{2}\Delta\epsilon}\right\}^2 = \left(\frac{\Delta l}{\Delta s}\right)^2 + \left(\frac{\Delta m}{\Delta s}\right)^2 + \left(\frac{\Delta n}{\Delta s}\right)^2;$$

therefore proceeding to the limit, we obtain

$$\left(\frac{d\epsilon}{ds}\right)^2 = \left(\frac{dl}{ds}\right)^2 + \left(\frac{dm}{ds}\right)^2 + \left(\frac{dn}{ds}\right)^2,$$

hence, if s be the independent variable and ρ be the radius of the circle of curvature,

$$\frac{1}{\rho^2} = \left(\frac{d^2x}{ds^2}\right)^2 + \left(\frac{d^2y}{ds^2}\right)^2 + \left(\frac{d^2z}{ds^2}\right)^2.$$

If any other variable be the independent variable

$$\frac{dl}{ds} = \frac{d\frac{dx}{ds}}{ds} = \frac{ds\,d^2x - dx\,d^2s}{(ds)^3},$$

and $$\frac{1}{\rho^2} = \frac{(ds\,d^2x - dx\,d^2s)^2 + (ds\,d^2y - dy\,d^2s)^2 + (ds\,d^2z - dz\,d^2s)^2}{(ds)^6},$$

and since $$(dx)^2 + (dy)^2 + (dz)^2 = (\overline{ds}|)^2;$$

$$\therefore\ dx\,d^2x + dy\,d^2y + dz\,d^2z = ds\,d^2s,$$

hence the numerator $$= ds^2\,(\overline{d^2x}|^2 + \overline{d^2y}|^2 + \overline{d^2z}|^2)$$
$$- 2ds\,d^2s\,.\,ds\,d^2s + \overline{ds}|^2\ \overline{d^2s^2}|;$$

$$\therefore\ \frac{1}{\rho^2} = \frac{\overline{d^2x}|^2 + \overline{d^2y}|^2 + \overline{d^2z}|^2 - \overline{d^2s}|^2}{\overline{ds}|^4}.$$

The equations of the principal normal at P are

$$\frac{\xi - x}{\frac{dl}{ds}} = \frac{\eta - y}{\frac{dm}{ds}} = \frac{\zeta - z}{\frac{dn}{ds}} = \frac{r}{\frac{1}{\rho}},$$

and if ξ, η, ζ be the co-ordinates of the center of curvature

$$r = \rho;$$

$$\therefore\ \xi = x + \rho^2\frac{dl}{ds},\ \text{\&c.}$$

525. Another method of determining these quantities is that depending upon the property that the center of curvature is the point in which the line of intersection of consecutive normal planes meets the osculating plane, mentioned in Art. (507).

The equation of the normal plane at P is

$$(\xi - x)\, dx + (\eta - y)\, dy + (\zeta - z)\, dz = 0 \qquad (1),$$

and from the consecutive normal plane at Q we have

$$(\xi - x)\, d^2x + (\eta - y)\, d^2y + (\zeta - z)\, d^2z = ds^2 \qquad (2).$$

The equation of the osculating plane is

$$(\xi - x)\,(dy\, d^2z - dz\, d^2y) + (\eta - y)\,(dz\, d^2x - dx\, d^2z)$$
$$+ (\zeta - z)\,(dx\, d^2y - dy\, d^2x) = 0.$$

Employing undetermined multipliers A and B, we obtain the equations

$$(\xi - x)\,\{A\, dx + d^2x + B\,(dy\, d^2z - dz\, d^2y)\} = ds^2,$$
$$A\, dy + d^2y + B\,(dz\, d^2x - dx\, d^2z) = 0,$$
$$A\, dz + d^2z + B\,(dx\, d^2y - dy\, d^2x) = 0,$$

and multiplying these equations successively by dx, dy, dz and by d^2x, d^2y, d^2z the coefficient of B vanishes in both cases, therefore

$$(\xi - x)\,(A ds^2 + ds\, d^2s) = ds^2 dx,$$
$$(\xi - x)\,(A ds\, d^2s + \overline{d^2x}|^2 + \overline{d^2y}|^2 + \overline{d^2z}|^2) = ds^2\, d^2x;$$
$$\therefore\ (\xi - x)\,(\overline{d^2x}|^2 + \overline{d^2y}|^2 + \overline{d^2z}|^2 - \overline{d^2s}|^2) = ds\,(ds\, d^2x - dx\, d^2s) \quad (3).$$

Again, multiplying (1) and (2) by d^2s and ds, and subtracting, we have

$$(\xi - x)\,(ds\, d^2x - dx\, d^2s) + \ldots\ldots = \overline{ds}|^3,$$

therefore by equations corresponding to (3),

$$\rho^2\,(\overline{d^2x}|^2 + \overline{d^2y}|^2 + \overline{d^2z}|^2 - \overline{d^2s}|^2) = \overline{ds}|^4,$$

which results are the same as those given in the preceding article.

526. *To calculate the rate of torsion at any point of a curve in space.*

Let (x, y, z) be the point P, and l, m, n the direction cosines of the normal to the osculating plane, and let $\Delta\tau$ be the angle between the two osculating planes at P and an adjacent point Q, where $PQ = \Delta s$; then, exactly as in Art. (524), we obtain

$$\left(\frac{d\tau}{ds}\right)^2 = \left(\frac{dl}{ds}\right)^2 + \left(\frac{dm}{ds}\right)^2 + \left(\frac{dn}{ds}\right)^2;$$

and $$\frac{l}{dy\,d^2z - dz\,d^2y} = \frac{m}{dz\,d^2x - dx\,d^2z} = \frac{n}{dx\,d^2y - dy\,d^2x},$$

or $$\frac{l}{u} = \frac{m}{v} = \frac{n}{w} = \frac{1}{R};$$

$$\therefore\ dl = \frac{du}{R} - \frac{u\,dR}{R^2},$$

$$dm = \frac{dv}{R} - \frac{v\,dR}{R^2},$$

and $$dn = \frac{dw}{R} - \frac{w\,dR}{R^2}.$$

Also $$R^2 = u^2 + v^2 + w^2;$$

$$\therefore\ R\,dR = u\,du + v\,dv + w\,dw;$$

$$\therefore\ \overline{dl}|^2 + \overline{dm}|^2 + \overline{dn}|^2 = \frac{\overline{du}|^2 + \overline{dv}|^2 + \overline{dw}|^2}{R^2} - \frac{2R\,\overline{dR}|^2}{R^3} + \frac{R^2\,\overline{dR}|^2}{R^4}$$

$$= \frac{R^2\left(\overline{du}|^2 + \overline{dv}|^2 + \overline{dw}|^2\right) - (R\,dR)^2}{R^4}$$

$$= \frac{(v\,dw - w\,dv)^2 + (w\,du - u\,dw)^2 + (u\,dv - v\,du)^2}{R^4},$$

and $$u\,dv - v\,du = u\,(dz\,d^3x - dx\,d^3z) - v\,(dy\,d^3z - dz\,d^3y)$$

$$= (u\,d^3x + v\,d^3y + w\,d^3z)\,dz;$$

$$\therefore\ \left(\frac{d\tau}{ds}\right)^2 = \frac{(u\,d^3x + v\,d^3y + w\,d^3z)^2}{(u^2 + v^2 + w^2)^2} = \frac{1}{\sigma^2},$$

where σ is the radius of torsion.

527. We may observe, although it is not necessary to investigate the conditions, that $\frac{d\tau}{ds}$ may be zero identically, which would correspond to the case of a plane curve; or, that it may change sign in passing through a point, which will be what is called a point of *inflected torsion*, if it pass through zero, or a *cuspidal* point if it pass through infinity; or, that it may pass through zero, without changing sign, in which case it is called a point of *suspended torsion;* or, it may pass through infinity without changing sign, in which case it is a point of *infinite torsion.*

Again, $\frac{d\epsilon}{ds}$ may be zero identically, in which case the line is rectilinear; it may change sign through zero or infinity, in which case there is a point of inflexion or a cusp; or it may be zero or infinity without changing sign, in which case there is a point of suspended or infinite curvature.

528. It is convenient to solve problems relating to the distances of tangents, binormals, &c. at points very near to one another, and others of this kind, by taking as the axes of x, y, z the tangent at any point of the curve, the binormal, and the principal normal. If the axes be so chosen, and the co-ordinate be expressed in ascending powers of s, the arc reckoned from the point of contact, observing that at the point of contact

$$\frac{dx}{ds} = 1, \quad \frac{dy}{ds} = 0 = \frac{dz}{ds},$$

$$\text{and} \quad \frac{d^2x}{ds^2} = 0 = \frac{d^2y}{ds^2},$$

we have

$$x = s + as^3 + \ldots$$

$$y = bs^3 + \ldots$$

$$z = cs^2 + \ldots$$

and since

$$\left(\frac{dx}{ds}\right)^2 + \left(\frac{dy}{ds}\right)^2 + \left(\frac{dz}{ds}\right)^2 \equiv 1,$$

$$(3bs^2 + \ldots)^2 + (2cs + \ldots)^2 \equiv -6as^2 - \ldots$$

and hence

$$a = -\frac{2c^2}{3};$$

therefore, if s be very small,

$$x = s - \frac{2c^2}{3}s^3, \quad y = bs^3, \quad \text{and} \quad z = cs^2.$$

$$\frac{dx}{ds} = 1 - 2c^2s^2, \quad \frac{dy}{ds} = 3bs^2, \quad \frac{dz}{ds} = 2cs;$$

$$\frac{d^2x}{ds^2} = -4c^2s, \quad \frac{d^2y}{ds^2} = 6bs, \quad \frac{d^2z}{ds^2} = 2c;$$

$$l = \frac{dy}{ds}\frac{d^2z}{ds^2} - \frac{dz}{ds}\frac{d^2y}{ds^2} = -6bcs^2,$$

$$m = \frac{dz}{ds}\frac{d^2x}{ds^2} - \frac{dx}{ds}\frac{d^2z}{ds^2} = -2c,$$

$$n = \frac{dx}{ds}\frac{d^2y}{ds^2} - \frac{dy}{ds}\frac{d^2x}{ds^2} = 6bs.$$

Let ϵ, τ be the angles contained between the tangents and osculating planes respectively, at the origin and at a point P, whose distance from the origin, measured along the curve, is s.

Therefore $\sin^2\epsilon = 1 - \left(\frac{dx}{ds}\right)^2 = 4c^2s^2$ ultimately;

$$\therefore\ \epsilon = 2cs \text{ ultimately.}$$

Also $\sin^2\tau = 1 - \frac{m^2}{l^2+m^2+n^2} = \frac{l^2+n^2}{l^2+m^2+n^2} = \frac{36b^2s^2}{4c^2};$

$$\therefore\ \tau = \frac{3bs}{c} \text{ ultimately.}$$

Hence, if ρ and σ be the radii of curvature and torsion,

$$c = \text{limit}\,\frac{\epsilon}{2s} = \frac{1}{2\rho},$$

$$b = \frac{c}{3}\,\text{limit}\,\frac{\tau}{s} = \frac{1}{6\rho\sigma},$$

and the equations of the curve near the origin assume the forms.

$$x = s - \frac{s^3}{6\rho^2}, \quad y = \frac{s^3}{6\rho\sigma}, \quad z = \frac{s^2}{2\rho}.$$

529. *To find the shortest distance between two consecutive tangents.*

Employing the axes chosen in the last article, the distance between the tangent at P and the axis of x is the perpendicular from the origin upon the projection of the tangent at P on the plane of yz, and the equation of this projection is

$$(\eta - y)\frac{dz}{ds} - (\zeta - z)\frac{dy}{ds} = 0,$$

which reduces to

$$2\sigma\eta - \zeta s + \frac{s^3}{6\rho} = 0;$$

therefore δ, the distance between two consecutive tangents, is ultimately equal to $\dfrac{s^3}{12\rho\sigma}$.

530. *The distance, from the osculating plane at any point, of a point near the former, is twice the distance of the tangents.*

The distance of P from the osculating plane at O

$$= y = \frac{s^3}{6\rho\sigma} = 2\delta.$$

531. *The angle between the line of intersection of osculating planes at two consecutive points and the tangent at one of the points is equal to half the angle of contingence.*

The equation of the osculating plane being

$$l(\xi - x) + m(\eta - y) + n(\zeta - z) = 0,$$

the line of intersection with the osculating plane at O has the equation

$$l(\xi - x) + n(\zeta - z) - my = 0,$$

and the inclination to the axis of x is ultimately

$$-\frac{l}{n} = cs = \tfrac{1}{2}\epsilon.$$

532. *To find the angle between the osculating plane at any point and the tangent at an adjacent point.*

The angle required $= \dfrac{dy}{ds} = \dfrac{s^2}{2\rho\sigma}$ ultimately.

533. *To find the angle between two consecutive principal normals and their shortest distance.*

The equations of a principal normal corresponding to a distance s from O are ultimately

$$\frac{\xi - s}{-\frac{s}{\rho^2}} = \frac{\eta}{\frac{s}{\rho\sigma}} = \frac{\zeta}{\frac{1}{\rho}}.$$

The direction cosines are proportional to $-\frac{s}{\rho}$, $\frac{s}{\sigma}$ and 1, therefore the angle between this normal and that at the origin

$$= \sqrt{1 - \frac{1}{1 + \frac{s^2}{\rho^2} + \frac{s^2}{\sigma^2}}} = s\sqrt{\frac{1}{\rho^2} + \frac{1}{\sigma^2}}.$$

The shortest distance is equal to the perpendicular from O upon the projection on the plane of xy, whose equation is

$$\rho(\xi - s) + \sigma\eta = 0;$$

therefore the shortest distance is

$$\frac{\rho s}{\sqrt{\rho^2 + \sigma^2}}.$$

PROBLEMS.

1. PROVE that the locus of the center of absolute curvature of a helix is another helix, coaxial with the former.

2. Prove that the equation of the polar surface to the helix is

$$x \cos\phi + y \sin\phi + a \tan^2 \alpha = 0,$$

where $$x^2 + y^2 = \tan^2\alpha\,\{a^2 \tan^2\alpha + (z - a \tan\alpha\phi)^2\},$$

and that its edge of regression is a helix of the same inclination on a cylinder whose radius is $a \tan^2 \alpha$.

3. Prove that the area of the hemispherical spiral in which the latitude λ and longitude l are connected by the equation

$$4\lambda + l = 2\pi,$$

contained between the spiral and the base of the hemisphere is to the area of the hemisphere as $2 : \pi$.

4. A curve is formed by the intersection of a hemisphere, and a cylinder whose base is the circle described on a radius of the base

of the hemisphere as diameter, prove that the area of the hemisphere included between the curve, the meridian touching the cylinder and a quadrant of the base of the hemisphere, is equal to the square on the radius of the hemisphere.

5. Prove that the volume contained between the cylinder, the hemisphere, the meridian plane touching the cylinder, and the base of the hemisphere is $\frac{2}{9}$ths of the cube of the radius of the hemisphere.

6. A hemisphere is pierced by a cylinder, whose circular base touches the base of the hemisphere, the diameter of the base of the cylinder being less than the radius of the hemisphere. Prove that the area of the cylinder included between the hemisphere and its base is equal to the rectangle contained by the diameter of the cylinder and the chord of the base of the hemisphere which touches the base of the cylinder and is parallel to the common tangent of the bases.

7. If ρ, σ, R be the radii of curvature, torsion and spherical curvature of a curve at a point whose distance measured from a fixed point along the curve is s, prove that

$$\rho^2 + \sigma^2 \left(\frac{d\rho}{ds}\right)^2 = R^2.$$

8. When the polar surface of a curve is developed into a plane, prove that the curve itself degenerates into a point on the plane, and if r, p be the radius vector and perpendicular on the tangent to the developed edge of regression of the polar surface drawn from this point, prove that

$$\sqrt{r^2 - p^2} = \sigma \frac{d\rho}{ds}.$$

9. Prove that the angle between the shortest distance of two consecutive tangents at two points, and the binormal at one is equal to half the corresponding angle of tension.

10. Prove that the angle between the chord joining two consecutive points and the tangent at one of them is half the angle of contingence.

11. Prove that the angle between the radius of the osculating sphere, and the edge of regression of the polar surface is equal to the angle between the radius of the osculating circle and the locus of the center of curvature.

CHAPTER XXIII.

CURVATURE OF SURFACES. LINES OF CURVATURE.

Def. Two surfaces are said to have a complete contact of the nth order at a common point, when the sections of the surfaces, made by any plane passing through that point, have a contact of the nth order.

534. *To find the conditions necessary in order that two surfaces whose equations are given may have a complete contact of the nth order at a given point.*

If P be the given point, and if PQ, PQ' be equal arcs measured along the curves in which any plane λ, μ, ν intersects the two surfaces, these curves will have a contact of the nth order, if the limiting value of $\dfrac{QQ'}{PQ^{n+1}}$ be finite.

Hence the values of

$$\frac{dx}{ds},\ \frac{dy}{ds},\ \frac{dz}{ds},\ \ldots\ldots\ldots\ \frac{d^nx}{ds^n},\ \frac{d^ny}{ds^n},\ \frac{d^nz}{ds^n},$$

must be respectively equal in the two curves. But the values of these differential coefficients are found from the equations

$$\frac{dz}{ds} = \frac{dz}{dx}\frac{dx}{ds} + \frac{dz}{dy}\frac{dy}{ds}, \qquad \lambda\frac{dx}{ds} + \mu\frac{dy}{ds} + \nu\frac{dz}{ds} = 0,$$

$$\left(\frac{dx}{ds}\right)^2 + \left(\frac{dy}{ds}\right)^2 + \left(\frac{dz}{ds}\right)^2 = 1,$$

and the equations formed by differentiating these. We shall, therefore, obtain them as functions of

$$\lambda,\ \mu,\ \nu,\ \frac{dz}{dx},\ \frac{dz}{dy},\ \frac{d^2z}{dx^2},\ \frac{d^2z}{dx\,dy},\ \frac{d^2z}{dy^2},\ \ldots\ldots\ \frac{d^nz}{dy^n}.$$

In order then that these may be respectively equal in the two curves *for all values of* $\lambda : \mu : \nu$, we must have the values of

$$\frac{dz}{dx} \cdots\cdots\cdots \frac{d^n z}{dy^n}$$

respectively equal in the two surfaces at the given point. These are, therefore, the conditions required.

We may observe that if two surfaces have a complete contact of the nth order at a common point, there will be $n+1$ directions in which a plane can be drawn meeting them in curves which have a contact of the $(n+1)^{th}$ order. For, taking the origin at the common point, and referring them to their common tangent plane as that of xy, the equations of the two surfaces will each be of the form

$$z = rx^2 + 2sxy + ty^2 + \ldots + A_1 x^{n+1} + A_2 x^n y + \ldots + A_{n+1} y^{n+1} + \ldots,$$

the coefficients being the same in the two surfaces for all terms of not more than n dimensions in x and y.

If then we take a section of the surface by one of the $n+1$ planes, determined by the equation

$$A_1 x^{n+1} + A_2 x^n y + \ldots + A_{n+1} y^{n+1} = A_1' x^{n+1} + \ldots,$$

the difference of the ordinates of the two curves in such a plane will in the neighbourhood of the origin, be of the order of smallness $(n+2)$, and the curves will have a contact of the $(n+1)^{th}$ order. Hence if two surfaces have a common tangent plane, two normal sections can be drawn which have a contact of the second order. If two surfaces have a complete contact of the second order, three normal sections can be drawn which have a contact of the third order.

If the surfaces be conicoids, it may easily be found that two of these three sections are the real or impossible generating lines through the point, and that the third is always real, having the same tangent lines as their remaining curve of intersection, which is then a plane curve passing through the point of contact.

535. *To determine the number of constants which must be involved in the general equation of a surface, in order that it may be made to have a complete contact of the* n^{th} *order with a given surface, at a given point.*

Substituting the values of x and y at the given point, we must have the corresponding values of

$$z, \frac{dz}{dx}, \frac{dz}{dy}, \dots\dots \frac{d^n z}{dy^n}$$

the same in the two surfaces, which will supply

$$1 + 2 + 3 + \dots + (n + 1) \equiv \frac{(n + 1)(n + 2)}{2}$$

relations among the constants which must be satisfied in order that the proposed contact may take place.

Thus, when $n = 2$, we require six disposable constants, whence we cannot, in general, find a sphere which has a complete contact of the second order with a surface at any point. Since, however, there are four constants in the general equation of a sphere, our six conditions for contact of the second order will lead to two relations among the co-ordinates of the point, which together with the equation of the surface will determine in general a definite number of points at which a sphere can be drawn as required. Such points are called umbilici.

The general equation of a conicoid, referred to axes parallel to its principal axes, involves six constants, and we can therefore always determine a conicoid, with axes in any proposed directions, which shall have a contact of the second order with a given surface at any point.

536. Also the general equation of a conicoid, having a given point at the extremity of one of its principal axes, involves six constants; and we can, therefore, always determine a conicoid which shall, at the extremity of one of its axes, have a complete contact of the second order with a given surface at a given point.

When $n = 3$, the number of conditions required is ten, and we cannot therefore, in general, determine a conicoid which has a contact of the third order at a given point of a surface. Two conicoids cannot have a contact of the third order at any point without complete coincidence. We can, however, find, at any point of a conicoid, any number of conicoids such that all *normal* sections through the point have contact of the third order. The general form of the equation of such conicoids, referred to the tangent plane and normal at the given point, will be

$$z = ax^2 + by^2 + a'yz + b'zx + c'xy + Cz^2,$$

the constants a, b, a', b', c' being the same for all, and C arbitrary.

537. *To investigate the relations between the curvatures of the different normal sections of a given surface at a given point.*

Let the surface be referred to the tangent plane and normal at the given point; its equation may then be written in the form

$$z = ax^2 + by^2 + cz^2 + 2a'yz + 2b'zx + 2c'xy$$
$$+ \text{ terms of higher dimensions.}$$

Also, let ρ be the radius of curvature of any normal section, inclined at an angle θ to the plane of zx. Then, if (x, y, z) be a point near the origin, we shall have

$$\frac{1}{\rho} = 2 \text{ lt.} \frac{z}{x^2 + y^2} = 2 \text{ lt.} \frac{ax^2 + by^2 + 2c'xy}{x^2 + y^2},$$

$$\text{or } \frac{1}{2\rho} = a \cos^2 \theta + 2c' \sin \theta \cos \theta + a \sin^2 \theta,$$

whence it appears that the radius of curvature is proportional to the square of the diameter of the conic

$$ax^2 + 2c'xy + by^2 = 1,$$

which is parallel to the tangent line through which the normal section is drawn.

It follows from this, that there will generally be a maximum and a minimum value of the radius of curvature of a normal section at a given point, corresponding respectively to the transverse and conjugate axis of this conic. If ρ_1, ρ_2 be these principal radii, and ρ, ρ' the radii of curvature of any other normal sections at right angles to each other, we shall have

$$\frac{1}{\rho} + \frac{1}{\rho'} = \frac{1}{\rho_1} + \frac{1}{\rho_2},$$

or, *the sum of the curvatures of any two normal sections at right angles to each other, at any point of a given surface, is constant;* which is "Euler's Theorem."

It appears also from the above, that the section made by a plane parallel and near to the tangent plane at any point which is not a singular point, is ultimately a conic; this conic, from the properties here proved, is called the "*Indicatrix*" of the surface at the given point.

The *indicatrix* will be an hyperbola, ellipse, or two parallel straight lines, according as the two tangents to the intersection of the surface by its tangent plane are real, impossible, or coincident; and the point is called a hyperbolic, elliptic, or parabolic point accordingly. The criterion distinguishing these different cases is

$$U^2(u'^2 - vw) + \dots + 2VW(uu' - v'w') + \dots > < = 0,$$

writing U for $\dfrac{dF}{dx}$, u for $\dfrac{d^2F}{dx^2}$, u' for $\dfrac{d^2F}{dydz}$, &c. See Art. (289).

If the equation of the surface be given in the form $z = \phi(x, y)$, the corresponding criterion is

$$s^2 - rt > < = 0.$$

At a hyperbolic point, the radius of curvature, being proportional to the square of the corresponding diameter of the indicatrix, will change sign on passing through the section whose tangent lines are parallel to the asymptotes. This indicates that the corresponding radii of curvature will be drawn on opposite sides of the tangent plane. The radius of curvature will be infinite for the limiting sections, as might have been inferred from the fact that in those directions, three consecutive points of the surface lie in a straight line, Art. (286). At an elliptic point, the whole of the surface in the neighbourhood of the point lies on the same side of the tangent plane, and only touches it in one point. At a parabolic point, the surface in the neighbourhood of the point also lies wholly on one side of the tangent plane, but it has contact with it at points near the given one. In this case, the radius of curvature is of constant sign, and becomes infinite in only one direction. It has a minimum value for the normal section at right angles to this direction, and if ρ_1 be this minimum value, ρ the radius of curvature of a section inclined at an angle θ to this, we shall have

$$\rho = \rho_1 \sec^2 \theta.$$

The reader will observe that every point of a developable surface is a parabolic point, and every point of a skew surface a hyperbolic point. Arts. (295, 298).

538. *To determine the radius of curvature of a given oblique section at any point of a surface.*

Let O be the given point, Ox the tangent line to the given oblique section, Oz the normal at O, ϕ the angle between the plane of section and the plane zOx; through N any point on

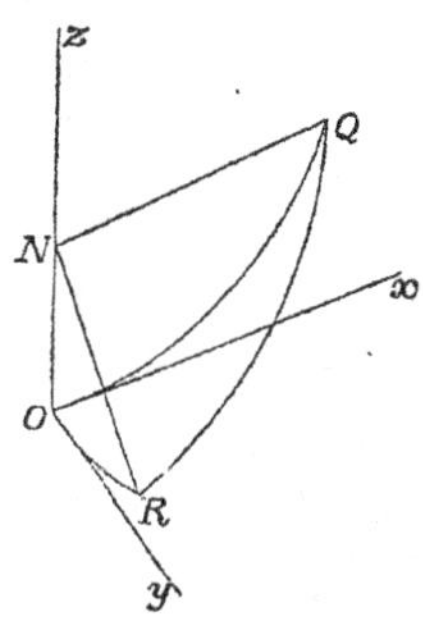

the normal draw NQ parallel to Ox to meet the surface in Q, and through QN draw a plane parallel to the given section cutting the surface in the curve QR. Then in the limit RQ will coincide with the given section, and the required radius of curvature will be the limit of the radius of curvature of RQ at R. Let r be this limit, ρ the radius of curvature of the normal section having the same tangent line, that is, of OQ.

Then $$2r = \text{lt.}\frac{QN^2}{NR}, \quad 2\rho = \text{lt.}\frac{QN^2}{ON},$$

$$\frac{r}{\rho} = \text{lt.}\frac{ON}{NR} = \cos ONR = \cos\phi,$$

since the distance from O of the projection of R on ON is a quantity which ultimately vanishes compared with ON.

Hence, *the radius of curvature of any oblique section is equal to the radius of curvature of the normal section through the same tangent line, multiplied by the cosine of the angle between the planes,* which is "Meunier's Theorem."

Combining this result with those of the last article, we see that the curvatures of all sections of a surface at any point are completely determined when we know the directions of the principal sections, and the magnitude of the principal radii of curvature at that point.

539. *To determine the radius of curvature of a normal section of a surface through a given tangent line at a given point, in terms of the co-ordinates of the point.*

Let (x, y, z) be the given point, $F(x, y, z) = 0$ the equation of the surface, then the equations of the normal will be

$$\frac{x'-x}{U} = \frac{y'-y}{V} = \frac{z'-z}{W}.$$

Let $x+dx$, $y+dy$, $z+dz$ be co-ordinates of a consecutive point along a normal section, the direction cosines of whose tangent line are λ, μ, ν.

The distance of this point from the normal is

$$\sqrt{dx^2+dy^2+dz^2-\frac{(Udx+Vdy+Wdz)^2}{U^2+V^2+W^2}} \equiv m,$$

and its distance from the tangent plane is

$$\frac{Udx+Vdy+Wdz}{\sqrt{U^2+V^2+W^2}} \equiv n;$$

but, by the condition that the second point lies on the surface, we have, neglecting small quantities of a higher order than the second,

$$Udx+Vdy+Wdz+\tfrac{1}{2}(udx^2+\ldots+2u'dydz+\ldots)=0.$$

Now, if ρ be the required radius of curvature,

$$\rho = \tfrac{1}{2}\,\text{lt.}\,\frac{m^2}{n} = \pm\,\text{lt.}\,\frac{(dx^2+dy^2+dz^2)\sqrt{U^2+V^2+W^2}}{udx^2+\ldots+2u'dydz+\ldots};$$

or, since $$\frac{dx}{\lambda}=\frac{dy}{\mu}=\frac{dz}{\nu},$$

$$\rho=\frac{\pm\sqrt{U^2+V^2+W^2}}{u\lambda^2+v\mu^2+w\nu^2+2u'\mu\nu+2v'\nu\lambda+2w'\lambda\mu}.$$

We shall have the conditions

$$U\lambda+V\mu+W\nu=0, \quad \lambda^2+\mu^2+\nu^2=1,$$

and the problem of finding the directions of the principal sections, and the magnitudes of the principal radii of curvature, is the same as that of finding the magnitude and direction of the principal axes of the section of the conicoid

$$ux^2+\ldots+2u'yz+\ldots=1,$$

made by the plane $\quad Ux+Vy+Wz=0.$

540. *To determine the sections of principal curvature, and the radii of principal curvature, at any point of a given surface, in terms of the co-ordinates of the point.*

Writing the equations of the last article, we have to make

$$u\lambda^2+v\mu^2+w\nu^2+2u'\mu\nu+2v'\nu\lambda+2w'\lambda\mu\equiv\frac{1}{r}$$

a maximum or minimum, by variation of λ, μ, ν, subject to the conditions

$$\lambda^2+\mu^2+\nu^2=1, \quad U\lambda+V\mu+W\nu=0.$$

Differentiating, and using undetermined multipliers A, B, we have

$$u\lambda+w'\mu+v'\nu+A\lambda+BU=0,$$
$$w'\lambda+v\mu+u'\nu+A\mu+BV=0,$$
$$v'\lambda+u'\mu+w\nu+A\nu+BW=0,$$

whence, multiplying by λ, μ, ν, and adding, we have

$$\frac{1}{r}+A=0;$$

also, we obtain

$$\frac{\lambda}{U\{(v+A).(w+A)-u'^2\}+V\{u'v'-w'(w+A)\}+W\{w'u'-v'(v+A)\}}$$
$$=\frac{\mu}{\ldots\ldots\ldots\ldots}=\frac{\nu}{\ldots\ldots\ldots\ldots},$$

which leads to the quadratic equation in r,

$$U^2\left\{\left(v-\frac{1}{r}\right)\left(w-\frac{1}{r}\right)-u'^2\right\}+\ldots\ldots$$

$$+2VW\left\{v'w'-u'\left(u-\frac{1}{r}\right)\right\}+\ldots\ldots=0;$$

and since $\frac{1}{r}=\frac{\sqrt{U^2+V^2+W^2}}{\rho}$, this equation gives the values of the principal radii of curvature, and the values of $\lambda:\mu:\nu$, corresponding to each root, are then given by the preceding system of equations.

It may readily be shewn from these equations that the principal sections are at right angles, for if A_1, A_2 be the two values of A, and $(\lambda_1, \mu_1, \nu_1)$, $(\lambda_2, \mu_2, \nu_2)$ the corresponding direction-cosines, then multiplying the system

$$(u+A_1)\,\lambda_1+w'\mu_1+v'\nu_1+BU=0,$$
&c.

by λ_2, μ_2, ν_2 respectively, and adding, we obtain

$$\lambda_1(u\lambda_2+w'\mu_2+v'\nu_2)+\ldots+A_1(\lambda_1\lambda_2+\mu_1\mu_2+\nu_1\nu_2)=0.$$

Treating the other system similarly, we obtain

$$\lambda_2(u\lambda_1+w'\mu_1+v'\nu_1)+\ldots+A_2(\lambda_1\lambda_2+\mu_1\mu_2+\nu_1\nu_2)=0,$$

whence, subtracting,

$$(A_1-A_2)(\lambda_1\lambda_2+\mu_1\mu_2+\nu_1\nu_2)=0,$$

which shews that, except when the principal radii of curvature are equal, the sections are at right angles. When the principal radii are equal, all the radii of curvature of normal sections are equal, and a sphere can be described having a complete contact of the second order, or the point is an umbilicus. The principal sections are of course then indeterminate.

541. *To determine the conditions for an umbilicus.*

At an umbilicus, r retains a constant value for all values of λ, μ, ν subject to the given conditions; or

$$\left(u-\frac{1}{r}\right)\lambda^2+\left(v-\frac{1}{r}\right)\mu^2+\left(w-\frac{1}{r}\right)\nu^2+2u'\mu\nu+2v'\nu\lambda+2w'\lambda\mu=0,$$

for all values of λ, μ, ν, satisfying the equation

$$U\lambda + V\mu + W\nu = 0.$$

The left-hand member of the latter equation must therefore be a factor of the left-hand member of the former, and the other factor will therefore be

$$\frac{\lambda}{U}\left(u - \frac{1}{r}\right) + \frac{\mu}{V}\left(v - \frac{1}{r}\right) + \frac{\nu}{W}\left(w - \frac{1}{r}\right).$$

Multiplying the two, and equating coefficients,

$$\frac{V}{W}\left(w - \frac{1}{r}\right) + \frac{W}{V}\left(v - \frac{1}{r}\right) = 2u',$$

$$\frac{W}{U}\left(u - \frac{1}{r}\right) + \frac{U}{W}\left(w - \frac{1}{r}\right) = 2v',$$

$$\frac{U}{V}\left(v - \frac{1}{r}\right) + \frac{V}{U}\left(u - \frac{1}{r}\right) = 2w',$$

which, on eliminating r, lead to the two conditions

$$\frac{W^2v + V^2w - 2VWu'}{V^2 + W^2} = \frac{U^2w + W^2u - 2WUv'}{W^2 + U^2}$$

$$= \frac{V^2u + U^2v - 2UVw'}{U^2 + V^2}.$$

These two equations, together with the equation of the surface, will, in general, determine a definite number of points, among which are included all the umbilici. It may happen that a common factor exists, so that the three equations are satisfied by the co-ordinates of any point lying on a certain curve. Such a curve is called *a line of spherical curvature.*

It should also be observed that U, V, W have been assumed to be *finite* in the foregoing investigation. Should one of them, as U, vanish, we must have, in the same manner, $V\mu + W\nu$ a factor, and must therefore have

$$\left(u - \frac{1}{r}\right)\lambda^2 + \ldots + 2u'\mu\nu + \ldots$$

$$\equiv (V\mu + W\nu)\left\{k\lambda + \left(v - \frac{1}{r}\right)\frac{\mu}{V} + \left(w - \frac{1}{r}\right)\frac{\nu}{W}\right\}.$$

This identity gives

$$u=\frac{1}{r},\quad 2u'=\left(v-\frac{1}{r}\right)\frac{W}{V}+\left(w-\frac{1}{r}\right)\frac{V}{W},\quad 2v'=kW,\quad 2w'=kV,$$

$$\text{or}\ Vv'=Ww',\quad 2u'=(v-u)\frac{W}{V}+(w-u)\frac{V}{W},$$

which with $U=0$, and the equation of the surface, give *four* relations between the co-ordinates, and these will not, in general, be simultaneously true of any point on the surface. Solutions of the equations for an umbilicus which make U, V, or W vanish, must therefore be excluded.

542. *To determine the number of umbilici on a surface of the nth degree.*

If the equations for an umbilicus be written in the form

$$\frac{P}{P'}=\frac{Q}{Q'}=\frac{R}{R'},$$

the degree of P', Q', R' is $2(n-1)$, and of P, Q, R is $3n-4$. The degree of the surfaces

$$QR'-Q'R=0,\quad RP'-R'P=0,$$

is therefore $5n-6$, and the degree of their curve of intersection is $(5n-6)^2$. But the curve $R=0$, $R'=0$ is part of their intersection, and does not lie on the surface $PQ'-P'Q=0$. The degree of the curve

$$\frac{P}{P'}=\frac{Q}{Q'}=\frac{R}{R'}$$

is therefore $(5n-6)^2-2(n-1)(3n-4)\equiv 19n^2-46n+28$.

But this curve includes three curves similar to

$$U=0,\quad u=\frac{W^2v+V^2w-2VWu'}{V^2+W^2},$$

which do not meet the surface in umbilici. The degree of this curve is

$$(n-1)(3n-4),$$

and the degree of the curve, which meets the surface in umbilici only, is therefore

$$19n^2-46n+28-3(n-1)(3n-4)\equiv 10n^2-25n+16.$$

The whole number of real and impossible umbilici is therefore

$$n(10n^2 - 25n + 16).$$

Thus, in a conicoid, this number is 12, four in each of the principal planes; but not more than one system is real, and if the surface be a ruled surface, none are real.

There can never be real umbilici on a ruled surface of any degree whatever, as every point of a ruled surface is either parabolic or hyperbolic.

Lines of Curvature.

543. DEF. *If on any surface a curve be traced whose tangent line at any point lies in one of the planes of principal curvature at that point, such a curve is called a line of curvature.*

Since there are two principal planes of curvature at each point, there will be two lines of curvature passing through every point, and these will cross each other at right angles.

544. *To find the differential equations of the lines of curvature on any surface.*

Referring to the equations of Art. (540), we see that λ, μ, ν are the direction-cosines of the tangents to the planes of principal curvature at any point, or are the direction-cosines of the tangents to the lines of curvature through that point.

Hence, if (x, y, z) be the point, $(x + dx, y + dy, z + dz)$ a consecutive point on a line of curvature, we shall have

$$\frac{dx}{\lambda} = \frac{dy}{\mu} = \frac{dz}{\nu},$$

which, since λ, μ, ν are determined in terms of x, y, z, are the differential equations of the lines of curvature.

Since λ, μ, ν were determined so as to satisfy the condition

$$U\lambda + V\mu + W\nu = 0,$$

we shall have

$$Udx + Vdy + Wdz = 0,$$

and one integral of these equations will be the original equation of the surface. Any other integral we may find, will involve one arbitrary constant, which may be determined, so as to make the line of curvature pass through any proposed point on the surface.

It appears from the above, that all the surfaces represented by the equation

$$\phi(x, y, z) = c,$$

for different values of c, will have the same differential equations to their lines of curvature.

545. *The normals to any surface, at two consecutive points of any line of curvature, intersect each other.*

Let (x, y, z), $(x + dx, y + dy, z + dz)$ be two consecutive points on the surface, the normals at which intersect. The equations of these normals will be

$$\frac{x'-x}{U} = \frac{y'-y}{V} = \frac{z'-z}{W} \equiv k, \qquad (1),$$

$$\frac{x'-x-dx}{U + udx + w'dy + v'dz} = \ldots\ldots = \ldots\ldots \equiv k', \qquad (2),$$

whence, we have the three equations

$$(k'-k)\,U + dx + k'(udx + w'dy + v'dz) = 0,$$

$$\ldots\ldots\ldots\ldots\ldots\ldots\ldots\ldots\ldots\ldots\ldots\ldots$$

which are precisely the same equations as those of Art. (540), A being replaced by $\frac{1}{k'}$, and B by $1 - \frac{k}{k'}$; whence, we shall have

$$\frac{dx}{\lambda} = \frac{dy}{\mu} = \frac{dz}{\nu},$$

λ, μ, ν having the values there determined, or the consecutive point must be taken along a line of curvature.

The condition for the intersection of the two normals (1) and (2) gives at once

$$dx\,\{V(v'dx + u'dy + wdz) - W(w'dx + vdy + u'dz)\} + \ldots = 0,$$

or $(Vv' - Ww')\,dx^2 + \ldots + \{U(v-w) - Vw' + Wv'\}\,dy\,dz + \ldots = 0,$

which, combined with the equation

$$Udx + Vdy + Wdz = 0,$$

will give the direction of the tangents to the lines of curvature at any point.

These equations are then also the differential equations of the lines of curvature on the surface.

A line of curvature of any surface may therefore be defined to be a curve such that the normals to the surface, at any two consecutive points of the curve, intersect each other.

546. The property of lines of curvature proved in the last article, may be easily shewn geometrically by means of the indicatrix at the point.

For, let O be the point, xOy the tangent plane at the point, and draw a plane parallel and indefinitely near to xy, meeting

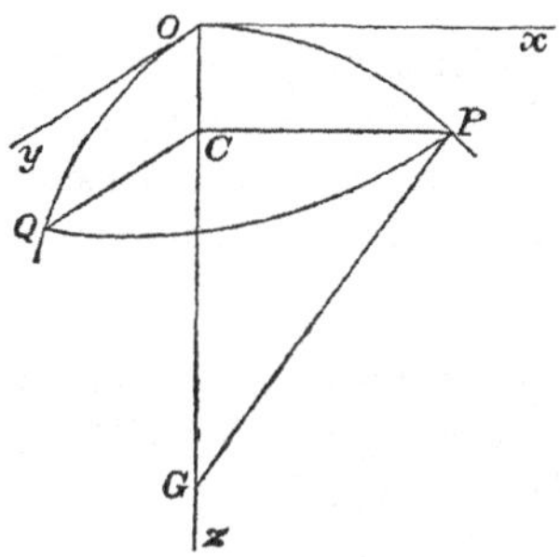

the surface in the curve PQ, and the normal at P in C. This curve will then, ultimately, be, in general, a conic whose center is C; Art. (537).

Let P be a point on this conic at which the normal to the surface intersects Oz, the normal at O. The tangent to the conic at P is then perpendicular to Oz and to the normal at P, and since these intersect, it is also perpendicular to CP, which lies in the same plane with them. The point P is, therefore, at the extremity of one of the axes of the conic: or POz is one of the sections of principal curvature at O. Art. (537).

Hence OP coincides with one of the lines of curvature at P, and there are therefore two directions, at right angles to each other, in which we may proceed from any given point to a consecutive point, at which the normal to the surface intersects the normal at the given point.

At an umbilicus, the indicatrix becomes a circle, and these directions will be indeterminate. In this case the shortest distance between consecutive normals vanishes, to the second order of smallness, in whatever direction the consecutive points be taken. By taking account of the terms of a higher order, we can determine three directions, two of which may be impossible, in which the shortest distance between consecutive normals vanishes, to the third order of smallness. The equations giving these directions will be

$$Udx + Vdy + Wdz = 0,$$

$$(VW' - V'W)\,dx + \ldots\ldots = 0,$$

U', V', W' being the values of U, V, W at a consecutive point, including the terms of two dimensions in dx, dy, dz. In the second of these equations, the terms of one dimension vanish identically, those of two dimensions vanish at an umbilicus, and there remain only the terms of three dimensions, which, combined with the first equation, give us a cubic for the determination of the ratios $dx : dy : dz$.

547. The foregoing equations, for determining the principal curvatures, undergo a considerable simplification, if the equation of the surface is of the form

$$\phi_1(x) + \phi_2(y) + \phi_3(z) = 0.$$

We shall then have u', v', w' all zero; the equation giving the length of the radius of curvature of any normal section, whose tangent line is (λ, μ, ν), will be

$$\frac{1}{r} \equiv \frac{\sqrt{U^2 + V^2 + W^2}}{\rho} = u\lambda^2 + v\mu^2 + w\nu^2;$$

the quadratic equation for the principal radii of curvature will be

$$\frac{U^2}{u - \frac{1}{r}} + \frac{V^2}{v - \frac{1}{r}} + \frac{W^2}{w - \frac{1}{r}} = 0;$$

the differential equations of the lines of curvature will be

$$Udx + Vdy + Wdz = 0,$$

$$U(v - w)\,dy\,dz + V(w - u)\,dz\,dx + W(u - v)\,dx\,dy = 0;$$

and the conditions for an umbilicus are

$$\frac{V^2w+W^2v}{V^2+W^2}=\frac{W^2u+U^2w}{W^2+U^2}=\frac{U^2v+V^2u}{U^2+V^2}.$$

548. *To obtain equations for the determination of the curvatures of a surface at any point, when the equation of the surface gives one of the co-ordinates explicitly in terms of the other two.*

These equations could of course be deduced from those already obtained, but the results being considerably shorter, though not generally so easy to deal with in practice, we give a separate investigation.

Let the equation of the surface be

$$z=\phi(x, y),$$

and let the first and second differential coefficients of z be denoted by p, q, r, s, t. Then if (x, y, z), $(x+dx, y+dy, z+dz)$ be consecutive points, λ, μ, ν the direction-cosines of the corresponding tangent line, the distances of the second point from the normal, and tangent plane, at (x, y, z), are respectively

$$\sqrt{dx^2+dy^2+dz^2-\frac{(dz-p\,dx-q\,dy)^2}{1+p^2+q^2}}\equiv D,$$

$$\frac{dz-p\,dx+q\,dy}{\sqrt{1+p^2+q^2}}\equiv P.$$

Also $dz=p\,dx+q\,dy+\frac{1}{2}(r\,dx^2+2s\,dx\,dy+t\,dy^2)+\dots$

Hence, if ρ be the radius of curvature of the corresponding normal section,

$$\rho=\tfrac{1}{2}\text{lt.}\frac{D^2}{P}=\text{lt.}\frac{dx^2+dy^2+(p\,dx+q\,dy)^2}{r\,dx^2+2s\,dx\,dy+t\,dy^2}\sqrt{1+p^2+q^2},$$

or, $$\frac{1}{R}\equiv\frac{\sqrt{1+p^2+q^2}}{\rho}=\frac{r+2s\,m+tm^2}{1+p^2+2pqm+(1+q^2)\,m^2},$$

m being the value of $\dfrac{dy}{dx}$ for the particular tangent line.

The equation for determining the principal radii of curvature will be found by making this equation in m have equal roots, and will therefore be

$$(1+q^2-Rt)(1+p^2-Rr)=(pq-Rs)^2,$$

or, $$(rt-s^2)R^2-\{(1+p^2)t-2pqs+(1+q^2)r\}R+1+p^2+q^2=0.$$

Also, when R is a maximum or minimum, m must satisfy the equation

$$\{1+p^2+2pqm+(1+q^2)m^2\}(s+tm)=(r+2sm+tm^2)\{pq+(1+q^2)m\},$$

or, $$m^2\{pqt-s(1+q^2)\}+m\{t(1+p^2)-r(1+q^2)\}+(1+p^2)s-pqr=0.$$

Replacing m by $\frac{dy}{dx}$, this is an equation which must be satisfied at all points on a line of curvature, and since it only involves x and y, it is the differential equation of the projections of the lines of curvature on the plane of xy.

The conditions for an umbilicus may be determined from either of the properties, that R is independent of m, or that the equation for m is indeterminate, at such a point. They will be

$$\frac{1+p^2}{r}=\frac{pq}{s}=\frac{1+q^2}{t}.$$

They may also be found from the condition that the equation in R may have equal roots, which may easily be expressed as the sum of two squares, and for real values of the differential coefficients, these must separately vanish, and we obtain the same equations as before.

549. *The lines of curvature on a conicoid are its curves of intersection with confocal surfaces.*

Let the equation of the surface be

$$\frac{x^2}{a}+\frac{y^2}{b}+\frac{z^2}{c}=1; \qquad (1),$$

the differential equations of the lines of curvature will then be (Art. 547),

$$\frac{x}{a}\,dx+\frac{y}{b}\,dy+\frac{z}{c}\,dz=0,$$

$$x(b-c)\,dy\,dz+y(c-a)\,dz\,dx+z(a-b)\,dx\,dy=0.$$

But for the curve of intersection of (1) with a confocal surface,

$$\frac{x^2}{a+k}+\frac{y^2}{b+k}+\frac{z^2}{c+k}=1, \qquad (2),$$

we have $\dfrac{xdx}{a(b-c)(a+k)}=\dfrac{ydy}{b(c-a)(b+k)}=\dfrac{zdz}{c(a-b)(c+k)}$,

$$=\frac{\dfrac{x^2}{a(a+k)}+\dfrac{y^2}{b(b+k)}+\dfrac{z^2}{c(c+k)}}{\dfrac{x(b-c)}{dx}+\dfrac{y(c-a)}{dy}+\dfrac{z(a-b)}{dz}},$$

and subtracting (2) from (1), we have

$$\frac{x^2}{a(a+k)}+\frac{y^2}{b(b+k)}+\frac{z^2}{c(c+k)}=0$$

at all points of the curve of intersection. Hence, we must have, at all points on such a curve,

$$\frac{x(b-c)}{dx}+\frac{y(c-a)}{dy}+\frac{z(a-b)}{dz}=0,$$

or, $\quad x(b-c)\,dy\,dz+y(c-a)\,dz\,dx+z(a-b)\,dx\,dy=0,$

and the differential equations of the curve of intersection of the given conicoid with any confocal surface coincide with the differential equations of the lines of curvature; and the equations of such curves, involving an arbitrary constant k, are therefore the complete integral of the differential equations of the lines of curvature.

Having given any one point (x', y', z'), we shall have the quadratic equation

$$\frac{x'^2}{a(a+k)}+\frac{y'^2}{b(b+k)}+\frac{z'^2}{c(c+k)}=0,$$

for determining k, and to each value we shall have a corresponding line of curvature passing through the point (x', y', z').

We may also solve the equations by obtaining the differential equation of the projection on the plane of xy. This equation is

$$\{x(b-c)\,dy+y(c-a)\,dx\}\left(\frac{x}{a}\,dx+\frac{y}{b}\,dy\right)$$

$$=(a-b)\left(1-\frac{x^2}{a}-\frac{y^2}{b}\right)dx\,dy,$$

or,

$$xy\frac{b-c}{b}\left(\frac{dy}{dx}\right)^2+\left\{\frac{a-c}{a}x^2+\frac{c-b}{b}\cdot y^2-(a-b)\right\}\frac{dy}{dx}+xy\cdot\frac{c-a}{a}=0,$$

or, $$xy\frac{a(b-c)}{b(a-c)}\left(\frac{dy}{dx}\right)^2+\left\{x^2-\frac{a(b-c)}{b(a-c)}y^2-\frac{a(a-b)}{a-c}\right\}\frac{dy}{dx}-xy=0,$$

of which the integral is (Boole's *Differ. Equations*, page 135)

$$\left(x^2+\frac{y^2}{C}\right)\frac{a-c}{a(a-b)}=\frac{1}{1-C\frac{a(b-c)}{b(a-c)}},$$

that is, a conic

$$\frac{x^2}{a'}+\frac{y^2}{b'}=1,$$

a', b' being connected by the equation

$$\frac{a'}{a}(a-c)-\frac{b'}{b}(b-c)=a-b.$$

Now the equation of the projection of the curve of intersection of the confocal conicoids (1) and (2) is

$$\frac{x^2(a-c)}{a(a+k)}+\frac{y^2(b-c)}{b(b+k)}=1,$$

a conic whose semi-axes are connected by the same relation. This conic in fact coincides with the former, if we give k such a value that

$$C\frac{a(b-c)}{b(a-c)}\equiv\frac{b+k}{a+k}.$$

The relation between the axes of this conic shews that it touches four real or impossible straight lines, whose equation, the singular solution of the differential equation, is

$$\left\{x^2-\frac{a(b-c)}{b(a-c)}y^2-\frac{a(a-b)}{a-c}\right\}^2=-4\frac{a(b-c)}{b(a-c)}x^2y^2.$$

If a, b, c be positive, and in descending order, these lines will only be possible in the plane of zx, and it will be seen that they are the tangent lines to the principal section in that plane at the umbilici.

If two of the three a, b, c, be negative, there will in like manner be only one system of possible straight lines; and if one only be negative, all the systems are impossible.

The general result is, that the lines of curvature passing through any point on a given conicoid are the intersections of the given surface with the two confocal conicoids passing through the proposed point. This has only been shewn for central conicoids, but the paraboloids are of course included as a limiting case.

550. The preceding results may also be proved in the following manner. We will only take the case of the ellipsoid.

Let P be any point on the ellipsoid, PCP' the diameter through P, CL the radius parallel to the tangent line at P to

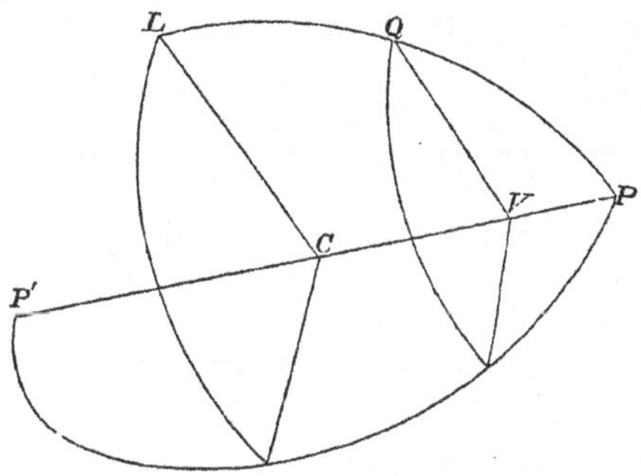

a normal section whose radius of curvature is required, PQL the central section having the same tangent line. Through Q, a point near P at a distance ϖ from the tangent plane at P, draw a plane parallel to the tangent plane, meeting CP in V. Let p be the perpendicular distance from C to the tangent plane at P. The required radius of curvature is then

$$\tfrac{1}{2}\,\text{lt.}\,\frac{PQ^2}{\varpi} = \tfrac{1}{2}\,\text{lt.}\,\frac{QV^2}{\varpi};$$

$$\text{and}\quad \frac{QV^2}{CL^2} = \frac{PV \,.\, VP'}{CP^2}, \quad \frac{\varpi}{p} = \frac{PV}{CP},$$

$$\text{whence}\quad \frac{QV^2}{\varpi} = \frac{CL^2 \,.\, VP'}{p \,.\, CP}, \text{ and lt. } \frac{QV^2}{\varpi} = 2\,\frac{CL^2}{p},$$

or the required radius of curvature is $\frac{CL^2}{p}$. The tangents to the sections of principal curvature at P are therefore parallel to the axes of the central section conjugate to CP, and if α, β be the semi-axes of the section, the values of the principal radii of curvature are $\frac{\alpha^2}{p}$, $\frac{\beta^2}{p}$ respectively.

Now, let the equation of the ellipsoid referred to its axes be

$$\frac{x^2}{a^2}+\frac{y^2}{b^2}+\frac{z^2}{c^2}=1,$$

and let x', y', z' be the co-ordinates of P. The equation of the conjugate central section is

$$\frac{xx'}{a^2}+\frac{yy'}{b^2}+\frac{zz'}{c^2}=0,$$

and the equation, giving the values of its semi-axes, is

$$\frac{\frac{x'^2}{a^2}}{a^2-r^2}+\frac{\frac{y'^2}{b^2}}{b^2-r^2}+\frac{\frac{z'^2}{c^2}}{c^2-r^2}=0,\quad \text{Art. (250)};$$

and the equation giving the values of the principal radii of curvature is therefore

$$\frac{x'^2}{a^2(a^2-p\rho)}+\frac{y'^2}{b^2(b^2-p\rho)}+\frac{z'^2}{c^2(c^2-p\rho)}=0,$$

and, since

$$\frac{x'^2}{a^2}+\frac{y'^2}{b^2}+\frac{z'^2}{c^2}=1,$$

we shall also have

$$\frac{x'^2}{a^2-p\rho}+\frac{y'^2}{b^2-p\rho}+\frac{z'^2}{c^2-p\rho}=1.$$

551. *If three confocal conicoids (A), (B), (C) intersect in a point P, the centers of principal curvature of (A) at P are the poles with respect to (B) and (C) of the tangent plane to (A) at P.*

Let the equation of (A) be

$$\frac{x^2}{a^2}+\frac{y^2}{b^2}+\frac{z^2}{c^2}=1,$$

and let (x', y', z') be the co-ordinates of P. The equation of a confocal surface through P will be

$$\frac{x^2}{a^2+k}+\frac{y^2}{b^2+k}+\frac{z^2}{c^2+k}=1,$$

k being given by the equation

$$\frac{x'^2}{a^2+k}+\frac{y'^2}{b^2+k}+\frac{z'^2}{c^2+k}=1,$$

and, comparing these with the equations of the last article, we see that the two values of k are the two values of $p\rho$ with the sign reversed.

Now, if X, Y, Z be the co-ordinates of one of the centers of principal curvature, corresponding to a radius of curvature ρ, we shall have

$$\frac{X-x'}{\frac{x'}{a^2}}=\frac{Y-y'}{\frac{y'}{b^2}}=\frac{Z-z'}{\frac{z'}{c^2}}$$

$$=-\frac{\sqrt{(X-x')^2+(Y-y')^2+(Z-z')^2}}{\sqrt{\frac{x'^2}{a^4}+\frac{y'^2}{b^4}+\frac{z'^2}{c^4}}}=-\frac{\rho}{\frac{1}{p}}\equiv k;$$

whence, $X=x'\left(1+\frac{k}{a^2}\right), \quad Y=y'\left(1+\frac{k}{b^2}\right), \quad Z=z'\left(1+\frac{k}{c^2}\right).$

But these are the co-ordinates of the pole of the plane

$$\frac{xx'}{a^2}+\frac{yy'}{b^2}+\frac{zz'}{c^2}=1,$$

with respect to the conicoid

$$\frac{x^2}{a^2+k}+\frac{y^2}{b^2+k}+\frac{z^2}{c^2+k}=1.$$

The result stated is therefore true for central conicoids, and therefore generally. This proposition is due to Mr Salmon.

552. *At any point in a line of curvature of a central conicoid, the rectangle contained by the central radius parallel to the tangent line at the point, and the perpendicular from the center on the tangent plane at the point, is constant.*

Let the equations of the line of curvature be

$$\frac{x^2}{a^2}+\frac{y^2}{b^2}+\frac{z^2}{c^2}=1, \qquad (1),$$

$$\frac{x^2}{a^2+k}+\frac{y^2}{b^2+k}+\frac{z^2}{c^2+k}=1, \qquad (2),$$

then the radius conjugate to the central section of (1) through the tangent line at (x, y, z) to the line of curvature will be parallel to the normal to (2) at that point. For, it is one axis of the section parallel to the tangent plane at (x, y, z), and, therefore, perpendicular to the other axis which is parallel to the tangent line. It is also perpendicular to the normal at (x, y, z), and therefore to the plane containing the tangent line and the normal, or to the tangent plane to (2). Hence its direction-cosines are proportional to

$$\frac{x}{a^2+k},\ \frac{y}{b^2+k},\ \frac{z}{c^2+k},$$

and its length will therefore be

$$\sqrt{\frac{\frac{x^2}{(a^2+k)^2}+\frac{y^2}{(b^2+k)^2}+\frac{z^2}{(c^2+k)^2}}{\frac{x^2}{a^2(a^2+k)^2}+\frac{y^2}{b^2(b^2+k)^2}+\frac{z^2}{c^2(c^2+k)^2}}},$$

which, by virtue of equations (1) and (2), is equal to $2\sqrt{-k}$. But, if α, β be the two semi-axes of the central section parallel to the tangent plane at (x, y, z), and p the perpendicular upon the tangent plane, $\alpha\beta p = abc$. Now, of the two α, β, one is parallel to the tangent line at (x, y, z), and the other has been proved to be constant. Hence, if d be the radius parallel to the tangent line, we have

$$pd=\frac{2abc}{\sqrt{-k}},$$

which is a constant quantity throughout the same line of curvature.

553. *To apply the general equations to find the umbilici of a central conicoid.*

It is obvious, from the preceding articles, that the umbilici are the points at which the tangent planes are parallel to circular sections of the conicoid, which sufficiently determines them. But, as an example, we may use the equations of Art. (541), to find the umbilici of the conicoid

$$\frac{x^2}{a^2}+\frac{y^2}{b^2}+\frac{z^2}{c^2}=1.$$

In this case

$$U=\frac{2x}{a^2},\quad V=\frac{2y}{b^2},\quad W=\frac{2z}{c^2},\quad u=\frac{2}{a^2},\quad u'=0,\ \&c.$$

The equations therefore become

$$\frac{\dfrac{y^2}{b^4c^2}+\dfrac{z^2}{c^4b^2}}{\dfrac{y^2}{b^4}+\dfrac{z^2}{c^4}}=\frac{\dfrac{z^2}{c^4a^2}+\dfrac{x^2}{a^4c^2}}{\dfrac{z^2}{c^4}+\dfrac{x^2}{a^4}}=\frac{\dfrac{x^2}{a^4b^2}+\dfrac{y^2}{b^4a^2}}{\dfrac{x^2}{a^4}+\dfrac{y^2}{b^4}}=\frac{1}{k^2},$$

or,

$$\frac{y^2}{b^4}\left(\frac{1}{c^2}-\frac{1}{k^2}\right)=-\frac{z^2}{c^4}\left(\frac{1}{b^2}-\frac{1}{k^2}\right),$$

$$\frac{z^2}{c^4}\left(\frac{1}{a^2}-\frac{1}{k^2}\right)=-\frac{x^2}{a^4}\left(\frac{1}{c^2}-\frac{1}{k^2}\right),$$

$$\frac{x^2}{a^4}\left(\frac{1}{b^2}-\frac{1}{k^2}\right)=-\frac{y^2}{b^4}\left(\frac{1}{a^2}-\frac{1}{k^2}\right);$$

which are satisfied by the three systems, $x=0$, $k^2=a^2$; $y=0$, $k^2=b^2$; or $z=0$, $k^2=c^2$; only one of which however leads to real values of all the co-ordinates; the second, if a, b, c be in order of magnitude, which gives

$$y=0,\quad \frac{\dfrac{z^2}{c^2}}{b^2-c^2}=\frac{\dfrac{x^2}{a^2}}{a^2-b^2}=\frac{1}{a^2-c^2},$$

results which are the same as those obtained from the circular sections.

554. As an example of finding general expressions for the radii of curvature of families of surfaces, we may take the following.

A surface is generated by the motion of a straight line which always intersects the axis of x, to prove that the radii of curvature at any point on the axis of x are

$$\frac{dx}{d\phi}\frac{\cos\theta \pm 1}{\sin\theta},$$

x being the distance of the point from the origin, θ the angle which the corresponding generator makes with the axis of x, and ϕ that which its projection upon yz makes with the axis of y.

Let P, P' be contiguous points on the axis of x, at distances x, $x+dx$, from the origin, θ, ϕ; $\theta+d\theta$, $\phi+d\phi$; the angles for the corresponding generating lines: Q a point on the generator through P' at a distance λ from P'.

The tangent plane at P contains the axis of x, and the generator through P, and its equation is therefore

$$y\sin\phi - z\cos\phi = 0;$$

the co-ordinates of Q are

$$x+dx+\lambda\cos(\theta+d\theta),\quad \lambda\sin(\theta+d\theta)\cos(\phi+d\phi),$$
$$\lambda\sin(\theta+d\theta)\sin(\phi+d\phi),$$

and we have

$$PQ^2 = dx^2 + 2dx\,\lambda\cos(\theta+d\theta)+\lambda^2.$$

But the distance (p) of Q, from the tangent plane at P, is

$$\lambda\sin(\theta+d\theta)\sin d\phi,$$

whence, ρ being the radius of curvature of the normal section through PQ, we have

$$2\rho = \text{lt.}\frac{PQ^2}{p} = \frac{\left(\dfrac{dx}{d\phi}\right)^2 + 2\dfrac{dx}{d\phi}\mu\cos\theta+\mu^2}{\mu\sin\theta},$$

denoting the limit of $\dfrac{\lambda}{d\phi}$ by μ.

The equation giving the principal radii of curvature is, therefore,

$$\left(\frac{dx}{d\phi}\right)^2 = \left(\frac{dx}{d\phi}\cos\theta - \rho\sin\theta\right)^2,$$

$$\text{or,} \quad \frac{dx}{d\phi}(\cos\theta \pm 1) = \rho \sin\theta,$$

giving the values for ρ as stated.

If the surface be a right conoid, whose axis is the axis of x, these values become $\pm \frac{dx}{d\phi}$, which agrees with the fact that at any point on the axis of such a surface, the asymptotes of the indicatrix, being parallel to the generators, are at right angles to each other, and the principal radii of curvature must therefore be equal and of opposite signs.

555. *To find the osculating plane of a line of curvature at any point of a surface.*

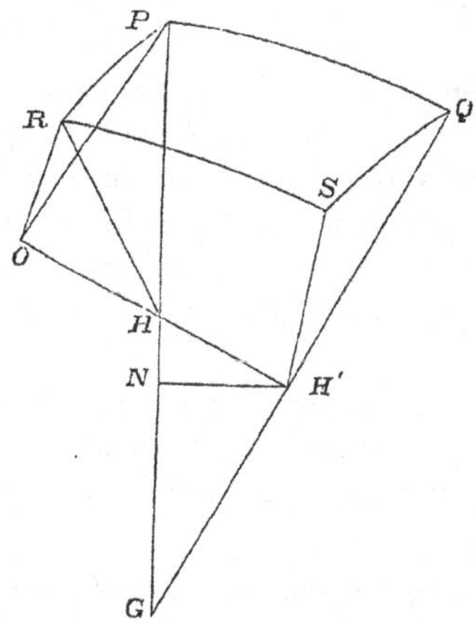

Let PQ, PR be small arcs of lines of curvature drawn through P, a point in the surface, RS, QS lines of curvature through R, Q respectively; and let PHG, $QH'G$, RH, SH' be normals to the surface at P, Q, R, S, so that PH, QH' are ultimately the radii of curvature of the principal normal sections PR, QS, and PG that of PQ; let these be R', $R' + dR'$, and R, dR' being the increment of R' due to a change ds along the principal section PQ.

The tangent to PR at P is perpendicular to the plane PH', and therefore to HH', and the tangent at R is for a similar reason perpendicular to HH', which is therefore parallel to the binormal at P to the line PR, and determines the osculating plane POR.

If ϕ be the inclination of the principal normal section to the osculating plane of PR, $OHP = \frac{\pi}{2} - \phi$.

Draw $H'N$ perpendicular to PG,

$$\tan\phi = \text{lim.}\ \frac{HN}{H'N} = \text{lim.}\ \frac{HN}{PQ} \cdot \frac{PQ}{H'N}$$

$$= \frac{dR'}{ds} \cdot \frac{R}{R - R'}.$$

Cor. In the case of a surface of revolution, since R' is the same for all points in the circular line of curvature supposed to correspond to R, $\frac{dR'}{ds} = 0$, and the osculating plane coincides with the normal plane.

Gauss' Theorems on Curvature.

556. If a portion of a surface be cut off by a closed curve, the *total curvature* of this portion is defined by Gauss to be the area of the surface of a sphere whose radius is unity, cut off by a cone whose vertex is the center, and whose generating lines are parallel to the normals to the surface at every point of the bounding closed curve.

The *measure of curvature* of a surface at any point is the ratio of the total curvature of any elementary portion of the surface including the point to the area of that elementary portion.

557. *To shew that Gauss' measure of curvature is a proper measure, and to calculate its value.*

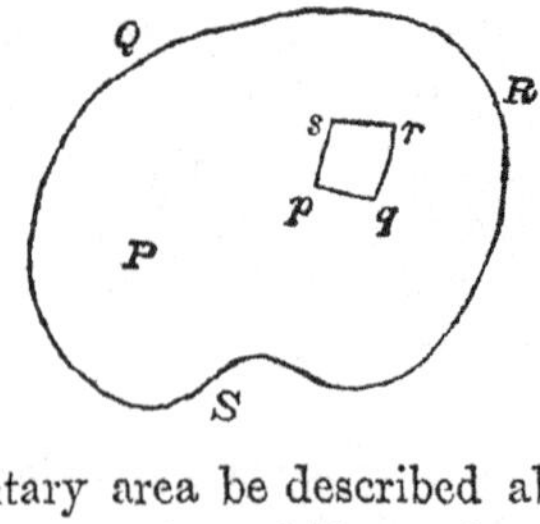

Let any elementary area be described about the point P of the surface, and let a series of lines of curvature divide this area into sub-elementary portions such as $pqrs$, and let ρ, ρ' be the principal radii of curvature at p in the directions pq, ps, the portion of the area on the unit sphere corresponding to pr will be on sides equal to $\frac{pq}{\rho}$, $\frac{ps}{\rho'}$, and area $pr = pq \cdot ps$ ultimately.

Gauss' measure of curvature gives the value $\lim. \dfrac{\Sigma \frac{pr}{\rho\rho'}}{\Sigma (pr)}$.

But $\frac{1}{\rho\rho'}$ lies between $\frac{1}{RR'}(1+\epsilon_1)$ and $\frac{1}{RR'}(1+\epsilon_2)$,
where ϵ_1, ϵ_2 vanishes in the limit;

$$\therefore \text{ the measure of curvature is } \frac{1}{RR'},$$

which is independent of the form of the elementary portion.

COR. The measure of curvature at any point of a surface $z = f(x, y)$ is therefore $\dfrac{rt - s^2}{(1+p^2+q^2)^2}$. Art. (548).

558. *To shew that the measure of curvature is not altered by any deformation of the surface supposed inextensible.*

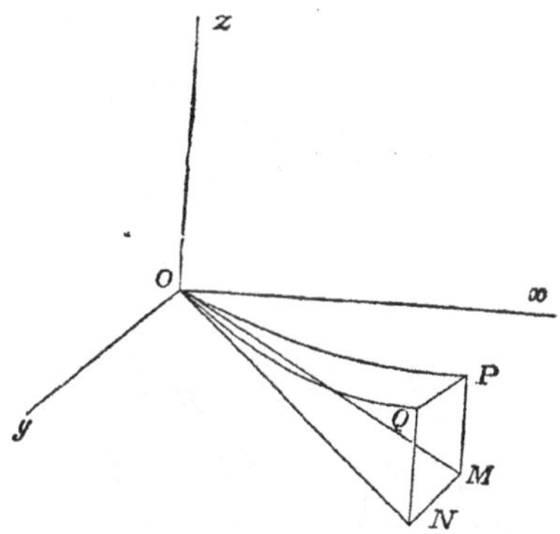

Let a surface be referred to axes such that Oz is a normal at O, and the planes of zx, zy are principal normal sections; and let R, R' be the radii of these sections.

Let a curve be described whose geodesic distances from O are constant and each equal to s;

OP, OQ two of the distances whose planes are inclined to zx at angles θ, $\theta + d\theta$;

OM, ON the traces on xy being r, $r+dr$, the distances of P, Q from xy being z, $z+dz$, and $x\equiv lr$, $y\equiv mr$;

$$PQ^2 = d\sigma^2 = r^2\,\overline{d\theta}|^2 + \overline{dr}|^2 + \overline{dz}|^2.$$

Now
$$z = \frac{x^2}{2R} + \frac{y^2}{2R'} + \&c.$$
$$= \frac{r^2}{2}\left(\frac{l^2}{R} + \frac{m^2}{R'}\right);$$

let ρ be the radius of curvature of OP;

$$\therefore\ r = \rho \sin\frac{s}{\rho} = s - \frac{s^3}{6\rho^2},$$

$$z = \rho \operatorname{vers}\frac{s}{\rho} = \frac{s^2}{2\rho};$$

and
$$\frac{1}{\rho} = \frac{l^2}{R} + \frac{m^2}{R'},$$

$$d\frac{1}{\rho} = -2\left(\frac{1}{R} - \frac{1}{R'}\right) lm\, d\theta,$$

$$dr = \frac{2s^2}{3}\left(\frac{1}{R} - \frac{1}{R'}\right) lm\, d\theta,$$

$$dz = s^2\left(\frac{1}{R} - \frac{1}{R'}\right) lm\, d\theta,$$

$$rd\theta = s\left\{1 - \frac{s^2}{6}\left(\frac{l^2}{R} + \frac{m^2}{R'}\right)^2\right\} d\theta;$$

$$\therefore\ d\sigma^2 = s^2\left\{1 - \frac{s^2}{3}\left(\frac{l^2}{R} + \frac{m^2}{R'}\right)^2 + s^2\left(\frac{1}{R} - \frac{1}{R'}\right)^2 l^2m^2\right\}\overline{d\theta}|^2,$$

$$\text{and } \sigma = \int_0^{2\pi} s\,d\theta\left\{1 - \frac{s^2}{6}\left(\frac{l^2}{R} + \frac{m^2}{R'}\right)^2 + \frac{s^2}{2}\left(\frac{1}{R} - \frac{1}{R'}\right)^2 l^2m^2\right\},$$

but
$$\int_0^{\frac{\pi}{2}} l^4\,d\theta = \frac{3\,.\,1}{4\,.\,2}\,\frac{\pi}{2} = \int_0^{\frac{\pi}{2}} m^4\,d\theta;$$

and
$$\int_0^{\frac{\pi}{2}} l^2m^2\,d\theta = \frac{1}{2}\,\frac{\pi}{2} - \frac{3\,.\,1}{4\,.\,2}\,\frac{\pi}{2} = \frac{1}{8}\,\frac{\pi}{2}.$$

Hence, the perimeter

$$\sigma = 2\pi s\left\{1 - \frac{s^2}{16}\left(\frac{1}{R^2} + \frac{1}{R'^2}\right) + \frac{2}{3RR'} - \left(\frac{1}{R} - \frac{1}{R'}\right)^2\right\}$$

$$= 2\pi s\left(1 - \frac{s^2}{6RR'}\right);$$

$$\text{the area} = \int_0^s \sigma ds = \pi s^2\left(1 - \frac{s^2}{12RR'}\right).$$

Hence, since in any deformation the distances of all points measured along the surface remain unaltered, in order that this may be the case about every point, the perimeters of geodesic circles must be ultimately unaltered: therefore the measure of curvature at any point remains unchanged.

The student, who is interested in the subject of deformation of surfaces, instantaneous and permanent lines of bending, will find a most ingenious article by Maxwell in the *Cambridge Philosophical Transactions*, Vol. IX. Part 4, No. 19.

559. Certain properties of the principal radii of curvature may be conveniently investigated by considering the angle between the two tangents to the curve in which a surface is intersected by the tangent plane at any point. In these directions three consecutive points lie in a straight line, and the radius of curvature of a normal section through one of these tangents is therefore infinite. Hence, if θ be the angle which one of these tangents makes with the tangents to a section of principal curvature, we shall have

$$0 = \frac{\cos^2\theta}{\rho} + \frac{\sin^2\theta}{\rho'},$$

ρ, ρ' being the algebraic magnitudes of the radii of principal curvature. Thus, for points at which the radii of principal curvature are equal in magnitude and opposite in sign, we shall have $\tan^2\theta = 1$, and the tangents to the curve of intersection will therefore also be at right angles. As an example of this method we shall take the following. To prove that, in the surface

$$x(x^2 + y^2 + z^2) = 2a(x^2 + y^2),$$

at all points lying on the plane $x=a$, the radii of curvature will be equal in magnitude, and of opposite signs. This, by what has been said, will be true, if we can prove that the two straight lines, drawn through any such point, to meet the surface in three consecutive points, are at right angles to each other.

Let $$\frac{x'-a}{l}=\frac{y'-y}{m}=\frac{z'-z}{n}=r$$

be a straight line which meets the surface in three consecutive points. The equation

$$(a+lr)(z+nr)^2=(a-lr)\{(a+lr)^2+(y+mr)^2\}$$

must then have all its roots equal to zero.

This gives us the equations

$$z^2=a^2+y^2,\quad lz^2+2naz=2a(al+my)-l(a^2+y^2)$$
$$2nlz+n^2a=a(l^2+m^2)-2l(la+my),$$

of which the two latter become, by reason of the first,

$$ly^2-may+naz=0,$$
$$l^2a-m^2a+n^2a+2nlz+2lmy=0.$$

The equation giving the two values of $n:l$, is

$$ay^2(al^2+an^2+2znl)-(ly^2+naz)^2+2ly^2(ly^2+naz)=0,$$

whence, if (l_1, m_1, n_1), (l_2, m_2, n_2) be the directions of the two lines, we have

$$\frac{n_1n_2}{l_1l_2}=\frac{y^2(a^2+y^2)}{a^2(y^2-z^2)}=\frac{-y^2z^2}{a^4},$$

and similarly, eliminating n, we obtain

$$\frac{m_1m_2}{l_1l_2}=\frac{a^2z^2+y^4-2y^2z^2}{a^2(y^2-z^2)}=\frac{y^2z^2-(y^2-z^2)^2}{a^4}=\frac{y^2z^2-a^4}{a^4};$$

whence, finally,

$$1+\frac{m_1m_2}{l_1l_2}+\frac{n_1n_2}{l_1l_2}=0,$$

or the two lines are at right angles. Hence at all points of the surface lying in the plane $x=a$, the radii of curvature are equal and of opposite sign.

560. DUPIN'S THEOREM. *If there be three series of surfaces such that any two surfaces of different series intersect everywhere at right angles, the curves of intersection are lines of curvature on both surfaces.*

Let the origin be a point of intersection of three surfaces, one of each series, and the tangents to their lines of intersection the axes. The equations of the three surfaces may then be written

$$x + ay^2 + 2byz + cz^2 + \ldots = 0, \qquad (1).$$

$$y + a'z^2 + 2b'zx + c'x^2 + \ldots = 0, \qquad (2).$$

$$z + a''x^2 + 2b''xy + c''y^2 + \ldots = 0, \qquad (3).$$

At a consecutive point on the curve of intersection of (2) and (3), we have

$$y = 0, \quad z = 0, \quad x = x',$$

and the equations of the tangent planes are, ultimately,

$$x \,.\, 2c'x' + y + z \,.\, 2b'x' = 0,$$

$$x \,.\, 2a''x' + y \,.\, 2b''x' + z = 0,$$

and since these also are at right angles,

$$4a''c'x'^2 + 2b''x' + b'x' = 0,$$

or, taking the limit, $\qquad b' + b'' = 0.$

Similarly, $b'' + b = 0$, $b + b' = 0$, which lead us to

$$b = 0, \quad b' = 0, \quad b'' = 0,$$

or the axes are tangents to the lines of curvature on each surface.

Hence, the tangent lines, at any point of intersection of three surfaces, to their curves of intersection, are tangents to the lines of curvature of the surfaces through that point, and consequently, their curves of intersection must coincide with the lines of curvature.

561. As an example of three series of surfaces, satisfying the conditions of Dupin's theorem, we will take the locus of the points of contact, of tangent planes drawn to a series of confocal conicoids, from fixed points on the axes. For each fixed point,

we shall obtain a surface locus, and for each axis a series of such surfaces. We will now prove that these surfaces intersect everywhere at right angles.

Let $(\alpha, 0, 0)$, $(0, \beta, 0)$, $(0, 0, \gamma)$ be the co-ordinates of three points from which tangent planes are drawn to the series of conicoids, confocal with the ellipsoid

$$\frac{x^2}{a^2}+\frac{y^2}{b^2}+\frac{z^2}{c^2}=1.$$

The equations of the surfaces may readily be found to be

$$\frac{x}{\alpha}+\frac{y^2}{\alpha x-a^2+b^2}+\frac{z^2}{\alpha x-a^2+c^2}=1, \qquad (1),$$

$$\frac{x^2}{\beta y+a^2-b^2}+\frac{y}{\beta}+\frac{z^2}{\beta y+c^2-b^2}=1, \qquad (2),$$

and

$$\frac{x^2}{\gamma z+a^2-c^2}+\frac{y^2}{\gamma z+b^2-c^2}+\frac{z}{\gamma}=1. \qquad (3).$$

The points of real intersection of (1) and (2) may, by subtracting their equations, be found to lie in the plane

$$\alpha x-a^2=\beta y-b^2,$$

which, combined with (1), gives the equation

$$\frac{x}{\alpha}(\alpha x-a^2+c^2)+\frac{y}{\beta}(\beta y-b^2+c^2)+z^2=\alpha x-a^2+c^2,$$

the equation of a sphere. The curves of intersection are then circles.

Let the equation of any one of the confocal conicoids be

$$\frac{x^2}{a^2+k}+\frac{y^2}{b^2+k}+\frac{z^2}{c^2+k}=1,$$

and let the tangent plane at (x', y', z') pass through the straight line

$$\frac{x}{\alpha}+\frac{y}{\beta}=1, \quad z=0;$$

the point (x', y', z') will then lie on the intersection of (1) and (2). This gives us

$$\alpha x' = a^2 + k, \ \beta y' = b^2 + k, \ z'^2 = (c^2 + k)\left\{1 - \frac{a^2 + k}{\alpha^2} - \frac{b^2 + k}{\beta^2}\right\} \quad (4);$$

and if (x', y', z') be the point of intersection of (1), (2), (3), we shall have

$$\alpha x' - a^2 = \beta y' - b^2 = \gamma z' - c^2 = k,$$

k being determined by the equation $\dfrac{a^2 + k}{\alpha^2} + \dfrac{b^2 + k}{\beta^2} + \dfrac{c^2 + k}{\gamma^2} = 1.$

Let (λ, μ, ν) be proportional to the direction-cosines of the tangent line at this point to the intersection of (1) and (2). Then, by (4),

$$\alpha\lambda = \beta\mu = \frac{2\nu z'}{1 - \dfrac{a^2 + k}{\alpha^2} - \dfrac{b^2 + k}{\beta^2} - (c^2 + k)\left(\dfrac{1}{\alpha^2} + \dfrac{1}{\beta^2}\right)} = \frac{2\dfrac{\nu}{\gamma}}{\dfrac{1}{\gamma^2} - \dfrac{1}{\alpha^2} - \dfrac{1}{\beta^2}}.$$

Similarly, if (λ', μ', ν') be the direction of the tangent at this point to the intersection of (1) and (3),

$$\alpha\lambda' = \frac{\dfrac{2\mu'}{\beta}}{\dfrac{1}{\beta^2} - \dfrac{1}{\alpha^2} - \dfrac{1}{\gamma^2}} = \gamma\nu',$$

and these will be at right angles, if

$$\frac{1}{\alpha^2} + \frac{1}{2}\left(\frac{1}{\beta^2} - \frac{1}{\alpha^2} - \frac{1}{\gamma^2}\right) + \frac{1}{2}\left(\frac{1}{\gamma^2} - \frac{1}{\alpha^2} - \frac{1}{\beta^2}\right) = 0,$$

which is obviously true.

The tangent lines to the three circles of intersection are then mutually at right angles, and therefore the planes, each of which contains two of these tangent lines, are mutually at right angles, and these are the tangent planes to the surfaces at the point. Hence, at any point which is the intersection of three such surfaces, the tangent planes are mutually at right angles. But, by varying γ, α and β remaining constant, we can make this point coincide with any proposed point on the intersection of (1) and (2), and the surfaces will therefore cut each other orthogonally at all points on their curve of intersection.

The angle between the curve of intersection of (1) and (2), and the tangent plane to (1) at the point in question will be found to be $\tan^{-1}\dfrac{\beta}{\alpha}$, a value which, not depending on γ, will be constant throughout the curves, as is always the case when a line of curvature is a plane curve.

The lines of curvature on any surface (1) will then be two systems of circles, whose planes are parallel to the axes of z and y respectively, and pass each through one of two fixed points on the axis of x.

XXIV.

1. THE principal radii of curvature, at the points of the surface, $a^2x^2 = z^2(x^2+y^2)$, where $x=y=z$, are given by the equation

$$2\rho^2 - 11\sqrt{3}a\rho + 18a^2 = 0.$$

2. Prove that, in the surface

$$x^2(y-z) + ayz = 0,$$

the radii of principal curvature, at points where it is met by the cone

$$(x^2 + 6yz)\,yz = (y-z)^4,$$

are equal, and opposite.

3. Deduce the conditions for an umbilicus from the equation giving the radii of curvature, by making the roots of the equation equal.

The only surface of revolution, such that the radii of curvature are at every point equal and opposite, is that produced by the revolution of a catenary about its directrix.

4. Prove that the only real points on the surface

$$x(x^2+y^2+z^2) = 2a(x^2+y^2),$$

at which the radii of curvature are equal and of opposite signs, are those lying on the hyperbola

$$x = a,\quad z^2 - y^2 = a^2.$$

5. A surface is generated by the revolution of a parabola about its directrix; shew that the principal radii of curvature at any point are to each other in a constant ratio.

6. If ρ, ρ' be the principal radii of curvature at a point of a surface, where the normal's direction-cosines are l, m, n, prove that

$$\frac{1}{\rho}+\frac{1}{\rho'}=\frac{dl}{dx}+\frac{dm}{dy}+\frac{dn}{dz}.$$

7. Shew how to find the lines on a surface at which one of the principal curvatures vanishes. How does the form of the surface alter in passing across such a line? Is the line in question a line of curvature?

8. The locus of the centers of curvature of all plane sections of a given surface at a given point is the surface whose equation, referred to the normal and the principal tangents, as axes, is

$$z\,(x^2+y^2)=(x^2+y^2+z^2)\left(\frac{x^2}{\rho_2}+\frac{y^2}{\rho_1}\right).$$

9. Shew that the projection, on the plane of xy, of the indicatrix at any point of the surface $z=(e^y+e^{-y})\cos x$ is a rectangular hyperbola.

10. If at any point of a surface ρ_1, ρ_2 be radii of curvature of normal sections at right angles to each other, and R_1, R_2 be principal radii of curvature, the sections corresponding to ρ_1, ρ_2 being inclined at an angle α, prove that

$$\frac{\cos^2\alpha}{\rho_1}-\frac{\sin^2\alpha}{\rho_2}=\frac{\cos 2\alpha}{R_1},$$

and

$$\frac{\sin^2\alpha}{\rho_1}-\frac{\cos^2\alpha}{\rho_2}=-\frac{\cos 2\alpha}{R_2}.$$

11. Find the umbilici on the surface whose equation is

$$\frac{x^3}{a^3}+\frac{y^3}{b^3}+\frac{z^3}{c^3}=1,$$

and shew that the radius of curvature there is $\frac{1}{2}(a^6+b^6+c^6)^{\frac{1}{6}}$.

12. A surface is generated by the motion of a variable circle which always intersects the axis of x, and is parallel to the plane of yz. If r be the radius of the circle at a point on the axis of x, and θ the inclination of the diameter through that point to the axis of z, prove that the principal radii of curvature at the point are given by the equation $\rho^2 r+p^2(\rho-r)=0$, where p is the value of $\frac{dx}{d\theta}$ at the point.

13. A surface is generated by a straight line which always intersects a given circle, and the straight line through the center of the circle normal to its plane, prove that the principal radii of curvature of the surface, at any point on the circle, are given by the equation

$$\rho^2\left(\frac{d\theta}{d\phi}\right)^2 - a\rho\cos\theta - a^2 = 0,$$

a being the radius of the circle, θ the angle which the generator at the point makes with the fixed line, and ϕ the angle which the radius of the circle through the point makes with a fixed radius.

14. Find the differential equation of surfaces possessing the property, that the projections on a fixed plane of their lines of curvature cross each other everywhere at right angles. Prove that it is satisfied by surfaces of revolution whose axes are perpendicular to the fixed plane; and obtain the general solution.

15. Two surfaces touch each other at the point P; if the principal curvatures of the first surface at P be denoted by $a \pm b$, those of the second by $a' \pm b'$; and if ϖ be the angle between the principal planes to which $a+b$, $a'+b'$ refer, δ the angle between the two branches at P of the curve of intersection of the surfaces, shew that

$$\cos^2\delta = \frac{a^2 - 2aa' + a'^2}{b^2 - 2bb'\cos 2\varpi + b'^2}.$$

16. The measure of curvature at any point of an ellipsoid is proportional to p^4, p being the perpendicular from the center on the tangent plane.

17. The measure of curvature at any point of the paraboloid

$$\frac{y^2}{\rho} + \frac{z^2}{\rho'} = x$$

varies as $\left(\frac{p}{z}\right)^4$, p being the perpendicular from the origin on the tangent plane.

18. The planes drawn through the center of an ellipsoid, parallel to the tangent planes at points along a line of curvature, envelope a cone which intersects the ellipsoid in a spherical conic.

19. If R be the radius of absolute curvature at any point of a line of curvature on an ellipsoid, r, r' the radii of curvature of the normal sections of the ellipsoid, and the confocal hyperboloid which contains the line of curvature, through the tangent to the line of curvature, prove that

$$\frac{1}{R^2} = \frac{1}{r^2} + \frac{1}{r'^2}.$$

20. If P be any point on a surface, a, b the principal radii of curvature at P, $PQ = s$ an indefinitely small arc taken along a normal section making an angle θ with the principal section to which a refers, prove that if D be the minimum distance between the normals at P and Q, c the distance from P of their point of nearest approach,

$$D^2 = \frac{\sin^2\theta\cos^2\theta\,(a-b)^2 s^2}{a^2\sin^2\theta + b^2\cos^2\theta}, \quad c = \frac{ab\,(a\sin^2\theta + b\cos^2\theta)}{a^2\sin^2\theta + b^2\cos^2\theta}.$$

Prove that the maximum value of the ratio $\dfrac{D}{s}$ is $\left(\dfrac{a-b}{a \pm b}\right)^2$, the upper or lower sign being taken, according as the point is elliptic or hyperbolic. Prove also, that, in the case of the ellipsoid, this maximum value is greatest at the extremities of the mean axis.

21. Prove that the surface generated by normals drawn to a surface, at points whose distance from the normal at a fixed point P is constant and small, will intersect a normal plane making an angle θ with one of the principal planes at P, in two parabolas, whose radii of curvature are

$$\frac{\rho_2(\rho_2-\rho_1)}{\rho_1\sin^2\theta}, \quad \frac{\rho_1(\rho_1-\rho_2)}{\rho_2\cos^2\theta},$$

ρ_1, ρ_2 being the principal radii of curvature.

22. If a line of curvature be a plane curve, its plane cuts the tangent planes to the surface, at points lying along it, at a constant angle.

23. If one series of lines of curvature on a surface be plane curves, lying in parallel planes, the other series will also be plane curves.

24. If R be the radius of absolute curvature at any point of the curve, which is the intersection of two surfaces A, B, and r, r' be the radii of curvature of the sections of the surfaces A, B made by the tangent planes to B, A respectively, and θ be the angle between the tangent planes, prove that

$$\frac{1}{R^2} = \frac{1}{r^2} - \frac{2\cos\theta}{rr'} + \frac{1}{r'^2}.$$

25. If a plane curve be given by the equations

$$\frac{x}{a} = \cos\theta + \log_e\tan\frac{\theta}{2}, \quad \frac{y}{a} = \sin\theta;$$

the surface, produced by the revolution of this curve about the axis of x, will have its measure of curvature constant.

26. In a surface, generated as in (13), if $\phi = \log \tan \frac{\theta}{2}$, the measure of curvature will be the same at corresponding points on the axis of z and on the circle.

27. On an umbilical conicoid, the projections of the lines of curvature on the planes of circular section by lines parallel to an axis, form a series of confocal conics, the foci of which are the projections of the umbilici.

28. Prove that the three surfaces

$$yz = ax,$$

$$\sqrt{x^2 + y^2} + \sqrt{x^2 + z^2} = b,$$

$$\sqrt{x^2 + y^2} - \sqrt{x^2 + z^2} = c,$$

intersect each other always at right angles; and hence prove that, on a hyperbolic paraboloid, whose principal sections are equal parabolas, the sum or the difference of the distances of any point on a line of curvature, from the two generators through the vertex, is constant.

29. In the helicoid, whose equation is $y = x \tan \frac{z}{a}$, the lines of curvature are the intersections of the helicoid with the surfaces represented by the equation

$$\frac{2\sqrt{x^2 + y^2}}{a} = ce^{\frac{z}{a}} + \frac{1}{c}e^{-\frac{z}{a}},$$

for different values of c.

Also, prove that the principal radii of curvature are, at every point, constant, equal in magnitude, but of opposite signs.

CHAPTER XXIV.

GEODESIC LINES.

562. Def. A *geodesic* line of a given surface, between two given points on it, is a line of maximum or minimum length. Any infinitesimal arc of such a line will manifestly be the minimum line between its extremities, but if the two given points be at a finite distance, a geodesic passing through them may be either a maximum or minimum, and there may be an infinite number of such maxima and minima.

563. *The osculating plane at any point of a geodesic on any surface contains the normal to the surface.*

The distance between two indefinitely near points will be the least possible, when the curvature of the line joining them is the least, or when the radius of curvature is the greatest. Now the curvature of any curve on a surface will be, at any point, the same as that of the section of the surface made by the osculating plane at that point, since the two curves will have three coincident points. Also, of all sections, having a common tangent line, the normal section is that of least curvature, by Meunier's Theorem (Art. 538). Hence, the osculating plane of a geodesic, at any point, must be a normal section.

This also appears from the consideration of a stretched weightless string, joining any two points on a surface. This will manifestly assume the form of the shortest line joining the points, and since the resultant of the tensions of two consecutive elements of the string is balanced by the normal reaction of the surface, the normal must lie in the plane of these elements, that is, in the osculating plane of the curve.

The equations of any geodesic line on the surface $F(x, y, z)=0$, may therefore be written in the form

$$\frac{\frac{d^2x}{ds^2}}{\frac{dF}{dx}}=\frac{\frac{d^2y}{ds^2}}{\frac{dF}{dy}}=\frac{\frac{d^2z}{ds^2}}{\frac{dF}{dz}}.$$

One integral of these equations is, of course, $F(x, y, z)=0$, and, if another can be found, it will involve two constants, to be determined so as to make the line pass through any two proposed points.

The form of the equations, connecting the constants of integration with the co-ordinates of the two proposed points, may be such that an infinite number of values can be given to the constants, to each of which will correspond a geodesic through the two given points. As an example of such a family of geodesics joining two given points, we will take the very simple case of a right circular cylinder.

564. *To investigate the equations of a geodesic, joining two given points on a right circular cylinder.*

Let the equation of the cylinder be

$$x^2+y^2=a^2,$$

and let the given points be $(a, 0)$; $(a \cos \alpha, a \sin \alpha, c)$, respectively.

The general equations of a geodesic give

$$\frac{\frac{d^2x}{ds^2}}{x}=\frac{\frac{d^2y}{ds^2}}{y}=\frac{\frac{d^2z}{ds^2}}{0}.$$

Hence $\frac{dz}{ds}$ is constant $\equiv \cos \beta$, whence $\left(\frac{dx}{ds}\right)^2+\left(\frac{dy}{ds}\right)^2=\sin^2 \beta$.

Eliminating x and s, the equation connecting y and z is

$$\sin \beta \frac{dz}{dy}=\frac{a}{\sqrt{a^2-y^2}},$$

therefore $$z \sin \beta = a \sin^{-1}\left(\frac{y}{a}\right)+C.$$

To determine C we shall have the equation

$$C + a \sin^{-1} 0 = 0,$$

and the equation of the geodesic will be, whatever possible value we assign C,

$$y = a \sin\left(\frac{z \sin \beta}{a}\right).$$

We shall then have, to determine β, the equation

$$\sin \alpha = \sin\left(\frac{c \sin \beta}{a}\right),$$

which admits of any solution of the form

$$\frac{c \sin \beta}{a} = r\pi + (-1)^r \alpha, \ r \text{ being any whole number;}$$

which give, for the geodesic, the equation

$$\frac{y}{a} = \sin \frac{z}{c} \{r\pi + (-1)^r \alpha\},$$

and there will be a different curve for every different value of r. Thus there will be an infinite number of geodesics joining any two proposed points on the cylinder.

This is obvious geometrically, for we can wind a string round the cylinder in either direction as many times as we please so as to start from one point, and pass through the other, and retain its form under tension. If α be a positive angle less than two right angles, the value $r = 2m$ will give a geodesic crossing every generating line m times, and a certain portion of them $m + 1$ times; and $r = 2m + 1$ will give a geodesic crossing every generating line m times, and a certain portion, supplementary to the former portion, $m + 1$ times. Two points describing geodesics of different systems will start in opposite directions from one of the points to proceed to the other.

565. *To investigate the equations of a geodesic, joining two given points on a cone of revolution.*

Let the equation of the cone be

$$y^2 + z^2 = x^2 \tan^2 \alpha,$$

and let the two given points be $(a,\ a \tan\alpha,\ 0)$; $(b,\ b\tan\alpha\cos\beta,\ b\tan\alpha\sin\beta)$; then, for the geodesic, we have

$$\frac{\dfrac{d^2x}{ds^2}}{-x\tan^2\alpha} = \frac{\dfrac{d^2y}{ds^2}}{y} = \frac{\dfrac{d^2z}{ds^2}}{z},$$

$$z\frac{d^2y}{ds^2} - y\frac{d^2z}{ds^2} = 0, \quad \text{or} \quad z\frac{dy}{ds} - y\frac{dz}{ds} = h.$$

If we take $y = r\cos\phi$, $z = r\sin\phi$, we have $r = x\tan\alpha$, and the above equation becomes

$$r^2\frac{d\phi}{ds} + h = 0.$$

Also, we have the equation

$$\left(\frac{dx}{ds}\right)^2 + r^2\left(\frac{d\phi}{ds}\right)^2 + \overline{\frac{dr}{ds}}\Big|^2 = 1, \quad \text{or} \quad \frac{1}{\sin^2\alpha}\left(\frac{dr}{ds}\right)^2 + r^2\left(\frac{d\phi}{ds}\right)^2 = 1,$$

whence
$$\frac{1}{\sin^2\alpha}\left(\frac{dr}{d\phi}\right)^2 + r^2 = \frac{r^4}{h^2},$$

the solution of which is

$$r = h\sec(\phi\sin\alpha + C),$$

h, C being constants, to be determined so as to make the geodesic pass through the two given points. Hence

$$a\tan\alpha = h\sec C,$$

$$b\tan\alpha = h\sec(\beta\sin\alpha + C),$$

and, eliminating h, C, we obtain, for the geodesic, the equation

$$\frac{r}{a}\sin\{(\beta - \phi)\sin\alpha\} + \frac{r}{b}\sin(\phi\sin\alpha) = \sin(\beta\sin\alpha)\tan\alpha.$$

Now, the two points on the cone remain unchanged, if for β we substitute $2m\pi + \beta$, m being any positive or negative integer, and to each such value will correspond a geodesic line, joining the two points. The number of these lines is however limited by the condition that the angle $(2m\pi + \beta)\sin\alpha$ must lie between the limits π and $-\pi$. For, if we take $(2m\pi + \beta)\sin\alpha = n\pi + \delta$, n

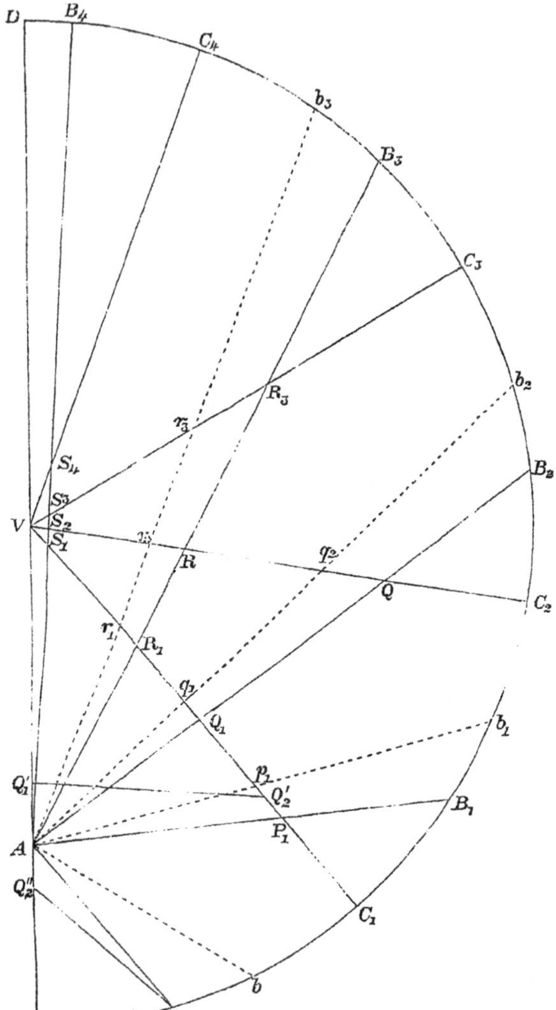

[*To face page* 453.]

being any whole number, then, if n be odd, we shall have, when $\phi \sin \alpha = \delta$, $r = -b$; and if n be even, when $\phi = \pi + \delta$, $r = -b$; or, in both cases, we shall have negative values of r, for a value of ϕ less than the one corresponding to our second point. Hence, r must have passed through the values 0 or ∞. But, it is manifest, that $r = 0$ does not satisfy the equation; hence r must have passed through the value ∞ before arriving at the second point, and there will be no corresponding geodesic of finite length. The number of finite geodesics joining the two points will therefore exceed by unity the sum of the integral parts of the two quantities

$$\frac{\frac{\pi}{\sin \alpha} - \beta}{2\pi}, \quad \frac{\frac{\pi}{\sin \alpha} + \beta}{2\pi},$$

and cannot differ by more than unity from the integral part of $\frac{1}{\sin \alpha}$. It has been assumed that β lies between 0 and 2π.

The form of the geodesic lines which can be drawn from any point A to any other B on a cone, are exhibited in the figure. With the radius VB and center V let a semicircle be described, whose base contains the point A, and let CVC_1 be the sector into which the portion of the cone terminated by the circle containing B would be developed, VA being less than VB. Let equal sectors be placed in order whose bases are C_1C_2, C_2C_3, C_3C_4, which in the figure are all the complete sectors which can lie within the limits of the semicircle.

Take $C_1B_1 = C_2B_2 = C_3B_3 = C_4B_4 = C_1b = C_2b_1 = C_3b_2 = CB$,

and join A by straight lines to all the points B and b, intersecting the radii VC_1, &c. in P, p, Q, q, R, r and S.

Take $VQ_1' = VQ_1$, $VQ_2'' = VQ_2' = VQ_2$, and join $Q_1'Q_2'$, $Q_2''B$.

One of the lines AB_2 crosses into three of the sectors, and if the second and third be placed upon the first the portions of the line corresponding will fall on $Q_1'Q_2'$ and $Q_2''B$, and if the cone be formed of the sector CVC_1, these portions will form a continuous line, being a geodesic line from A to B. Similarly for AB_1, AB_3, and AB_4.

In the same way, if the semicircle on the opposite side of CD were divided into sectors and a similar construction made, we should obtain other geodesic lines corresponding to the dotted lines Ab, Ab_1, &c. the figure being supposed turned about CD to the opposite side.

It is evident that the number of geodesic lines on each side will be limited by the consideration that each line AB must cross the sectors consecutively, in order that a continuous line may be obtained on the cone, corresponding to each. Hence, in the particular case of the figure, there will be five setting out on the one side of VC and four on the other.

Generally, if γ be the developed angle between the generating lines through A and B, and δ the complete developed angle of the cone, the number $r+1$ on one side is given by the inequality $\gamma + r\delta < \pi$, or the number is the integer next less than $\frac{\pi-\gamma}{\delta}+1$, the number on the other is that next less than

$$\frac{\pi-\delta+\gamma}{\delta}+1, \quad \text{or } \frac{\pi+\gamma}{\delta}.$$

566. *To prove that, throughout a geodesic on any surface of revolution, the distance of any point from the axis varies as the cosecant of the angle between the geodesic and the meridian.*

Let the axis of the surface of revolution be taken as the axis of z, and let the equation of the surface be $z = f(x^2+y^2)$.

Then, for any geodesic, we have

$$\frac{\dfrac{d^2x}{ds^2}}{\dfrac{dz}{dx}} = \frac{\dfrac{d^2y}{ds^2}}{\dfrac{dz}{dy}} = \frac{\dfrac{d^2z}{ds^2}}{-1}.$$

But from the form of the equation,

$$x\frac{dz}{dy} - y\frac{dz}{dx} = 0,$$

whence
$$x\frac{d^2y}{ds^2} - y\frac{d^2x}{ds^2} = 0, \quad x\frac{dy}{ds} - y\frac{dx}{ds} = h.$$

This may be transformed by the formulæ

$$x = \rho \cos \phi, \quad y = \rho \sin \phi,$$

when it takes the form $\rho^2 \dfrac{d\phi}{ds} = h$.

Now, $\rho \dfrac{d\phi}{ds}$ may be readily proved to be the sine of the angle between the curve in question, and the meridian of the surface of revolution. If this angle be denoted by I, we see that throughout any geodesic on a surface of revolution

$$\rho \propto \frac{1}{\sin I}.$$

567. *To prove that, throughout a geodesic on a central conicoid, pd is constant, p being the perpendicular on the tangent plane from the center, and d the central radius parallel to the geodesic.*

Let the equation of the conicoid be

$$\frac{x^2}{a} + \frac{y^2}{b} + \frac{z^2}{c} = 1,$$

and let (l, m, n) be the direction of the tangent line at any point of a geodesic. We shall then have the equations

$$\frac{\frac{dl}{ds}}{\frac{x}{a^2}} = \frac{\frac{dm}{ds}}{\frac{y}{b^2}} = \frac{\frac{dn}{ds}}{\frac{z}{c^2}} \equiv k, \qquad (1),$$

$$\frac{lx}{a^2} + \frac{my}{b^2} + \frac{nz}{c^2} = 0, \qquad (2),$$

$$l^2 + m^2 + n^2 = 1. \qquad (3).$$

Differentiating (2) we obtain

$$\frac{x}{a^2}\frac{dl}{ds} + \frac{y}{b^2}\frac{dm}{ds} + \frac{z}{c^2}\frac{dn}{ds} + \frac{l^2}{a^2} + \frac{m^2}{b^2} + \frac{n^2}{c^2} = 0,$$

whence, by (1), $k\left(\dfrac{x^2}{a^4} + \dfrac{y^2}{b^4} + \dfrac{z^2}{c^4}\right) + \dfrac{l^2}{a^2} + \dfrac{m^2}{b^2} + \dfrac{n^2}{c^2} = 0.$

Therefore $\left(l\dfrac{kx}{a^4}+m\dfrac{ky}{b^4}+n\dfrac{kz}{c^4}\right)\left(\dfrac{x^2}{a^4}+\dfrac{y^2}{b^4}+\dfrac{z^2}{c^4}\right)$

$$+\left(\frac{l^2}{a^2}+\frac{m^2}{b^2}+\frac{n^2}{c^2}\right)\left(\frac{lx}{a^4}+\frac{my}{b^4}+\frac{nz}{c^4}\right)=0,$$

or $$\left(\frac{l}{a^2}\frac{dl}{ds}+\frac{m}{b^2}\frac{dm}{ds}+\frac{n}{c^2}\frac{dn}{ds}\right)\left(\frac{x^2}{a^4}+\frac{y^2}{b^4}+\frac{z^2}{c^4}\right)$$

$$+\left(\frac{l^2}{a^2}+\frac{m^2}{b^2}+\frac{n^2}{c^2}\right)\left(\frac{x}{a^4}\frac{dx}{ds}+\frac{y}{b^4}\frac{dy}{ds}+\frac{z}{c^4}\frac{dz}{ds}\right)=0,$$

which gives the first integral

$$\left(\frac{l^2}{a^2}+\frac{m^2}{b^2}+\frac{n^2}{c^2}\right)\left(\frac{x^2}{a^4}+\frac{y^2}{b^4}+\frac{z^2}{c^4}\right)=C.$$

Also $$\frac{1}{p^2}=\frac{x^2}{a^4}+\frac{y^2}{b^4}+\frac{z^2}{c^4},\quad \frac{1}{d^2}=\frac{l^2}{a^2}+\frac{m^2}{b^4}+\frac{n^2}{c^2},$$

and therefore, finally, $\dfrac{1}{p^2d^2}=C$, or pd is constant.

The equations of a geodesic on a conicoid have not been integrated farther, but the property here proved to belong to them leads to many important results. The following geometrical proof of this property, and the corresponding one for lines of curvature, was published by Mr Joyce, in the *Quarterly Journal*, Vol. 5, page 265.

Consider any curve traced on a surface as the limit of an equilateral inscribed polygon, as in Chap. XXII. The tangent planes to the surface at the different points of the curve will be the limiting positions of planes passing through the sides of this polygon, and intersecting the surface in conics, of which these sides are diameters, and the normals to the surface will be the limiting positions of lines drawn through the middle points of the chords perpendicular to these planes. Let the line of intersection of two consecutive tangent planes be called the *conjugate line*. Then, if two consecutive normals to the surface make equal angles with the osculating plane containing the corresponding sides of the polygon, these sides must also make equal angles with the conjugate line, and the angles which they make with the same part of the conjugate line will be equal, or supplementary, according as the normals lie on the same side, or on opposite sides, of the osculating plane.

If the normals lie on the same side they will intersect, and the limit of the polygon will then be a line of curvature.

If the normals lie on opposite sides, then, in the limit, when they coincide, the normal to the surface, which is their common limit, must lie in the osculating plane, or the curve becomes a geodesic.

On any surface, therefore, consecutive sides of those polygons which become, in the limit, lines of curvature will make equal angles with the corresponding conjugate line; and, for those whose limits are geodesics, these angles are supplementary.

Now, let LM, MN be consecutive elements of such a polygon, on a conicoid whose centre is O, MP the corresponding conjugate line, Om the central radius vector parallel to the conjugate line, ml, mn chords of the ellipsoid parallel to ML, MN. Then

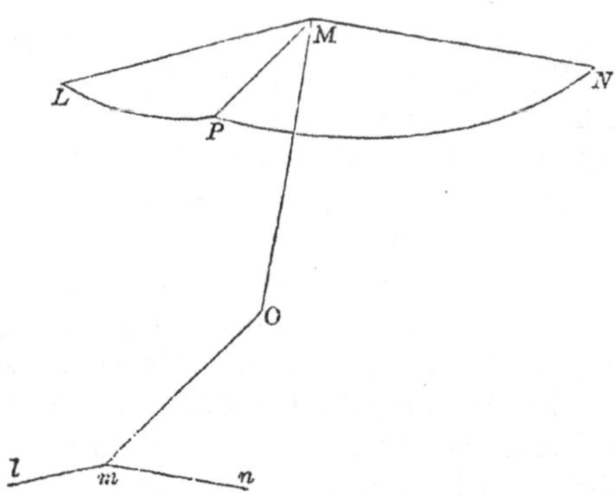

ml, mn will ultimately be tangents to the central section made by a plane parallel to the tangent plane to the surface, at a point on the curve. Also, since the angles Oml, Omn, are either equal or supplementary, the perpendiculars from O on these tangents will be equal. That is, if a plane be drawn through the center parallel to the tangent plane at any point on the curve, and a tangent line be drawn to this section parallel to the tangent line to the curve, the perpendicular from the center on this tangent line will be of constant length. Let p' be this length. Then $p'pd$ will be the volume of a parallelopiped enveloping the conicoid, and having its faces parallel to conjugate planes, and will therefore be constant.

Hence, since p' has been shewn to be constant, pd is constant.

568. *The constant pd has the same value for all geodesics passing through an umbilicus.*

For, at the umbilicus, p is common to all the geodesics, also d being parallel to a tangent line at an umbilicus, is a diameter of a central circular section, and is therefore equal to ab, whence $pd = \frac{ac}{b} b = ac$.

569. *The locus of a point on an ellipsoid, the sum or difference of whose geodesic distances from two adjacent umbilici is constant, is a line of curvature.*

For, let U, V be two adjacent umbilici, P any point on the locus, then at P, pd will be the same for the geodesics PU, PV, and p being the same, d will also be the same: that is, the central radii parallel to the tangents are equal. These radii must therefore make equal angles with the axes of the section in which they lie, which is parallel to the tangent plane at P. But these axes are parallel to the tangents to the lines of curvature at P. The geodesics PU, PV will therefore make equal angles with the lines of curvature at P. But if the sum of the geodesic distances PU, PV be constant, the locus of P will be some curve which always bisects the external angle between PU, PV, and if the difference be constant, the locus will always bisect the internal angle, by the same proof as is used for the corresponding property of plane conics. But the lines of curvature have been shewn to possess this property. The locus of P will therefore be one of these lines of curvature, according as the sum or the difference is constant.

570. *All geodesics joining two opposite umbilici are of equal length.*

Let P be any point on a line of curvature, U, V two umbilici for which $PU+PV$ is constant, U' the umbilicus opposite U. Then PU, PU', making equal vertically opposite angles, with the tangent to the line of curvature at P, will be parts of the same geodesic. Also the line of curvature will bisect the internal angle between PU', PV, and the difference of PU', PV will therefore be constant. But the sum of PU, PV is constant, and therefore the sum of PU, PU' is constant, or the geodesic UPU' is of constant length.

571. *The constant pd has the same value for all geodesics which touch the same line of curvature.*

For pd is constant throughout a line of curvature, and, at the point of contact with any geodesic, both p and d are the same for both curves.

572. *Two geodesic tangents, drawn to a line of curvature, will make equal angles with the lines of curvature at their intersection.*

For, pd will be the same for both, and therefore at the point of intersection, d will be the same for both, which, as in Art. (569), leads to the property enunciated.

It follows from this, by the same proof as for plane confocal conics, that if two tangents be drawn to a line of curvature, from a point lying on another line of curvature of the same system, that the sum of the tangents will exceed the intercepted arc by a constant quantity.

573. *The locus of the intersection of two geodesic tangents to two lines of curvature, at right angles to each other, lies on a sphere.*

Let the lines of curvature be given by the equations

$$\frac{x^2}{a^2-k_1^2}+\frac{y^2}{b^2-k_1^2}+\frac{z^2}{c^2-k_1^2}=1,$$

$$\frac{x^2}{a^2-k_2^2}+\frac{y^2}{b^2-k_2^2}+\frac{z^2}{c^2-k_2^2}=1,$$

then, along these lines, $pd=\frac{abc}{k_1}$, or, $\frac{abc}{k_2}$, Art. (552), and at the intersection of the two geodesics, pd has these values for both, and p is the same for both. Let a', b' be the semi-axes of the central section conjugate to their point of intersection, d, d' the central radii parallel to the tangents to the geodesics. These being at right angles, we have

$$\frac{1}{d^2}+\frac{1}{d'^2}=\frac{1}{a'^2}+\frac{1}{b'^2}=\frac{a'^2+b'^2}{a'^2b'^2}$$

$$\frac{1}{p^2d^2}+\frac{1}{p^2d'^2}=\frac{a'^2+b'^2}{a'^2b'^2p^2}=\frac{a^2+b^2+c^2-r^2}{a^2b^2c^2},$$

r being the central distance of the point of intersection.

Hence,

$$\frac{k_1^2+k_2^2}{a^2b^2c^2}=\frac{a^2+b^2+c^2-r^2}{a^2b^2c^2}, \text{ or } r^2=a^2+b^2+c^2-k_1^2-k_2^2,$$

whence r has a constant value, or the point of intersection lies on a sphere.

A particular case of this is, that the foot of the geodesic perpendicular from an umbilicus, on any tangent to a line of curvature, lies on a sphere, any geodesic through an umbilicus touching the limit of the lines of curvature, obtained by taking $k=b$.

574. Another form of the fundamental equation for geodesics may be found, in terms of the parameters of the lines of curvature at any point of the geodesic, and the angles which the geodesic makes with the lines of curvature.

Let θ, $\frac{\pi}{2}-\theta$ be these angles, and k_1, k_2 the parameters of the lines of curvature, then k_1, k_2 are the semi-axes of the central section conjugate to the point considered, and

$$\frac{1}{d^2}=\frac{\cos^2\theta}{k_1^2}+\frac{\sin^2\theta}{k_2^2}.$$

Also, if (x, y, z) be the point, the equation whose roots are k_1, k_2 is

$$\frac{x^2}{a^2(a^2-k^2)}+\frac{y^2}{b^2(b^2-k^2)}+\frac{z^2}{c^2(c^2-k^2)}=0,$$

whence $$k_1^2k_2^2=a^2b^2c^2\left(\frac{x^2}{a^4}+\frac{y^2}{b^4}+\frac{z^2}{c^4}\right)=\frac{a^2b^2c^2}{p^2}.$$

From these equations we have

$$\frac{a^2b^2c^2}{p^2d^2}=k_2^2\cos^2\theta+k_1^2\sin^2\theta,$$

or $$k_2^2\cos^2\theta+k_1^2\sin^2\theta=\text{const.}$$

If there be two geodesics at right angles to each other, for each of which pd is a given quantity, we shall have

$$k_2^2\cos^2\theta+k_1^2\sin^2\theta=\lambda^2,$$

$$k_2^2\sin^2\theta+k_1^2\cos^2\theta=\lambda'^2,$$

whence $$k_1^2+k_2^2=\lambda^2+\lambda'^2,$$

or $$a^2+b^2+c^2-x^2-y^2-z^2=\lambda^2+\lambda'^2,$$

and the point of intersection lies on the same sphere as that found in the last Article.

From each point of this locus, we can then draw two pairs of tangents at right angles to each other, one of each pair touching the corresponding line of curvature, and since only two geodesics can be drawn through each point (unless an umbilicus) having a given pd, we see that every geodesic, for which pd has a given value, will touch the corresponding line of curvature.

575. *If a point of an ellipsoid, and a line of curvature, be projected on a plane of circular section, by lines parallel to the greatest or least axis; the angle between the geodesic tangents drawn from the point to the line of curvature will be equal to the angle between the tangents from the projection of the point to the projection of the line of curvature.*

Let (x, y, z) be any point on the ellipsoid, (X, Y) the coordinates of its projection by a line parallel to the axis of x on a plane of circular section, the axis of Y coinciding with that of y, and the axis of X lying in the plane of circular section. We shall then have

$$y = Y, \quad z = X \sin \omega,$$

where 2ω is the angle between the planes of circular section.

Hence, we have

$$y^2 = Y^2, \quad z^2 = X^2 \frac{c^2 (a^2 - b^2)}{b^2 (a^2 - c^2)},$$

$$\frac{x^2}{a^2} = 1 - \frac{Y^2}{b^2} - \frac{X^2}{b^2} \cdot \frac{a^2 - b^2}{a^2 - c^2}.$$

The equation of the projection of the line of curvature, whose parameter is λ, will then be

$$\frac{1}{a^2 - \lambda^2} \left(1 - \frac{Y^2}{b^2} - \frac{X^2}{b^2} \frac{a^2 - b^2}{a^2 - c^2}\right)$$
$$+ \frac{Y^2}{b^2} \frac{1}{b^2 - \lambda^2} + \frac{X^2}{b^2 (c^2 - \lambda^2)} \frac{(a^2 - b^2)}{a^2 - c^2} = 0,$$

which reduces to

$$\frac{X^2}{\lambda^2 - c^2} + \frac{Y^2}{\lambda^2 - b^2} = \frac{b^2}{a^2 - b^2},$$

or the projection of a line of curvature is a conic having given foci, which are readily seen to be the projections of the umbilici.

Now, let (x', y', z') be a point from which geodesic tangents are drawn to the line of curvature λ, and let k_1, k_2 be the parameters of the lines of curvature through (x', y', z'). Then, if 2θ be the angle between the tangents, θ is given by the equation

$$k_1^2 \cos^2 \theta + k_2^2 \sin^2 \theta = \lambda^2.$$

Let (X', Y') be the projection of (x', y', z'), then k_1, k_2 are the roots of the equation

$$\frac{X'^2}{k^2 - c^2} + \frac{Y'^2}{k^2 - b^2} = \frac{b^2}{a^2 - b^2},$$

and if 2ϕ be the angle between the tangents from (X', Y') to the conic

$$\frac{X^2}{\lambda^2-c^2}+\frac{Y^2}{\lambda^2-b^2}=\frac{b^2}{a^2-b^2},$$

we shall have

$$\tan^2 2\phi=\frac{4\dfrac{b^2}{a^2-b^2}(\lambda^2-b^2)(\lambda^2-c^2)\left\{\dfrac{X'^2}{\lambda^2-c^2}+\dfrac{Y'^2}{\lambda^2-b^2}-\dfrac{b^2}{a^2-b^2}\right\}^2}{\left\{X'^2+Y'^2-(2\lambda^2-b^2-c^2)\dfrac{b^2}{a^2-b^2}\right\}^2}.$$

But, from the equation in k,

$$\frac{X'^2}{\lambda^2-c^2}+\frac{Y'^2}{\lambda^2-b^2}-\frac{b^2}{a^2-b^2}=-\frac{(\lambda^2-k_1^2)(\lambda^2-k_2^2)}{(\lambda^2-b^2)(\lambda^2-c^2)}\frac{b^2}{a^2-b^2},$$

and $$k_1^2+k_2^2=b^2+c^2+\frac{a^2-b^2}{b^2}(X'^2+Y'^2).$$

Hence $$\tan^2 2\phi=-4\frac{\dfrac{b^4}{(a^2-b^2)^2}(\lambda^2-k_1^2)(\lambda^2-k_2^2)}{\dfrac{b^4}{(a^2-b^2)^2}(k_1^2+k_2^2-2\lambda^2)^2}$$

$$\equiv-4\frac{(\lambda^2-k_1^2)(\lambda^2-k_2^2)}{(k_1^2+k_2^2-2\lambda^2)^2},$$

$$\sec^2 2\phi=\frac{(k_1^2+k_2^2)^2}{(k_1^2+k_2^2-2\lambda^2)^2},$$

$$k_1^2+k_2^2-2\lambda^2=\pm(k_1^2-k_2^2)\cos^2\phi,$$

which gives the equations

$$k_1^2\sin^2\phi+k_2^2\cos^2\phi=\lambda^2,$$

or $$k_1^2\cos^2\phi+k_2^2\sin^2\phi=\lambda^2,$$

which proves that 2ϕ is either equal to 2θ, or to $\pi-2\theta$, and in either case the acute angle between the geodesic tangents is equal to the acute angle between the tangents to the projection.

576. *If a geodesic be drawn through an umbilicus, inclined at a constant angle to a geodesic tangent to a fixed line of curvature, the point of intersection will lie on one of two spheres.*

For, the locus of the projection of the point will be the locus of the point of intersection of a straight line drawn through a focus with the tangent to a fixed conic, and will therefore be one of two circles. But, since

$$X^2 + Y^2 = \frac{b^2}{a^2 - b^2}(a^2 - x^2 - y^2 - z^2),$$

we see that if the projection lie on a circle in the plane of circular section, the corresponding point of the ellipsoid must lie on a sphere.

A particular case of this will be that, if two geodesics be drawn through two adjacent umbilici, inclined at a constant angle, their point of intersection will lie on a sphere.

577. *The tangent lines to a geodesic on a conicoid will all touch a fixed confocal conicoid.*

Let l, m, n be the direction-cosines of a tangent line to the conicoid,

$$\frac{x^2}{a^2} + \frac{y^2}{b^2} + \frac{z^2}{c^2} = 1,$$

at a point (x, y, z). This gives the condition

$$\frac{lx}{a^2} + \frac{my}{b^2} + \frac{nz}{c^2} = 0,$$

and the condition that this line may touch the confocal conicoid,

$$\frac{x^2}{a^2 - k^2} + \frac{y^2}{b^2 - k^2} + \frac{z^2}{c^2 - k^2} = 1,$$

is $$\left(\frac{l^2}{a^2 - k^2} + \frac{m^2}{b^2 - k^2} + \frac{n^2}{c^2 - k^2}\right)\left(\frac{x^2}{a^2 - k^2} + \frac{y^2}{b^2 - k^2} + \frac{z^2}{c^2 - k^2} - 1\right)$$
$$= \left(\frac{lx}{a^2 - k^2} + \frac{my}{b^2 - k^2} + \frac{nz}{c^2 - k^2}\right)^2,$$

which is, on reduction, a quadratic in k^2, one of whose roots will of course be zero.

Making the substitutions

$$2mn\,yz = b^2c^2\left(\frac{l^2x^2}{a^4} - \frac{m^2y^2}{b^4} - \frac{n^2z^2}{c^4}\right),$$

and the like, we obtain for the sum of the roots, and therefore for the second value of k^2, the expression

$$a^2b^2c^2\left(\frac{x^2}{a^4}+\frac{y^2}{b^4}+\frac{z^2}{c^4}\right)\left(\frac{l^2}{a^2}+\frac{m^2}{b^2}+\frac{n^2}{c^2}\right),$$

whence it follows that for the tangent lines to any fixed geodesic, this value of k^2 is constant.

This value is the same as the parameter of the line of curvature which is touched by the geodesic, or the tangent lines to a geodesic on a conicoid will all touch the confocal conicoid which contains the line of curvature touched by the geodesic.

When $k=b$, the corresponding pd is equal to ac, and the geodesic passes through an umbilicus. The confocal surface then reduces to the umbilical focal conic, or "all tangent lines to a geodesic, which passes through an umbilicus, will intersect the umbilical focal."

The plane of two consecutive tangents to the geodesic will be a tangent plane to the confocal, since it contains two consecutive tangent lines to that confocal.

Hence, the osculating planes of a geodesic will envelop a confocal conicoid. Also, since the osculating plane is normal to the surface on which the geodesic lies, it follows that the tangent planes drawn to two confocal conicoids through a common tangent line will be at right angles to each other.

Hence, the locus of the points of contact with the confocal surface will be a geodesic on that surface, for the osculating plane at any point of that locus will be a tangent plane to the first surface, and therefore at right angles to the tangent plane to the confocal, that is, will contain the normal to the confocal at the point of contact, and the curve will therefore be a geodesic.

These corresponding geodesics will touch the line of curvature, in which the confocals intersect, at the same point.

578. *The developable, enveloped by the tangent planes to a conicoid along a geodesic, will have its edge of regression on another conicoid.*

For the edge of regression is the locus of the intersection of three consecutive tangent planes, and any point on it will therefore be the pole of the plane passing through three consecutive points of the geodesic, that is, of an osculating plane of the geodesic. But the osculating plane is a tangent plane to a confocal conicoid, and the edge of regression will therefore lie on the reciprocal polar of this confocal with respect to the original, that is, will lie on another conicoid.

Also, the confocal will remain the same, for all geodesics which touch the same line of curvature; and therefore, the locus of the edge of regression corresponding to any such geodesic is a fixed conicoid.

For geodesics passing through an umbilicus, this fixed conicoid reduces to the hyperbola,

$$\frac{x^2(a^2-b^2)}{a^4}-\frac{z^2(b^2-c^2)}{c^4}=1.$$

579. *On the geodesic lines of the paraboloid.*

The equations may be integrated a first time for paraboloids in much the same way as for central conicoids, but we may more readily obtain the limiting form of the equation

$$\left(\frac{l^2}{a^2}+\frac{m^2}{b^2}+\frac{n^2}{c^2}\right)\left(\frac{x^2}{a^4}+\frac{y^2}{b^4}+\frac{z^2}{c^4}\right)=\frac{a^2b^2c^2}{k^2},$$

by transferring the origin to the extremity of the axis of x, and making the axes infinite, so as to reduce the equation of the conicoid to the form

$$\frac{y^2}{b}+\frac{z^2}{c}=2x.$$

The equation for the geodesic becomes, by this transformation,

$$\left(\frac{m^2}{b}+\frac{n^2}{c}\right)\left(1+\frac{y^2}{b^2}+\frac{z^2}{c^2}\right)=\frac{bc}{k};$$

b, c, k replacing the limits of $\dfrac{b^2}{a}$, $\dfrac{c^2}{a}$, $\dfrac{k^2}{a}$.

If k_1, k_2 be the parameters of the lines of curvature at a point (x, y, z), given by the equation

$$\frac{y^2}{b(b-k)}+\frac{z^2}{c(c-k)}+1=0,$$

and θ, $\frac{\pi}{2}+\theta$ be the angles which a geodesic, whose parameter is λ, makes with these lines of curvature, we shall have, by taking the limit of the equation of Art. (574),

$$k_2 \cos^2\theta + k_1 \sin^2\theta = \lambda.$$

Hence, for the locus of the intersection of geodesic tangents to two given lines of curvature, at right angles to each other, we shall have

$$k_1 + k_2 = \lambda + \lambda',$$

or

$$2x = b + c - \lambda - \lambda',$$

and the locus lies on a plane perpendicular to the axis. All the properties depending on the umbilici, and circular sections, will be true for the elliptic paraboloid. Thus, for one system of its lines of curvature the sum of the geodesic distances from the umbilici is constant on each line, and for the other system, the difference is constant.

580. It should be noticed that on any conicoid, generated by straight lines, each such straight line is of course a geodesic, but such geodesics are not included in the systems whose properties we have investigated. If (x, y, z) be a point on a fixed straight line, we shall have

$$\frac{d^2x}{ds^2} = 0, \quad \frac{d^2y}{ds^2} = 0, \quad \frac{d^2z}{ds^2} = 0,$$

and therefore the equations of a geodesic are satisfied, whatever be the forms of $\frac{dF}{dx}$, $\frac{dF}{dy}$, $\frac{dF}{dz}$. In our integration, it was tacitly assumed that $\frac{d^2x}{ds^2}$, &c., were finite.

581. *To find the radius of torsion at any point of any geodesic line on a surface.*

Take the point considered as origin, and let the surface be referred to the tangent plane, and planes of principal curvature, so that its equation assumes the form

$$2z = \frac{x^2}{\rho_1} + \frac{y^2}{\rho_2} + \ldots\ldots$$

ρ_1, ρ_2 being the principal radii of curvature.

Let a geodesic through the origin make an angle θ with the axis of x, and let $d\epsilon$ be the angle between two consecutive normal planes, $d\tau$ the angle between two consecutive osculating planes, to the geodesic. The normal to the surface at the consecutive point (x, y, z) on the geodesic lies in the osculating plane at that point. Hence, the normal at (x, y, z) may be made parallel to the normal at the origin, by turning it through the angles $d\tau$, $d\epsilon$ successively in planes at right angles to each other; or if λ be the angle between the two normals,

$$\cos\lambda = \cos d\tau \cos d\epsilon = 1 - \frac{d\tau^2 + d\epsilon^2}{2} \text{ ultimately.}$$

But $$\cos\lambda = \frac{1}{\sqrt{1 + \frac{x^2}{\rho_1^2} + \frac{y^2}{\rho_2^2}}} = 1 - \frac{1}{2}\left(\frac{x^2}{\rho_1^2} + \frac{y^2}{\rho_2^2}\right).$$

Hence, $$d\tau^2 + d\epsilon^2 = \frac{x^2}{\rho_1^2} + \frac{y^2}{\rho_2^2},$$

or, if ρ be the radius of absolute curvature of the geodesic, and σ the radius of torsion,

$$\frac{1}{\sigma^2} + \frac{1}{\rho^2} = \frac{\cos^2\theta}{\rho_1^2} + \frac{\sin^2\theta}{\rho_2^2},$$

but $\frac{1}{\rho} = \frac{\cos^2\theta}{\rho_1} + \frac{\sin^2\theta}{\rho_2}$, whence $\frac{1}{\sigma^2} = \sin^2\theta\cos^2\theta\left(\frac{1}{\rho_1} - \frac{1}{\rho_2}\right)^2$,

or $$\frac{1}{\sigma} = \left(\frac{1}{\rho_1} \sim \frac{1}{\rho_2}\right)\sin\theta\cos\theta.$$

The torsion of a geodesic consequently vanishes at a point where it touches a line of curvature, and the torsion of all geodesics passing through an umbilicus vanishes at the umbilicus. (Routh and Watson's *Cambridge Problems*, 1860.)

582. Def. If two points be taken on a curve traced on a surface, at a distance ds from each other; and if geodesic tangents be drawn to the curve at those points, the angle between them being du, the limiting value of $\frac{ds}{du}$ is the radius of geodesic curvature of the curve at the point.

If we suppose this indefinitely small part of the surface developed on the tangent plane at the intersection of the geodesics, the radius of geodesic curvature will be the radius of curvature of the corresponding plane curve, and du will be the projection, on the tangent plane, of the angle between the tangents to the curve in the osculating plane, whence, if ϕ be the angle between the osculating plane and the plane of normal section containing the tangent line to the curve, we shall have the angle of contingence of the curve $= \dfrac{du}{\sin\phi}$.

Hence, if ρ be the radius of absolute curvature,

$$\rho = \frac{ds}{du} \cdot \sin\phi,$$

and the radius of geodesic curvature is $\dfrac{\rho}{\sin\phi}$.

Hence, for geodesic lines, the radius of geodesic curvature is equal to the radius of absolute curvature.

XXV.

1. If a geodesic be drawn on a developable surface, and cut any generating line of the surface at an angle ψ, and at a distance t from the edge of regression measured along the generator, prove that

$$\frac{dt}{d\psi} + t\cot\psi = \rho,$$

ρ being the radius of curvature of the edge of regression at the point where the generator touches it.

2. On a right circular cone, whose vertical angle is $\frac{\pi}{3}$, two points are joined by a geodesic which completely surrounds the cone; prove that the two tangents to this geodesic at the double point are at right angles to each other.

3. If P be a point on a given line of curvature of a conicoid, PQ, PQ' geodesic tangents to another given line of curvature of the same system, PU, PV geodesics to two adjacent umbilici, prove that

$$\cos\frac{QPQ'}{2} : \cos\frac{UPV}{2} \text{ is a constant ratio.}$$

4. The angle between the geodesics passing through a point (x, y, z) on the ellipsoid

$$\frac{x^2}{a^2} + \frac{y^2}{b^2} + \frac{z^2}{c^2} = 1,$$

and through the umbilici, is given by the equation

$$\tan^2\theta = \frac{4(b^2 - c^2)(a^2 - b^2)y^2}{b^2(x^2 + y^2 + z^2 - a^2 + b^2 - c^2)^2}.$$

5. The angle between the geodesics passing through the point (x, y, z) of the paraboloid

$$\frac{y^2}{b} + \frac{z^2}{c} = 2x,$$

and through the umbilici, is given by the equation

$$\tan^2\theta = \frac{4(b - c)y^2}{b(b - c - 2x)^2},$$

b being greater than c.

6. The radius of absolute curvature of a geodesic of the ellipsoid

$$\frac{x^2}{a^2} + \frac{y^2}{b^2} + \frac{z^2}{c^2} = 1,$$

is $\frac{d^2}{p}$, p being the perpendicular from the center on the tangent plane, and d the central radius parallel to the tangent line.

7. If λ be the parameter of a geodesic on the conicoid

$$\frac{x^2}{a^2} + \frac{y^2}{b^2} + \frac{z^2}{c^2} = 1,$$

and R be the radius of torsion of the geodesic at a point (x, y, z), prove that

$$\frac{1}{R^2} = \frac{x^2p^6}{a^4b^4c^4}(a^2 + b^2 + c^2 - x^2 - y^2 - z^2 - \lambda^2) - \frac{p^4}{a^2b^2c^2},$$

p being the central distance of the tangent plane.

8. If two surfaces touch each other along a curve, and if the curve be a geodesic line on one surface, prove that it will also be a geodesic line on the other surface.

9. If P be a point on a geodesic line AP, drawn on a conoidal surface, s the length of AP, σ, N, O the projections of s, P, and the axis on any plane perpendicular to the axis, and p the projection of ON on the tangent to AP at P, then

$$\frac{dp}{d\sigma} = \frac{d\sigma}{ds}.$$

10. If a geodesic on any surface lie on a sphere, the radius of curvature of the geodesic at any point will be proportional to the perpendicular from the center of the sphere on the tangent plane to the surface.

11. The angle of geodesic contingence of a curve traced on a surface of revolution $= \dfrac{d(r\sin\theta)}{r\cos\theta}$ when θ is the angle made with a meridian, and r the radius of the corresponding parallel.

12. The sides of a geodesic triangle traced on a surface of revolution make, with the meridians which pass through the angular points, six angles; prove that the product of the sines of the three angles not adjacent is equal to the product of the three others.

13. The ratio of the radii of curvature and torsion in a geodesic line on a developable surface at any point is equal to the tangent of the inclination of the curve to the corresponding generating line.

14. Prove that if ψ be the inclination of a geodesic line to a generating line of the helicoid whose equation is

$$z = a\tan^{-1}\frac{y}{x},$$

and ϕ the inclination of the tangent plane at the corresponding point to the axis of z, $\sin\psi = C\cos\phi$.

15. If ABC be a geodesic triangle described on a conoidal helicoid, $\alpha_1, \alpha_2, \beta_1, \beta_2, \gamma_1, \gamma_2$ be the angles at which the sides cut the generating lines through A, B, C, prove that

$$\sin\alpha_1 \sin\beta_1 \sin\gamma_1 = \sin\alpha_2 \sin\beta_2 \sin\gamma_2.$$

16. A surface is generated by normals to a given curve, moving in such a manner that the angle between consecutive generating lines is equal to the arc intercepted, divided by a constant line a; prove that in any geodesic line on the surface, r being the distance from the curve measured on a generating line, ϕ the inclination to that line,

$$\frac{r^2}{a^2} = C^2\sin^2\phi - 1.$$

17. If a line of curvature on an ellipsoid be the intersection of the ellipsoid with a hyperboloid, one of whose principal sections is a rectangular hyperbola, no geodesic tangents at right angles can be drawn to it, except from the extremity of one of the axes.

THE END.

CAMBRIDGE: PRINTED AT THE UNIVERSITY PRESS.

Made in the USA
Monee, IL
13 September 2026

62976828R00282